Modern API Development with Spring and Spring Boot

Design highly scalable and maintainable APIs with REST, gRPC, GraphQL, and the reactive paradigm

Sourabh Sharma

BIRMINGHAM—MUMBAI

Modern API Development with Spring and Spring Boot

Copyright © 2021 Packt Publishing

Group Product Manager: Aaron Lazar

Publishing Product Manager: Alok Dhuri

Senior Editor: Rohit Singh

Content Development Editor: Kinnari Chohan

Technical Editor: Gaurav Gala

Copy Editor: Safis Editing

Project Coordinator: Francy Puthiry

Proofreader: Safis Editing

Indexer: Vinayak Purushotham

Production Designer: Alishon Mendonca

First published: May 2021

Production reference: 1280521

Published by Packt Publishing Ltd.
Livery Place
35 Livery Street
Birmingham
B3 2PB, UK.

ISBN 978-1-80056-247-9

www.packt.com

To my adored wife, Vanaja, and son, Sanmaya, for their unquestioning faith, support, and love.

To my parents, Mrs. Asha and Mr. Ramswaroop, for their blessings.

– Sourabh Sharma

Contributors

About the author

Sourabh Sharma works at Oracle as a lead technical member where he is responsible for developing and designing the key components of the blueprint solutions that are used by various Oracle products. He has over 18 years of experience delivering enterprise products and applications for leading companies. His expertise lies in conceptualizing, modeling, designing, and developing N-tier and cloud-based web applications as well as leading teams. He has vast experience in developing microservice-based solutions and implementing various types of workflow and orchestration engines. Sourabh believes in continuous learning and sharing knowledge through his books and training.

I would like to thank Kinnari, Eric Pirard, and Rohit for their hard work and critical review feedback, as well as Prajakta and Francy for their support. I would also like to thank Alok and Packt Publishing for providing me with the opportunity to publish this book.

About the reviewer

Eric Pirard has been a Java developer for a few years now. He's passionate about exploring new technologies that can help make developers' lives easy. He likes to help his friends and colleagues solve challenges in their projects and enjoys traveling and sports. There are so many things to do in addition to his exciting job that he looks for technology that can help him solve customer's problems as quickly as possible so he can spend more time with his family and friends. In short, Eric believes in enjoying every moment of life.

Table of Contents

2

Spring Concepts and REST APIs

3

API Specifications and Implementation

4
Writing Business Logic for APIs

5
Asynchronous API Design

Section 2: Security, UI, Testing, and Deployment

6

Security (Authorization and Authentication)

7

Designing a User Interface

8
Testing APIs

9
Deployment of Web Services

Section 3: gRPC, Logging, and Monitoring

10
gRPC Fundamentals

11
gRPC-based API Development and Testing

12
Logging and Tracing

Section 4: GraphQL

13

GraphQL Fundamentals

14

GraphQL API Development and Testing

Assessments

Other Books You May Enjoy

Index

Preface

The philosophy of API development has been evolving over the years to serve the modern needs of enterprise architecture, and developers need to know how to adapt to these modern API design principles. Apps are now developed with APIs that enable easy integration of a cloud environment and distributed systems. In this Spring book, you will discover various kinds of production-ready API implementations with the REST API, along with exploring async using the reactive paradigm, gRPC, and GraphQL.

You'll design evolving REST-based APIs supported by HATEOAS and ETags and develop reactive async non-blocking APIs. Next, you'll focus on how to secure REST APIs using Spring Security and how the developed APIs are consumed by the UI of the app. You will continue with testing, deploying, logging, and monitoring your developed APIs. In the concluding chapters, you'll explore more about developing APIs using gRPC and GraphQL, as well as designing modern scalable architecture with microservices. You'll also get up and running with a sample e-shopping app that will equip you with practical implementation knowledge of modern APIs.

By the end of this Spring Framework book, you'll be able to develop, test, and deploy highly scalable, maintainable, and developer-friendly APIs that can help your customers to transform their business.

Who this book is for

This book is for inexperienced Java programmers, comp science, or coding boot camp graduates who have knowledge of basic programming constructs, data structures, and algorithms in Java but lack the practical web development skills necessary to start working as a developer. Professionals who've recently joined a startup or a company and are tasked with creating real-world web APIs and services will also find this book helpful. This book is also a good resource for Java developers who are looking for a career move into web development to get started with the basics of web service development.

What this book covers

Chapter 1, RESTful Web Service Fundamentals, drives you through the fundamentals of RESTful APIs, or for short REST APIs, and their design paradigm. These basics will provide a solid platform to develop a RESTful web service. You will also learn the best practices while designing APIs. This chapter will also introduce the sample e-commerce app that will be used across the book while learning about the different aspects of API development.

Chapter 2, Spring Concepts and REST APIs, explores Spring fundamentals and features that are required to implement RESTful web services using the Spring Framework. This will provide the technical perspective required for developing a sample e-commerce app.

Chapter 3, API Specifications and Implementation, makes use of these two technologies to implement the REST APIs. We have chosen the design-first approach for implementation. You will make use of the OpenAPI specification to first design the APIs and later implement them. You will also learn how to handle the errors that occurred during the serving of the request. Here, the APIs of the sample e-commerce app will be designed and implemented for reference.

Chapter 4, Writing Business Logic for APIs, helps you implement the API's code in terms of business logic, along with data persistence. You will write services and repositories for implementation. You will also add hypermedia and ETag headers to API responses.

Chapter 5, Asynchronous API Design, covers asynchronous API design, where calls will be asynchronous and non-blocking. We'll develop these APIs using Spring WebFlux, which is itself based on Project Reactor (`https://projectreactor.io`). First, we'll walk through the reactive programming fundamentals and then migrate the existing e-commerce REST APIs (the previous chapter's code) to asynchronous (reactive) APIs to make things easier by correlating and comparing the existing (imperative) way and reactive way of programming.

Chapter 6, Security (Authorization and Authentication), explains how you can secure these REST endpoints using Spring Security. You'll implement token-based authentication and authorization for REST endpoints. Successful authentication will provide two types of tokens – a **JSON Web Token** (**JWT**) as an access token, and a refresh token in response. The JWT-based access token then will be used for accessing secured URLs. A refresh token will be used for requesting a new JWT if the existing JWT has expired. A valid request token can provide a new JWT to use. You'll associate users with roles such as Admin, User, and so on. These roles will be used as authorization to make sure that REST endpoints can only be accessed if the user holds a certain role. We'll also briefly discuss **Cross-Site Request Forgery** (**CSRF**) and **Cross-Origin Resource Sharing** (**CORS**).

Chapter 7, Designing a User Interface, concludes the end-to-end development and communication between different layers of the online shopping app. This UI app will be a **Single-Page Application (SPA)** that consists of interactive components such as Login, Product Listing, Product Detail, Cart, and Order Listing. By the end of the chapter, you will have learned about SPA and UI component development using React and consuming REST APIs using the browser's in-built Fetch API.

Chapter 8, Testing APIs, introduces manual and automated testing of APIs. You will learn about unit and integration test automation. After learning about automation in this chapter, you will be able to make both types of testing an integral part of the build. You will also set up the Java code coverage tool to calculate the different code coverage metrics.

Chapter 9, Deployment of Web Services, explains the fundamentals of containerization, Docker, and Kubernetes. You will then use this concept to containerize the sample e-commerce app using Docker. This container will then be deployed in a Kubernetes cluster. You are going to use Minikube for Kubernetes, which makes learning and Kubernetes-based development easier.

Chapter 10, gRPC Fundamentals, introduces the gRPC fundamentals.

Chapter 11, gRPC-based API Development and Testing, implements gRPC-based APIs. You will learn how to write a gRPC server and client, along with writing APIs based on gRPC. In the latter part of the chapter, you will be introduced to microservices and how they will help you to design modern, scalable architecture. Here, you will go through the implementation of two services – a gRPC server and a gRPC client.

Chapter 12, Logging and Tracing, explores the logging and monitoring tool called the **ELK (Elastic Search, Logstash, Kibana)** Stack and Zipkin. These tools will then be used to implement the distributed logging and tracing of the request/response of the API calls. Spring Sleuth will be used to inject the tracing information into API calls. You will learn how to publish and analyze the logging and tracing of different requests and logs related to responses. You will also use Zipkin for monitoring the performance of API calls.

Chapter 13, GraphQL Fundamentals, talks about the fundamentals of GraphQL – **schema definition language (SDL)**, queries, mutations, and subscriptions. This knowledge will help you in the next chapter, where you will implement an API based on GraphQL. During the course of this chapter, you will learn the basics of the GraphQL schema and solving the N+1 problem.

Chapter 14, GraphQL API Development and Testing, explains GraphQL-based API development and its testing. You will implement GraphQL-based APIs for a sample application in this chapter. A GraphQL server implementation will be developed based on the design-first approach.

To get the most out of this book

Software/ hardware covered in the book	OS requirements
Java 15	Windows, macOS and Linux (any)
Any Java IDE such as NetBeans, IntelliJ, or Eclipse	
Docker	
Kubernetes (Minikube)	
cURL, Postman, or any REST client	
An internet connection to clone the code and download the dependencies and Gradle	
Node.js 14.x with NPM 6.x	
Visual Studio Code	
React 17	
ELK Stack (7.12.1)	
Zipkin (2.23.2)	

Each chapter will contain special instructions to install the required tools if applicable.

If you are using the digital version of this book, we advise you to type the code yourself or access the code via the GitHub repository (link available in the next section). Doing so will help you avoid any potential errors related to the copying and pasting of code.

Knowledge of intermediate-level Java programming and beginner-level Spring is required to get the most out of this book, although this book also features a quick refresher on Spring fundamentals to help you. No prior knowledge of RESTful web services, gRPC, GraphQL, or microservices is expected.

Some of the code blocks included in the book are stripped for brevity. To run the entire code with all the dependencies, please refer to the GitHub files.

Download the example code files

You can download the example code files for this book from GitHub at `https://github.com/PacktPublishing/Modern-API-Development-with-Spring-and-Spring-Boot`. In case there's an update to the code, it will be updated on the existing GitHub repository.

We also have other code bundles from our rich catalog of books and videos available at `https://github.com/PacktPublishing/`. Check them out!

Download the color images

We also provide a PDF file that has color images of the screenshots/diagrams used in this book. You can download it here: `https://static.packt-cdn.com/downloads/9781800562479_ColorImages.pdf`.

Conventions used

There are a number of text conventions used throughout this book.

`Code in text`: Indicates code words in text, database table names, folder names, filenames, file extensions, pathnames, dummy URLs, user input, and Twitter handles. Here is an example: "Once the Gradle project is available, you can modify the `build.gradle` file to include the GDS dependencies and plugin."

A block of code is set as follows:

```
}
type Tag {
  id: String
  name: String
}
```

When we wish to draw your attention to a particular part of a code block, the relevant lines or items are set in bold:

```
generateJava {
  generateClient = true
  packageName = "com.packt.modern.api.generated"
  typeMapping = ["BigDecimal": "java.math.BigDecimal"]
}
```

Any command-line input or output is written as follows:

```
$ gradlew clean build
```

> **Tips or important notes**
> Appear like this.

Get in touch

Feedback from our readers is always welcome.

General feedback: If you have questions about any aspect of this book, mention the book title in the subject of your message and email us at customercare@packtpub.com.

Errata: Although we have taken every care to ensure the accuracy of our content, mistakes do happen. If you have found a mistake in this book, we would be grateful if you would report this to us. Please visit www.packtpub.com/support/errata, selecting your book, clicking on the Errata Submission Form link, and entering the details.

Piracy: If you come across any illegal copies of our works in any form on the Internet, we would be grateful if you would provide us with the location address or website name. Please contact us at copyright@packt.com with a link to the material.

If you are interested in becoming an author: If there is a topic that you have expertise in and you are interested in either writing or contributing to a book, please visit authors.packtpub.com.

Reviews

Please leave a review. Once you have read and used this book, why not leave a review on the site that you purchased it from? Potential readers can then see and use your unbiased opinion to make purchase decisions, we at Packt can understand what you think about our products, and our authors can see your feedback on their book. Thank you!

For more information about Packt, please visit packt.com.

Section 1: RESTful Web Services

In this section, you will develop and test the production-ready and evolving REST-based APIs supported by HATEOAS and ETAGs. API specs will be written using OpenAPI specifications (Swagger). You will learn the fundamentals of reactive API development using Spring WebFlux. By the end of this section, you will be familiar with the fundamentals of REST, best practices, and how to write evolving APIs. After completing this section, you will be able to develop sync and reactive async non-blocking APIs.

This section comprises the following chapters:

- *Chapter 1, RESTful Web Service Fundamentals*
- *Chapter 2, Spring Concepts and REST APIs*
- *Chapter 3, API Specifications and Implementation*
- *Chapter 4, Writing Business Logic for APIs*
- *Chapter 5, Asynchronous API Design*

1
RESTful Web Service Fundamentals

In this chapter, you will go through the fundamentals of RESTful APIs, or REST APIs for short, and their design paradigms. We will take a brief look at the history of REST, learn how resources are formed, and understand methods and status codes before we move on to explore **Hypermedia As The Engine Of Application State** (**HATEOAS**). These basics should provide a solid platform for you to develop a RESTful web service. You will also learn the best practices for designing **Application Programming Interfaces** (**APIs**).

This chapter will also introduce a sample e-commerce app, which will be used throughout the book as you learn about the different aspects of API development. In this chapter, we will cover the following topics:

- Introducing REST APIs
- Handling resources and **Uniform Resource Identifiers** (**URIs**)
- Exploring **Hypertext Transfer Protocol** (**HTTP**) methods and status codes
- Learning HATEOAS
- Best practices for designing REST APIs
- Overview of an e-commerce app (our sample app)

Technical requirements

This chapter does not require any specific software. However, knowledge of HTTP is necessary.

Introducing REST APIs

An API is the means by which a piece of code communicates with another piece of code. You might have already written an API for your code or used one in your programs; for example, in Java libraries for collection, input/output, or streams that provide a variety of APIs to perform specific tasks.

Java's SDK APIs allow one part of a program to communicate with another part of a program. You can write a function and then expose it with public access modifiers so that other classes can use it. That function signature is an API for that class. However, APIs that are exposed using these classes or libraries only allow internal communication inside a single application or individual service. So, what happens when two or more applications (or services) want to communicate with each other? In other words, you would like to integrate two or more services. This is where system-wide APIs help us.

Historically, there were different ways to integrate one application with another – RPC, **Simple Object Access Protocol** (**SOAP**)-based services, and more. The integration of apps has become an integral part of software architectures, especially after the boom of the cloud and mobile phones. You now have social logins, such as Facebook, Google, and GitHub, which means you can develop your application even without writing an independent login module and get around security issues such as storing passwords in a secure way.

These social logins provide APIs using REST and GraphQL. Currently, REST is the most widely used, and it has become a standard for writing APIs for integration and web app consumption. We'll also discuss GraphQL in detail in the final chapters of this book (in *Chapter 13, GraphQL Fundamentals*, and *Chapter 14, GraphQL Development and Testing*).

REST stands for REpresentational State Transfer, which is a style of software architecture. Web services that adhere to the REST style are called RESTful web services. In the following sections, we will take a quick look at the history of REST to understand its fundamentals.

REST history

Before REST adoption, when the internet was just starting to become widely known and Yahoo and Hotmail were the popular mail and social messaging apps, there was no standard software architecture that offered a homogenous way to integrate with web applications. People were using SOAP-based web services, which, ironically, were not simple at all.

Then came the light. Roy Fielding, in his doctoral research, *Architectural Styles and the Design of Network-Based Software Architectures* (`https://www.ics.uci.edu/~fielding/pubs/dissertation/top.htm`), came up with REST in 2000. REST's architecture style allowed any server to communicate with any other server over the network. It simplified communication and made integration easier. REST was made to work on top of HTTP, which enables it to be used all over the web and in internal networks.

eBay was the first to exploit REST-based APIs. It introduced the REST API with selected partners in November 2000. Later, Amazon, Delicious (a site-bookmarking web app), and Flickr (the photo-sharing app) started providing REST-based APIs. In fact, **Amazon Web Services (AWS)** took advantage of Web 2.0 (with the invention of REST) and provided REST APIs to developers for AWS cloud consumption in 2006.

Later Facebook, Twitter, Google, and other companies started using it. Nowadays (in 2021), you will hardly find any web application developed without a REST API. Although, the GraphQL-based API for mobile apps is getting pretty close in terms of popularity.

REST fundamentals

REST works on top of the HTTP protocol. Each URI works as an API resource. Therefore, we should use nouns as endpoints instead of verbs. RPC-style endpoints use verbs, for example, `api/v1/getPersons`. In comparison, in REST, this endpoint could be simply written as `api/v1/persons`. You must be wondering, then, how can we differentiate between the different actions performed on a REST resource? This is where HTTP methods help us. We can make our HTTP methods act as a verb, for example, GET, DELETE, POST (for creating), PUT (for modifying), and PATCH (for partial updating). We'll discuss this in more detail later. For now, the `getPerson` RPC-style endpoint is translated into `GET api/v1/persons` in REST.

> **Note**
>
> The REST endpoint is a unique URI that represents a REST resource. For example, `https://demo.app/api/v1/persons` is a REST endpoint. Additionally, `/api/v1/persons` is the endpoint path and `persons` is the REST resource.

Here, there is client and server communication. Therefore, REST is based on the **Client-Server** concept. The client calls the REST API and the server responds with a response. REST allows a client (that is, a program, web service, or UI app) to talk to a remotely (or locally) running server (or web service) using HTTP requests and responses. The client sends to the web service with an API command wrapped in an HTTP request to the web. This HTTP request may contain a payload (or input) in the form of query parameters, headers, or request bodies. The called web service responds with a success/failure indicator and the response data wrapped inside the HTTP response. The HTTP status code normally denotes the status, and the response body contains the response data. For example, an HTTP status code of *200 OK* normally represents success.

From a REST perspective, an HTTP request is self-descriptive and has enough context for the server to process it. Therefore, REST calls are **stateless**. States are either managed on the client side or on the server side. A REST API does not maintain its state. It only transfers states from the server to the client or vice versa. Therefore, it is called REpresentation State Transfer, or REST for short.

It also makes use of HTTP cache control, which makes REST APIs **cacheable**. Therefore, the client can also cache the representation (that is, the HTTP response) because every representation is self-descriptive.

The following is a list of key concepts in REST:

- Resources and URIs
- HTTP methods

A sample REST call in plain text looks similar to the following:

```
GET /licenses HTTP/1.1
Host: api.github.com
```

Here, the `/licenses` path denotes the licenses resource. `GET` is an HTTP method. `1.1` at the end of the first line denotes the HTTP protocol version. The second line shares the host to call.

GitHub responds with a JSON object. The status is *200 OK* and the JSON object is wrapped in a response body, as follows:

```
HTTP/1.1 200 OK
date: Sun, 22 Sep 2020 18:01:22 GMT
content-type: application/json; charset=utf-8
server: GitHub.com
status: 200 OK
cache-control: public, max-age=60, s-maxage=60
vary: Accept, Accept-Encoding, Accept, X-Requested-With,
    Accept-Encoding
etag: W/"3cbb5a2e38ac6fc92b3d798667e828c7e3584af278aa3
    14f6eb1857bbf2593ba"
 … <bunch of other headers>
Accept-Ranges: bytes
Content-Length: 2507
X-GitHub-Request-Id: 1C03:5C22:640347:81F9C5:5F70D372
[
    {
        "key": "agpl-3.0",
        "name": "GNU Affero General Public License v3.0",
        "spdx_id": "AGPL-3.0",
        "url": "https://api.github.com/licenses/agpl-3.0",
        "node_id": "MDc6TGljZW5zZTE="
    },
    {
        "key": "apache-2.0",
        "name": "Apache License 2.0",
        "spdx_id": "Apache-2.0",
        "url": "https://api.github.com/licenses/
            apache-2.0",
        "node_id": "MDc6TGljZW5zZTI="
    },
    …
]
```

If you take note of the third line in this response, it tells you the value of the content type. It is good practice to have JSON as a content type for both the request and the response.

Handling resources and URIs

Every document on the **World Wide Web** (**WWW**) is represented as a resource in terms of HTTP. This resource is represented as a URI, which is an endpoint that represents a unique resource on a server.

Roy Fielding states that a URI is known by many names – a WWW address, a **Universal Document Identifier** (**UDI**), a URI, a **Uniform Resource Locator** (**URL**), and a **Uniform Resource Name** (**URN**).

So, what is a URI? A URI is a string (that is, a sequence of characters) that identifies a resource by its location, name, or both (in the WWW world). There are two types of URIs – URLs and URNs – as follows:

Figure 1.1 – The URI hierarchy

URLs are widely used and even known to non-developer users. URLs are not only restricted to HTTP; in fact, they are also used for many other protocols such as FTP, JDBC, and MAILTO. Therefore, a URL is an identifier that identifies the network location of a resource. We will go into more detail in the later sections.

The URI syntax

The URI syntax is as follows:

```
scheme:[//authority]path[?query][#fragment]
```

As per the syntax, the following is a list of components of a URI:

- **Scheme**: This refers to a non-empty sequence of characters followed by a colon (:). scheme starts with a letter and is followed by any combination of digits, letters, periods (.), hyphens (-), or plus characters (+).

  Scheme examples include HTTP, HTTPS, MAILTO, FILE, FTP, and more. URI schemes must be registered with the **Internet Assigned Numbers Authority** (**IANA**).

- **Authority**: This is an optional field and is preceded by `//`. It consists of the following optional subfields:

  a. **Userinfo**: This is a subcomponent that might contain a username and a password, which are both optional.

  b. **Host**: This is a subcomponent containing either an IP address or a registered host or domain name.

  c. **Port**: This is an optional subcomponent that is followed by a colon (`:`).

- **Path**: A path contains a sequence of segments separated by slash characters (`/`). In the preceding GitHub REST API example, `/licenses` is the path.

- **Query**: This is an optional component and is preceded by a question mark (`?`). The query component contains a query string of non-hierarchical data. Each parameter is separated by an ampersand (`&`) in the query component and parameter values are assigned using an equals (`=`) operator.

- **Fragment**: This is an optional field and is preceded by a hash (`#`). The fragment component includes a fragment identifier that gives direction to a secondary resource.

The following list contains examples of URIs:

- `www.packt.com`: This doesn't contain the scheme. It just contains the domain name. There is no port either, which means it points to the default port.

- `index.html`: This contains no scheme nor authority. It only contains the path.

- `https://www.packt.com/index.html`: This contains the scheme, authority, and path.

Here are some examples of different scheme URIs:

- `mailto:support@packt.com`
- `telnet://192.168.0.1:23/`
- `ldap://[2020:ab9::9]/c=AB?objectClass?obj`

From a REST perspective, the path component of a URI is very important because it represents the resource path and your API endpoint paths are formed based on it. For example, take a look at the following:

```
GET https://www.domain.com/api/v1/order/1
```

Here, `/api/v1/order/1` represents the path, and `GET` represents the HTTP method.

URLs

If you look closely, most of the URI examples mentioned earlier can also be called URLs. A URI is an identifier; on the other hand, a URL is not only an identifier, but it also tells you how to get to it.

As per **Request for Comments (RFC)**-3986 on URIs (`https://xml2rfc.tools.ietf.org/public/rfc/html/rfc3986.html`), the term URL refers to the subset of URIs that, in addition to identifying a resource, provide a means of locating the resource by describing its primary access mechanism (for example, its network "location").

A URL represents the full web address of a resource, including the protocol name (the scheme), the hostname port (in case the HTTP port is not 80; for HTTPS, the default port is 443), part of the authority component, the path, and optional query and fragment subcomponents.

URNs

URNs are not commonly used. They are also a type of URI that starts with a scheme – **urn**. The following URN example is directly taken from RFC-3986 for URIs (`https://xml2rfc.tools.ietf.org/public/rfc/html/rfc3986.html`):

```
urn:oasis:names:specification:docbook:dtd:xml:4.1.2
```

This example follows the `"urn:"` `<NID>` `":"` `<NSS>` syntax, where `<NID>` is the NAMESPACE IDENTIFIER, and `<NSS>` is the Namespace-specific String. We are not going to use URNs in our REST implementation. However, you can read more about them at RFC-2141 (`https://tools.ietf.org/html/rfc2141`).

As per RFC-3986 on URIs (`https://xml2rfc.tools.ietf.org/public/rfc/html/rfc3986.html`): The term URN has been used historically to refer to both URIs under the "urn" scheme RFC-2141, which are required to remain globally unique and persistent even when the resource ceases to exist or becomes unavailable, and to any other URI with the properties of a name.

Exploring HTTP methods and status codes

HTTP provides various HTTP methods. However, you are primarily going to use only five of them. To begin with, you want to have **Create, Read, Update, and Delete** (**CRUD**) operations associated with HTTP methods:

- POST: Create or search
- GET: Read

- PUT: Update
- DELETE: Delete
- PATCH: Partial update

Some organizations also provide the HEAD method for scenarios where you just want to retrieve the header responses from the REST endpoints. You can hit any GitHub API with the HEAD operation to retrieve only headers; for example, `curl --head https://api.github.com/users`.

> **Note**
> REST has no such requirement that specifies which method should be used for which operation. However, widely used industry guidelines and practices suggest following certain rules.

Let's discuss each method in detail in the following sections.

POST

The HTTP POST method is normally what you want to associate with creating resource operations. However, there are certain exceptions when you might want to use the POST method for read operations. However, it should be put into practice after a well-thought-out process. One such exception is the search operation where filter criteria have too many parameters that might cross the GET call's length limit.

A GET query string has a limit of 256 characters. Additionally, the GET HTTP method is limited to a maximum of 2,048 characters minus the number of characters in the actual path. On the other hand, the POST method is not limited by the size of the URL for submitting name and value pairs.

You may also want to use the POST method with HTTPS for a read call if the submitted input parameters contain any private or secure information.

For successful create operations, you can respond with the *201 Created* status, and for successful search or read operations, you should use the *200 OK* or *204 No Content* status codes, although the call is made using the POST HTTP method.

For failed operations, REST responses may have different error status codes based on the error type, which we will look at later in this section.

GET

The HTTP GET method is what you usually want to associate with read resource operations. Similarly, you must have observed the GitHub GET /licenses call that returns the available licenses in the GitHub system. Additionally, successful GET operations should be associated with the *200 OK* status code if the response contains data, or *204 No Content* if the response contains no data.

PUT

The HTTP PUT method is what you usually want to associate with update resource operations. Additionally, successful update operations should be associated with a *200 OK* status code if the response contains data, or *204 No Content* if the response contains no data. Some developers use the PUT HTTP method to replace existing resources. For example, GitHub API v3 uses PUT to replace the existing resource.

DELETE

The HTTP DELETE method is what you want to associate with delete resource operations. GitHub does not provide the DELETE operation on the licenses resource. However, if you assume it exists, it will certainly look very similar to DELETE / licenses/agpl-3.0. A successful delete call should delete the resource associate with the agpl-3.0 key. Additionally, successful DELETE operations should be associated with the *204 No Content* status code.

PATCH

The HTTP PATCH method is what you want to associate with partial update resource operations. Additionally, successful PATCH operations should be associated with a *200 OK* status code. PATCH is relatively new as compared to other HTTP operations. In fact, a few years ago, Spring did not have state-of-the-art support for this method for REST implementation due to the old Java HTTP library. However, currently, Spring provides built-in support for the PATCH method in REST implementation.

HTTP status codes

There are five categories of HTTP status codes, as follows:

- Informational responses (100–199)

- Successful responses (200–299)

- Redirects (300–399)

- Client errors (400–499)

- Server errors (500–599)

You can view a complete list of status codes at MDN Web Docs (`https://developer.mozilla.org/en-US/docs/Web/HTTP/Status`) or RFC-7231 (`https://tools.ietf.org/html/rfc7231`). However, you can find the most commonly used REST response status codes in the following table:

HTTP status code	Description
200 OK	For successful requests other than those already created.
201 Created	For successful creation requests.
202 Accepted	The request has been received but not yet acted upon. This is used when the server accepts the request, but the response cannot be sent immediately, for example, in batch processing.
204 No Content	For a successful operation that contains no data.
304 Not Modified	This is used for caching. The server responds to the client that the resource is not modified; therefore, the same cache resource can be used.
400 Bad Request	This is for a failed operation when input parameters are either incorrect or missing or the request itself is incomplete.
401 Unauthorized	This is for a failed operation due to unauthenticated requests. The specification says it's unauthorized, but semantically, it means unauthenticated.
403 Forbidden	This is for a failed operation when the invoker is not authorized to perform.
404 Not Found	This is for a failed operation when the requested resource doesn't exist.
405 Method Not Allowed	This is for a failed operation when the method is not allowed for the requested resource.
406 Not Acceptable	This is for a failed operation when the `Accept` header doesn't match. You can also use it when dependent operations are a barrier to the request process. For example, an order cannot be placed when a payment is rejected.
409 Conflict	This is for a failed operation when an attempt is made for a duplicate create operation.
429 Too Many Requests	This is for a failed operation when a user sends too many requests in a given amount of time ("rate limiting").
500 Internal Server Error	This is a failed operation due to a server error. This is a generic error.
502 Bad Gateway	This is for failed operations when the upstream server calls fail. For example, when an app calls a third-party payment service but the call fails.
503 Service Unavailable	This is for a failed operation when something unexpected has happened at the server, for example, an overload or a service fails.

Learning HATEOAS

With HATEOAS, RESTful web services provide information dynamically through hypermedia. Hypermedia is a part of the content that you receive from a REST call response. This hypermedia content contains links to different types of media such as text, images, and videos.

Hypermedia links can be contained either in HTTP headers or the response body. If you take a look at GitHub APIs, you will find that GitHub APIs provide hypermedia links in both headers and the response body. GitHub uses the header named "Link" to contain the paging-related links. Additionally, if you look at the responses of GitHub APIs, you'll also find other resource-related links with keys that have a postfix of `"url"`. Let's take a look at an example. We'll hit the `GET  /users` resource and analyze the response:

```
$ curl -v https://api.github.com/users
```

This will give you the following output:

```
HTTP/1.1 200 OK
date: Mon, 28 Sep 2020 05:49:56 GMT
content-type: application/json; charset=utf-8
server: GitHub.com
status: 200 OK
cache-control: public, max-age=60, s-maxage=60
vary: Accept, Accept-Encoding, Accept, X-Requested-With,
      Accept-Encoding
etag: W/"6308a6b7274db1f1ffa377aeeb5359a015f69fa6733298938
      9453c7f20336753"
x-github-media-type: github.v3; format=json
link: <https://api.github.com/users?since=46>; rel="next",
<https://api.github.com/users{?since}>; rel="first"
… <Some other headers>
…
[
  {
    "login": "mojombo",
    "id": 1,
    "node_id": "MDQ6VXNlcjE=",
    "avatar_url": "https://avatars0.githubusercontent.com/
                  u/1?v=4",
    "gravatar_id": "",
    "url": "https://api.github.com/users/mojombo",
    "html_url": "https://github.com/mojombo",
    "followers_url": "https://api.github.com/users/mojombo/
                     followers",
```

```
        "following_url": "https://api.github.com/users/mojombo/
                            following{/other_user}",
        "gists_url": "https://api.github.com/users/mojombo/
                        gists{/gist_id}",
        "starred_url": "https://api.github.com/users/mojombo
                          /starred{/owner}{/repo}",
        "subscriptions_url": "https://api.github.com/users/
                                mojombo/subscriptions",
        "organizations_url": "https://api.github.com/
                                users/mojombo/orgs",
        "repos_url": "https://api.github.com/users/
                        mojombo/repos",
        "events_url": "https://api.github.com/users/mojombo
                        /events{/privacy}",
        "received_events_url": "https://api.github.com/users
                                  /mojombo/received_events",
        "type": "User",
        "site_admin": false
    },
    {
        "login": "defunkt",
        "id": 2,
        "node_id": "MDQ6VXNlcjI=",
    ...
    ... <some more data>
    ]
```

In this code block, you'll find that the "Link" header contains the pagination information. Links to "next" page and "first" page are given as a part of the response. Additionally, you can find many URLs in the response body, such as "avatar_url" or "followers_url", which provide links to other hypermedia.

REST clients should possess a generic understanding of hypermedia. Then, REST clients can interact with RESTful web services without having any specific knowledge of how to interact with the server. You just call any static REST API endpoint and you will receive the dynamic links as a part of the response to interact further. REST allows clients to dynamically navigate to the appropriate resource by traversing the links. It empowers machines, as REST clients can navigate to different resources in a similar way to how humans look at a web page and click on any link. Put simply, the REST client makes use of these links to navigate.

HATEOAS is a very important concept of REST. It is one of the concepts that differentiate REST from RPC. Even Roy Fielding was so concerned with certain REST API implementations that he published the following blog on his website in 2008: `REST APIs must be hypertext-driven`.

You must be wondering what the difference between hypertext and hypermedia is. Essentially, hypermedia is just an extended version of hypertext. As Roy Fielding states:

> *"When I say hypertext, I mean the simultaneous presentation of information and controls such that the information becomes the affordance through which the user (or automaton) obtains choices and selects actions. Hypermedia is just an expansion on what text means to include temporal anchors within a media stream; most researchers have dropped the distinction.*
>
> *Hypertext does not need to be HTML on a browser. Machines can follow links when they understand the data format and relationship types."*

Best practices for designing REST APIs

It is too early to talk about the best practices for implementing APIs. APIs are designed first and implemented later. Therefore, you'll find design-related best practices mentioned in the next sections. You'll also find best practices for going forward during the course of your REST API implementation.

1. Use nouns and not verbs when naming a resource in the endpoint path

We previously discussed HTTP methods. HTTP methods use verbs. Therefore, it would be redundant to use verbs yourself, and it would make your call look like an RPC endpoint, for example, `GET /getlicenses`. In REST, we should always use the resource name because, according to REST, you transfer the states and not the instructions.

For example, let's take another look at the GitHub license API that retrieves the available licenses. It is GET /licenses. That is perfect. Let's assume that if you use verbs for this endpoint, then it will be GET /getlicenses. It will still work, but semantically, it doesn't follow REST because it conveys the processing instruction rather than state transfer. Therefore, only use resource names.

However, GitHub's public API only offers read operations on the licenses resource, out of all the CRUD operations. If we need to design the rest of the operations, their paths should look like following:

- POST /licenses: This is for creating a new license.
- PATCH /licenses/{license_key}: This is for partial updates. Here, the path has a parameter (that is, an identifier), which makes the path dynamic. Here, the license key is a unique value in the license collection and is being used as an identifier. Each license will have a unique key. This call should make the update in the given license. Please remember that GitHub uses PUT for the replacement of the resource.
- DELETE /licenses/{license_key}: This is for retrieving license information. You can try this with any license that you receive in the response of the GET /licenses call. One example is GET /licenses/agpl-3.0.

You can see how having a noun in the resource path with the HTTP methods sorts out any ambiguity.

2. Use the plural form for naming the collection resource in the endpoint path

If you observe the GitHub license API, you might find that a resource name is given in the plural form. It is a good practice to use the plural form if the resource represents a collection. Therefore, we can use /licenses instead of /license. A GET call returns the collection of licenses. A style call creates a new license in the existing license collection. For delete and patch calls, a license key is used to identify the specific license.

3. Use hypermedia (HATEOAS)

Hypermedia (that is, links to other resources) makes the REST client's job easier. There are two advantages if you provide explicit URL links in a response. First, the REST client is not required to construct the REST URLs on their own. Second, any upgrade in the endpoint path will be taken care of automatically and this, therefore, makes upgrades easier for clients and developers.

4. Always version your APIs

The versioning of APIs is key for future upgrades. Over time, APIs keep changing, and you may have customers who are still using an older version. Therefore, you need to support multiple versions of APIs.

There are different ways you can version your APIs, as follows:

- **Using headers**: The GitHub API uses this approach. You can add an `Accept` header that tells you which API version should serve the request; for example, consider the following:

  ```
  Accept: application/vnd.github.v3+json
  ```

  This approach gives you the advantage of setting the default version. If there is no `Accept` header, it should lead to the default version. However, if a REST client that uses a versioning header is not changed after a recent upgrade of APIs, it may lead to a functionality break. Therefore, it is recommended that you use a versioning header.

- **Using an endpoint path**: In this approach, you add a version in the endpoint path itself; for example, `https://demo.app/api/v1/persons`. Here, `v1` denotes that version 1 is being added to the path itself.

  You cannot set default versioning out of the box. However, you can overcome this limitation by using other methods, such as request forwarding. Clients always use the intended versions of the APIs in this approach.

Based on your preferences and views, you can choose either of the preceding approaches for versioning. However, the important point is that you should always use versioning.

5. Nested resources

Consider this very interesting question: how are you going to construct the endpoint for resources that are nested or have a certain relationship? Let's take a look at some examples of customer resources from an e-commerce perspective:

- `GET /customers/1/addresses`: This returns the collection of addresses for customer 1.

- `GET /customers/1/addresses/2`: This returns the second address of customer 1.

- `POST /customers/1/addresses`: This adds a new address to customer 1's addresses.

- `PUT /customers/1/address/2`: This replaces the second address of customer 1.

- `PATCH /customers/1/address/2`: This partially updates the second address of customer 1.

- `DELETE /customers/1/address/2`: This deletes the second address of customer 1.

So far so good. Now, can we can have an altogether separate addresses resource endpoint (`GET /addresses/2`)? It makes sense, and you can do that if there exists a relationship that requires it; for example, orders and payments. Instead of `/orders/1/payments/1`, you might prefer a separate `/payments/1` endpoint. In the microservice world, this makes more sense; for instance, you would have two separate RESTful web services for both orders and payments.

Now, if you combine this approach with hypermedia, it makes things easier. When you make a REST API request to customer 1, it will provide the customer 1 data and address links as hypermedia (that is, links). The same applies to orders. For orders, the payment link will be available as hypermedia.

However, in some cases, you might wish to have a complete response in a single request rather than using the hypermedia-provided URLs to fetch the related resource. This reduces your web hits. However, there is no thumb rule. For a flag operation, it makes sense to use the nested endpoint approach; for example, `PUT /gist/2/star` (which adds a star) and `DELETE /gist/2/star` (which undoes the star) in the case of the GitHub API.

Additionally, in some scenarios, you might not find a suitable resource name when multiple resources are involved, for example, in a search operation. In that case, you should use a `direct /search` endpoint. This is an exception.

6. Secure APIs

Securing your API is another expectation that requires diligent attention. Here are some recommendations:

- Always use HTTPS for encrypted communication.

- Always look for OWASP's top API security threats and vulnerabilities. These can be found on their website (`https://owasp.org/www-project-api-security/`) or the GitHub repository (`https://github.com/OWASP/API-Security`).

- Secure REST APIs should have authentication in place. REST APIs are stateless; therefore, REST APIs should not use cookies or sessions. Instead, these should be secure using JWT or OAuth 2.0-based tokens.

7. Documentation

Documentation should be easily accessible and up to date with the latest implementation with their respective versioning. It is always good to provide sample code and examples. It makes the developer's integration job easier.

A change log or a release log should list all of the impacted libraries, and if some APIs are deprecated, then replacement APIs or workarounds should be elaborated on inside the documentation.

8. Status codes

You might have already learned about status code in the *Exploring HTTP methods and status codes* section. Please follow the same guidelines discussed there.

9. Caching

HTTP already provides the caching mechanism. You just have to provide additional headers in the REST API response. Then, the REST client makes use of the validation to make sure whether to make a call or use the cached response. There are two ways to do it:

- **ETag**: ETag is a special header value that contains the hash or checksum value of the resource representation (that is, the response object). This value must change with respect to the response representation. It will remain the same if the resource response doesn't change.

Now, the client can send a request with another header field, called `If-None-Match`, which contains the ETag value. When the server receives this request and finds that the hash or checksum value of the resource representation value is different from `If-None-Match`, only then should it return the response with a new representation and this hash value in the ETag header. If it finds them to be equal, then the server should simply respond with a `304  (Not Modified)` status code.

- **Last-Modified**: This approach is identical to the ETag way. Instead of using the hash or checksum, it uses the timestamp value in RFC-1123 format (*Last-Modified: Wed, 21 Oct 2015 07:28:00 GMT*). It is less accurate than ETag and should only be used for the falling mechanism.

  Here, the client sends the `If-Modified-Since` header with the value received in the `Last-Modified` response header. The server compares the resource-modified timestamp value with the `If-Modified-Since` header value and sends a `304` status if there is a match; otherwise, it sends the response with a new `Last-Modified` header.

10. Rate limit

This is important if you want to prevent the overuse of APIs. The HTTP status code `429 Too Many Requests` is used when the rate limit goes over. Currently, there is no standard to communicate any warning to the client before the rate limit goes over. However, there is a popular way to communicate about it using response headers; these include the following:

- `X-Ratelimit-Limit`: The number of allowed requests in the current period
- `X-Ratelimit-Remaining`: The number of remaining requests in the current period
- `X-Ratelimit-Reset`: The number of seconds left in the current period
- `X-Ratelimit-Used`: The number of requests used in the current period

You can check the headers sent by the GitHub APIs. For example, they could look similar to the following:

- `X-Ratelimit-Limit`: 60
- `X-Ratelimit-Remaining`: 55
- `X-Ratelimit-Reset`: 1601299930
- `X-Ratelimit-Used`: 5

So far, we have discussed various concepts related to REST. Next, we will next move on to discuss our sample app.

An overview of the e-commerce app

The e-commerce app is a simple online shopping application. It provides the following features:

- A user can browse through the products.
- A user can add/remove/update the products in the cart.
- A user can place an order.
- A user can modify the shipping address.
- The application can only support a single currency.

E-commerce is a very popular domain. If we look at the features, we can divide the application into the following subdomains using bounded contexts:

- **Users**: This subdomain is related to users. We'll add the `users` RESTful web service, which provides REST APIs for user management.
- **Carts**: This subdomain is related to the cart. We'll add the `carts` RESTful web service, which provides REST APIs for cart management. Users can perform CRUD operations on cart items.
- **Products**: This subdomain is related to the products catalog. We'll add the `products` RESTful web service, which provides REST APIs to search and retrieve the products.
- **Orders**: This subdomain is related to orders. We'll add the `orders` RESTful web service, which provides REST APIs for users to place orders.
- **Payment**: This subdomain is related to payments. We'll add the `payment` RESTful web service, which provides REST APIs for payment processing.
- **Shipping**: This subdomain is related to shipping. We'll add the `shipping` RESTful web service, which provides REST APIs for order tracking and shipping.

Here's a visual representation of our app's architecture:

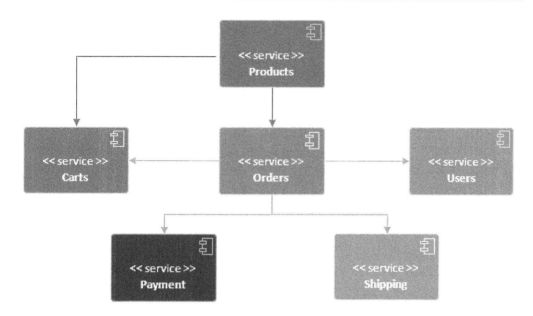

Figure 1.2 – The e-commerce app architecture

We'll implement a RESTful web service for each of the subdomains. We'll keep the implementation simple, and we will focus on learning these concepts throughout this book.

Summary

In this chapter, we learned the basic concepts of the REST architecture style. Now, you know how REST, which is based on HTTP, simplifies and makes integration easier. We also explored the different HTTP concepts that allow you to write REST APIs in a meaningful way. We also learned why HATEOAS is an integral part of REST implementation. Additionally, we learned the best practices for designing REST APIs. We also went through an overview of our e-commerce app. This sample app will be used throughout the book.

In the next chapter, you'll learn about the Spring Framework and its fundamentals.

Questions

1. Why have RESTful web services became so popular and, arguably, the industry standard?

2. What is the difference between RPC and REST?

3. How would you explain HATEOAS?

4. What error codes should be used for server-related issues?

5. Should verbs be used to form REST endpoints, and why?

Further reading

- *Architectural Styles and the Design of Network-based Software Architectures* can be found at `https://www.ics.uci.edu/~fielding/pubs/dissertation/top.htm`.

- The URI Generic Syntax (RFC-3986) can be found at `https://tools.ietf.org/html/rfc3986`.

- The URN Syntax (RFC-2141) can be found at `https://tools.ietf.org/html/rfc2141`.

- HTTP Response Status Codes – RFC 7231 can be found at `https://tools.ietf.org/html/rfc7231`.

- HTTP Response Status Codes – Mozilla Developer Network can be found at `https://developer.mozilla.org/en-US/docs/Web/HTTP/Status`.

- *REST APIs must be hypertext-driven* can be found at `https://roy.gbiv.com/untangled/2008/rest-apis-must-be-hypertext-driven`.

- RFC for the URI template can be found at `https://tools.ietf.org/html/rfc6570`.

- The OWASP API security project can be found at `https://owasp.org/www-project-api-security/` and `https://github.com/OWASP/API-Security`.

2
Spring Concepts and REST APIs

In the previous chapter, we learned about the REST architecture style. Before we go and implement RESTful web services using Spring and Spring Boot, we need to have a proper understanding of the basic Spring concepts. In this chapter, you will learn about the Spring fundamentals and features that are required to implement RESTful web services using the Spring Framework. This will provide the technical perspective required for developing the sample e-commerce app. If you are already aware of the Spring fundamentals required for implementing RESTful APIs, you can move on to the next chapter.

We'll cover the following topics as part of this chapter:

- Introduction to Spring
- Learning the basic concepts of the Spring Framework
- Working with the servlet dispatcher

Technical requirements

This chapter covers concepts and does not cover actual programs or code. However, you'll need basic Java knowledge.

Please visit the following link to download the code files: `https://github.com/PacktPublishing/Modern-API-Development-with-Spring-and-Spring-Boot/tree/main/Chapter02`

Introduction to Spring

Spring is a framework and is written in the Java language. It provides lots of modules, such as Spring Data, Spring Security, Spring Cloud, Spring Web, and so on. It is popular for building enterprise applications. Initially, it was looked at as a Java **Enterprise Edition (EE)** alternative. However, over the years, it has been preferred over Java EE. Spring supports **Dependency Injection (Inversion of Control)** and **Aspect-Oriented Programming** out of the box as its core. Apart from Java, Spring also supports other JVM languages such as Groovy and Kotlin.

With the introduction of Spring Boot, building web services have hit the nail right on the head as far as turnaround time for development is concerned. You hit the ground running immediately. This is huge and one of the reasons why Spring has become so popular lately.

Covering the Spring fundamentals itself requires a dedicated book. I'll try to be concise and cover all the features required for you to go ahead and grasp REST implementation knowledge in a granular way.

Before we proceed, you should go through the principle and design patterns – **Inversion of Control (IoC)**, **Dependency Injection (DI)**, and **Aspect-Oriented Programming (AOP)**.

The Inversion of Control pattern

Traditional CLI programs are typical methods for procedural programming implementation, where the flow is determined by the programmer and code runs sequentially, one piece after another. However, UI-based OS applications determine the flow of programs based on user inputs and events, which is dynamic.

Long ago, when mostly procedural ways of programming were a hit, you would have to look for a way to move the control of flow from a traditional procedural way (the programmer dictates the flow) to external sources such as a framework or components that determined the control flow of the program. This is what is called **IoC**. It is a very generic principle and part of most frameworks.

With the object-oriented programming approach, soon frameworks came up with IoC container pattern implementation that supports dependency injection.

The Dependency Injection pattern

Let's say you are writing a program that needs some data from a database. The program needs a database connection. You could use the JDBC database connection object. You could instantiate and assign the database connection object instantly in the program. Or, you could simply take the connection object as a constructor or a setter/factory method parameter. Then, the framework creates the connection object as per the configuration and assigns that object to your program at runtime. Here, the framework actually injects the connection object at runtime. This is called DI. Spring supports DI for class compositions.

> **Note:**
> The Spring Framework throws an error at runtime if any dependency is not available, or the proper object name is not marked when more than one type of object is available. In contrast to that, there are some frameworks that also check these dependencies at compile time, for example, Dagger.

DI is a type of IoC. IoC containers construct and maintain implementation objects. These types of objects (objects required by other objects – a form of dependency) are injected into objects that need them in a constructor, setter, or interface. This decouples the instantiation and allows dependency injection at runtime. Dependency injection can also be achieved using the Service Locator pattern. However, we'll stick to the IoC pattern approach.

We'll look at it more closely with a code example in the next section.

The Aspect-Oriented Programming paradigm

We have talked about procedural programming (for IoC) and object-oriented programming (for IoC and DI). Then comes AOP, another programming paradigm. AOP works in tandem with OOP. It's a good practice in OOP to handle only a single responsibility in a particular class – this principle is called the **Single Responsibility Principle** (applicable for modules/classes/methods). For example, if you are writing a `Gear` class in an automotive domain application, then the `Gear` class should only allow functions related to the gear object, or it should not be allowed to perform other functions such as braking. However, in programming models, often you need a feature/function that scatters across more than one class. In fact, sometimes, most classes use features such as logging or metrics.

Features such as logging, security, transaction management, and metrics are required across multiple classes/modules. The code of these features is scattered across multiple classes. In OOP, there is no way to abstract and encapsulate such features. This is where AOP helps you. These features (read aspects) are cross-cutting concerns that cut across multiple points in the object model. AOP provides a way to let you handle these aspects across multiple classes/modules.

AOP allows you to do the following:

- To abstract and encapsulate cross-cutting concerns.
- To add aspects behavior around your code.
- To make code modular for cross-cutting concerns to easily maintain and extend it.
- To focus on your business logic inside code. This makes code clean. Cross-cutting concerns are encapsulated and maintained separately.

Without AOP, it is very difficult and complex to achieve all of the preceding points.

Please note that this section helps you to understand IoC, DI, and AOP conceptually. In the next sections, you'll take a deep dive into the code implementation of these patterns and paradigms.

Now, we will go through the fundamentals of the Spring Framework and its basic building blocks.

Learning the basic concepts of the Spring Framework

The Spring Framework's backbone is the IoC container that is responsible for a bean's life cycle. In the Spring world, a Java object can be a **bean** if it is instantiated, assembled, and managed by the IoC container. You create *n*-number of beans, aka objects, for your application. A bean may have dependencies, that is, requiring other objects to work. The IoC container is responsible for injecting the object's dependencies when it creates that bean. In the Spring context, IoC is also known as **DI**.

In the following sections, we'll cover the following core Spring concepts:

- IoC containers
- Defining beans
- Configuring beans using Java

- How to code DI
- Writing code for AOP

> **Note:**
>
> You can refer to the Spring documentation (`https://docs.spring.io/spring-framework/docs/current/spring-framework-reference/`) for more information about the Spring Framework.

Let's get started!

IoC containers

The Spring Framework's IoC container core is defined in two packages: `org.springframework.beans` and `org.springframework.context`. **BeanFactory** (`org.springframework.beans.factory.BeanFactory`) and **ApplicationContext** (`org.springframework.context.ApplicationContext`) are two important interfaces that provide a basis for IoC containers. BeanFactory provides the configuration framework and basic functionality and takes care of bean instantiation and wiring. ApplicationContext can also take care of bean instantiation and wiring. However, it provides more enterprise-specific functionalities, as follows:

- Integrated life cycle management
- Automatic registration of `BeanPostProcessor` and `BeanFactoryPostProcessor`
- *Internationalization* (message resource handling) with easy access to `MessageSource`
- *Events publication* using a built-in `ApplicationEvent`
- Provides `WebApplicationContext`, an application layer specific context for web applications

`ApplicationContext` is a sub-interface of `BeanFactory`. Let's look at its class signature:

```
public interface ApplicationContext extends EnvironmentCapable,
ListableBeanFactory, HierarchicalBeanFactory, MessageSource,
ApplicationEventPublisher, ResourcePatternResolver
```

Here, `ListableBeanFactory` and `HierarchicalBeanFactory` are sub-interfaces of `BeanFactory`.

Spring recommends the use of `ApplicationContext` due to added features apart from state-of-the-art bean management.

ApplicationContext

Now, you know that the `ApplicationContext` interface represents the IoC container and manages the beans, you must be wondering how it gets to know about what beans to instantiate, assemble, and configure. From where does it get its instruction? The answer is configuration metadata. Configuration metadata allows you to express your application objects and interdependencies among those objects. Configuration metadata can be represented in three ways: through XML configuration, Java annotations, and Java Code. You write the business objects and provide the configuration metadata, and the Spring container generates a ready-to-use fully configured system as shown:

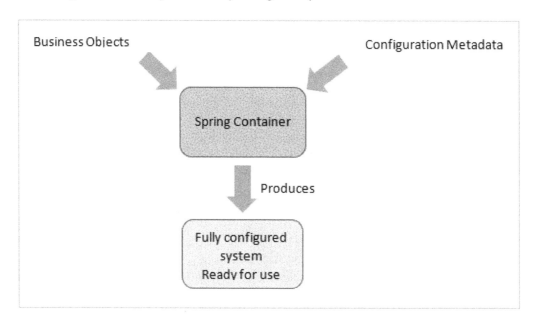

Figure 2.1 – Spring container

Let's see how you can define a Spring bean.

Defining beans

Beans are Java objects that are managed by IoC containers. The developer supplies the configuration metadata to the IoC container, which is then used by the container to construct, assemble, and manage the beans. Beans should have a unique identifier inside a container. A bean may have more than one identity using an **alias**.

You can define beans using XML, Java, and annotations. Let's declare a simple bean using Java-based configuration:

```
public class SampleBean {

    public void init() { // initialization logic }
    public void destroy() { // destruction logic }
    // bean code
}


public interface BeanInterface { // interface code }

public class BeanInterfaceImpl implements BeanInterface {
    // bean code
}


@Configuration
public class AppConfig {

    @Bean(initMethod = "init", destroyMethod = "destroy", name
        = {"sampleBean", "sb"}")
    @Description("Demonstrate a simple bean")
    public SampleBean sampleBean() {
        return new SampleBean();
    }
    @Bean
    public BeanInterface beanInterface() {
        return new BeanInterfaceImpl();
    }
}
```

Here, the bean is declared using the AppConfig class. @Configuration is a class-level annotation that shows that class contains code for configuration. @Bean is a method-level annotation that is used for defining the bean. You can also pass the bean's initialization and destruction life cycle method using the @Bean annotation attributes as shown in the previous code.

In general, a bean's name is a class name with its initial in lowercase. For example, the bean name of `BeanInterface` would be `beanInterface`. However, you can also use the name attribute to define a bean name and its aliases. `SampleBean` has two bean names: `sampleBean` and `sb`.

> **Note:**
> Default methods for destruction are close/shutdown public methods, which are called by a container automatically. However, if you wish to have a different method, that can be done as shown in the sample code. If you don't want a container to call the default destruction method, then you can assign an empty string to the `destroyMethod` attribute (`destroyMethod = ""`).

You can also create a bean using the interfaces shown in the previous code for the `BeanInterface` bean.

Note that the `@Bean` annotation should be inside the `@Component` annotation. The `@Component` annotation is a generic way to declare a bean. A class annotated with `@Configuration` lets the method works to return a bean annotated with `@bean`. `@Configuration` is meta-annotated with `@Component`, therefore the `@Bean` annotation works inside it. There are other annotations such as `@Controller`, `@Service`, and `@Repository`, which are also annotated with `@Component`.

The `@Description` annotation, as the name suggests, is used for describing a bean. When monitoring tools are used, these descriptions help to understand the beans at runtime.

@ComponentScan

The `@ComponentScan` annotation allows the autoscanning of beans. It takes a few arguments, such as base packages and their classes. The Spring container then looks into all the classes inside the base package and looks for beans. It scans all classes annotated with `@Component` or other annotations that are meta-annotated with `@Component`, such as `@Configuration`, `@Controller`, and so on.

By default, Spring Boot takes the default base package from the class, which has the `@ComponentClass` annotation. You can use the `basePackageClasses` attribute to identify which packages should be scanned.

Another way to scan more than two packages is by using the `basePackages` attribute. It allows you to scan more than one package.

If you want to use more than one @ComponentScan, then you can wrap them inside the
@ComponentScans annotation as shown:

```
@Configuration
@ComponentScans({
  @ComponentScan(basePackages = "com.packt.modern.api"),
  @ComponentScan(basePackageClasses = AppConfig.class)
})
class AppConfig { //code }
```

The bean's scope

The Spring container is responsible for creating the bean's instances. How instances will be
created by the Spring container are defined by scope. The default scope is singleton,
that is, only one instance would be created per IoC container and the same instance would
be injected. If you want to create a new instance each time it is requested, you can define
the prototype scope for the bean.

Singleton and prototype scopes are available for all Spring-based applications. There are
four more scopes available for web applications: request, session, application,
and websocket. For later scopes, the application context should be web-aware. Spring
Boot-based web applications are web-aware.

The following table contains all the scopes:

Scope	Usage
singleton	Creates a new instance per IoC container. Default scope.
prototype	Creates a new instance for each injection (for collaborated beans).
request	Only for web-aware context. A single bean instance would be created for each HTTP request and valid throughout the HTTP request life cycle.
session	Only for web-aware context. A single bean instance would be created for each HTTP session and valid throughout the HTTP session life cycle.
application	Only for web-aware context. A single instance would be created for application scope, that is, valid throughout the life cycle of the servlet-context.
websocket	Only for web-aware context. A single instance would be created for each WebSocket session.

Let's see how we can define the `singleton` and `prototype` scopes in code:

```
@Configuration
public class AppConfig {

    // no scope is defined so default singleton scope is
    // applied.
    // If you want to define it explicitly, you can do that
    // using
    // @Scope(value = ConfigurableBeanFactory.SCOPE_SINGLETON)
    // OR
    // @Scope(ConfigurableBeanFactory.SCOPE_SINGLETON)
    // Here, ConfigurableBeanFactory.SCOPE_SINGLETON is string
    // constant, which
    // value is "singleton". You can use the string also,
    // better to avoid it.
    @Bean
    public SingletonBean singletonBean() {
        return new SingletonBean();
    }


    @Bean
    @Scope(ConfigurableBeanFactory.SCOPE_PROTOTYPE)
    public PrototypeBean prototypeBean() {
        return new PrototypeBean();
    }


    @Bean
    @Scope(value = WebApplicationContext.SCOPE_REQUEST,
            proxyMode = ScopedProxyMode.TARGET_CLASS)
    // You need a proxyMode attribute because when web-aware
    // context is
    // instantiated, you don't have any HTTP request.
    // Therefore,
    // Spring injects the proxy as a dependency and
    // instantiate the bean when HTTP request is invoked.
    // OR, in short you can write below which is a shortcut for
```

```
    // above
    @RequestScope
    public ReqScopedBean requestScopedBean() {
        return new ReqScopedBean();
    }
}
```

Similarly, you can create a web-aware context-related bean as follows:

```
@Configuration
public class AppConfig {

    @Bean
    @Scope(value = WebApplicationContext.SCOPE_REQUEST,
            proxyMode = ScopedProxyMode.TARGET_CLASS)
    // You need a proxyMode attribute because when web-aware
    // context is
    // instantiated, you don't have any HTTP request.
    // Therefore,
    // Spring injects the proxy as a dependency and
    // instantiate the bean when HTTP request is invoked.
    // OR, in short you can write below which is a shortcut for
    // above
    @SessionScope
    public ReqScopedBean requestScopedBean() {
        return new ReqScopedBean();
    }


    @ApplicationScope
    public ReqScopedBean requestScopedBean() {
        return new ReqScopedBean();
    }


    // here "scopeName" is alias for value
    // interestingly, no shortcut. Also hard coded value for
    // websocket
    @Scope(scopeName = "websocket",
```

```
            proxyMode = ScopedProxyMode.TARGET_CLASS)
    public ReqScopedBean requestScopedBean() {
        return new ReqScopedBean();
    }

}
```

Bean is covered here succinctly. However, you can explore more about it in the official Spring documentation (https://docs.spring.io/spring-framework/docs/current/spring-framework-reference/).

Configuring beans using Java

Before Spring 3, you could only define beans using Java. Spring 3 introduced the @Configuration, @Bean, @import, and @DependsOn annotations to configure and define Spring beans using Java.

You have already learned about the @Configuration and @Bean annotation in the *Defining beans* section. Now, you will explore how to use the @import and @DependsOn annotations.

The @Import annotation is more useful when you develop an application without using autoconfiguration.

The @Import annotation

This is used for modularizing configurations when you have more than one configuration class. You can import the bean's definitions from other configuration classes. It is useful when you instantiate the context manually. Spring Boot uses autoconfiguration, therefore you don't need to use @Import. However, if you want to instantiate the context manually, then you would have to use @Import to modularize the configurations.

Let's say the configuration class FooConfig contains FooBean and the configuration class BarConfig contains BarBean. The BarConfig class also imports FooConfig using @Import:

```
@Configuration
public class FooConfig {

    @Bean
    public FooBean fooBean() {
        return new FooBean();
```

```
        }
}


@Configuration
@Import(FooConfig.class)
public class BarConfig {

    @Bean
    public BarBean barBean() {
        return new BarBean();
    }
}
```

Now, while instantiating the container (context), you can just supply `BarConfig` to load both `FooBean` and `BarBean` definitions in the Spring container as shown:

```
public static void main(String[] args) {
    ApplicationContext appContext = new
        AnnotationConfigApplicationContext(BarConfig.class);

    // now both FooBean and BarBean beans will be available...
    FooBean fooBean = appContext.getBean(FooBean.class);
    BarBean barBean = appContext.getBean(BarBean.class);
}
```

The @DependsOn annotation

The Spring container manages the bean initialization order. What if you have a bean that depends on another bean? You want to make sure that the dependent bean is initialized before the bean that needs it. `@DependsOn` helps you to achieve this when you configure beans using Java (not through XML).

You get the exception `NoSuchBeanDefinitionException` if a bean's initialization order is messed up and because of that the Spring container does not find the dependency.

Let's assume we have a bean named called `BazBean` that depends on the beans `FooBean` and `BarBean`. You can make use of the `@DependsOn` annotation to maintain the initializing order. The Spring container will follow the instructions and initialize both the `FooBean` and `BarBean` beans before creating `BazBean`. Here's how the code will look:

```
@Configuration
public class AppConfig {

    @Bean
    public FooBean fooBean() {
        return new FooBean();
    }

    @Bean
    public BarBean barBean () {
        return new BarBean ();
    }

    @Bean
    @DependsOn({"fooBean","barBean"})
    public BazBean bazBean (){
        return new BazBean ();
    }

}
```

How to code DI

Have a look at the following example. `CartService` has a dependency on `CartRepository`. `CartRepository` instantiation has been done inside the `CartService` constructor:

```
public class CartService {

    private CartRepository repository;

    public CartService() {
```

```
            this.repository = new CartRepositoryImpl();
    }
}
```

We can decouple this dependency in the following way:

```
public class CartService {

    private CartRepository repository;

    public CartService(CartRepository repository) {
        this.repository = repository;
    }
}
```

If you create a bean of the `CartRepository` implementation, you can easily inject the `CartRepository` bean using configuration metadata. Before that, let's have a look at the Spring container again.

You have seen how `ApplicationContext` can be initialized in *The @Import annotation* subsection of this chapter. When it gets created, it takes all the metadata from the bean's configuration. `@Import` allows you to have multiple configurations.

Each bean can have its dependencies, that is, a bean may need other objects to work (compositions) as in the `CartService` example. These dependencies can be defined using constructors, setter methods, or properties. These dependent objects (part of constructors, setter method arguments, or class properties) are injected by the Spring container (`ApplicationContext`) using the bean's definition and its scope. We'll look into each of these ways to define the DI.

> **Note**
>
> **DI** makes a class independent of its dependencies. Therefore, you might not have to change a class because of a change in dependency or it would become rare. That means you can change the dependency without changing the class. It also simplifies unit testing by mocking or spying on dependencies.

Using a constructor to define a dependency

Now, you'll see how you can inject `CartRepository` into the `CartService` constructor. A way to inject a dependency using a constructor is as follows:

```
@Configuration
public class AppConfig {

    @Bean
    public CartRepository cartRepository() {
        return new CartRepositoryImpl();
    }

    @Bean
    public CartService cartService() {
        return new CartService(cartRepository());
    }
}
```

Using a setter method to define a dependency

Now, let's change the `CartService` class. Instead of having a constructor, use the setter method to instantiate the dependency:

```
public class CartService {

    private CartRepository repository;

    public void setCartRepository(CartRepository repository) {
        this.repository = repository;
    }
}
```

Now, you can use the following configuration to inject the dependency:

```
@Configuration
public class AppConfig {

    @Bean
    public CartRepository cartRepository() {
```

```
        return new CartRepositoryImpl();
    }


    @Bean
    public CartService cartService() {
        CartService service = new CartService();
        Service.setCartService(cartRepository());
        Return service;

    }

}
```

> **Note**
>
> Spring recommends using constructor-based dependency injection over setter method or class property-based dependency injection. However, for an opt-in dependency, you should use setter method-based dependency injection cautiously.

Using a class property to define a dependency

Spring also provides an out-of-the-box solution for injecting a dependency using the @Autowired annotation. It makes code look cleaner. Have a look at the following example:

```
@Service
public class CartService {


    @Autowired
    private CartRepository repository;

}
```

The Spring container will take care of injecting the CartRepository bean. You'll learn more about @Autowired in the next section.

Configuring a bean's metadata using annotations

The Spring Framework provides lots of annotations to configure the metadata for beans. However, we'll focus on the most commonly used annotations: @Autowired, @Qualifier, @Inject, @Resource, @Primary, and @Value.

How to use @Autowired

The @Autowired annotation allows you to define the configuration part in a bean's class itself instead of writing a separate configuration class annotated with @Configuration. The @Autowired annotation can be applied to a field (as we saw in the class property-based dependency injection example), constructor, setter, or any method.

The Spring container makes use of reflections to inject the beans annotated with @Autowired. This also makes it more costly than other injection approaches.

Please make a note that applying @Autowired to class members will only work if there is no constructor or setter method to inject the dependent bean.

Here is a code example of the @Autowired way of injecting dependencies:

```
@Component
public class CartService {
    private CartRepository repository;
    private ARepository aRepository;
    private BRepository bRepository;
    private CRepository cRepository;


    @Autowired // member(field) based auto wiring
    private AnyBean anyBean;


    @Autowired // constructor based autowired
    public CartService(CartRepository cartRepository) {
        this.repository = repository;
    }


    @Autowired // Setter based auto wiring
    public void setARepository(ARepository aRepository) {
        this.aRepository = aRepository;
    }


    @Autowired // method based auto wiring
    public void xMethod(BRepository bRepository, CRepository
cRepository)
    {
        this.bRepository = bRepository;
```

```
            this.cRepository = cRepository;
    }
}
```

`@Autowired` works based on reflection. However, to remove the ambiguity, matching beans are found and injected using type matching, qualifier matching, or name matching in the same order of precedence. These are applicable to both field and setter method injections.

Match by type

The following example works because match by type takes precedence. It finds the `CartService` bean and injects it into `CartController`:

```
@Configuration
public class AppConfig {
    @Bean
    public CartRepository cartRepository() {
        return new CartRepositoryImpl();
    }    @Bean
    public CartService cartService() {
        CartService service = new CartService();
        Service.setCartService(cartRepository());
        Return service;
    }
}


@Controller
public class CartController {    @Autowired
    private CartService service;
}
```

Match by qualifier

Let's assume there is more than one bean of a given type. Then, the Spring container won't be able to determine the correct bean by type:

```
@Configuration
public class AppConfig {    @Bean
    public CartService cartService1() {
```

```
            return new CartServiceImpl1();
     }


     @Bean
     public CartService cartService2() {
            return new CartServiceImpl2();
     }
}


@Controller
public class CartController {      @Autowired
     private CartService service1;


     @Autowired
     private CartService service2;
}
```

This example, when run, will return `NoUniqueBeanDefinitionException`. To sort this out, we can make use of the `@Qualifier` annotation.

If you look closely, you'll find that the configuration class has two beans identified by their method names: `cartService1` and `cartService2`. Or, you can also make use of the value attribute of the `@Bean` annotation to give a name/alias. Now, you can use these names to assign these two different beans the same type using the `@Qualifier` annotation as shown:

```
@Controller
public class CartController {      @Autowired
     @Qualifier("cartService1")
     private CartService service1;


     @Autowired
     @Qualifier("cartService2")
     private CartService service2;
}
```

Match by name

Let's define a service using the @Service annotation, which is a type of @Component. Let's assume we have a component scan in place:

```
@Service(value="cartServc")
public class CartService {
    // code
}
```

```
@Controller
public class CartController {    @Autowired
    private CartService cartServc;
}
```

This code works because the field name of CartController for CartService is the same as was given to the value attribute of @Service annotation. If you change the field name from cartServc to something else, it will fail with NoUniqueBeanDefinitionException.

There are other annotations: @Inject (JSR-330 - https://jcp.org/en/jsr/detail?id=330) and @Resource (JSR-250- https://jcp.org/en/jsr/detail?id=250). @Inject also requires the javax.inject library. @Resource and @Inject are similar to @Autowired and can be used for injecting dependencies. Both @Autowired and @Inject have the same execution path precedence (by type, by qualifier, and by name in the same order). However, the @Resource execution path preference is by name (first preference), by type, and by qualifier (last preference).

What is the purpose of @Primary?

In the previous subsection, we saw that @Qualifier helps you to resolve which type should be used when multiple beans are available for injection. The @Primary annotation allows you to set one of the type's beans as the default. Bean annotation with @Primary will be injected into autowired fields:

```
@Configuration
public class AppConfig {    @Bean
    @Primary
    public CartService cartService1() {
        return new CartServiceImpl1();
```

```
    }

    @Bean
    public CartService cartService2() {
        return new CartServiceImpl2();
    }
}

@Controller
public class CartController {    @Autowired
    private CartService service;
}
```

In this example, the bean marked with @Primary will be used to inject the dependency into the CartController class for CartService.

When we can use @Value?

Spring supports the use of external property files: <xyz>.properties or <xyz>.yml. Now, you want to use the value of any property in your code. You can achieve this using the @Value annotation. Let's have a look at the sample code:

```
@Configuration
@PropertySource("classpath:application.properties")
public class AppConfig {}

@Controller
public class CartController {    @Value("${default.currency}")
    String defaultCurrency;
}
```

The defaultCurrency field would take its value from the default.currency field defined in the application.properties file. If you are using Spring Boot, you don't need to use @PropertySource. You just need to place application.yml or properties file under the src/main/resources directory.

Writing code for AOP

We discussed AOP previously, in the *Introduction to Spring* section. In simple terms, it is a programming paradigm that solves cross-cutting concerns such as logging, transactions, security, and so on. These cross-cutting concerns are known as aspects in AOP. It allows you to modularize your code and place cross-cutting concerns in a central place.

The following code captures the time taken by the method to execute:

```
class Test
    public void performSomeTask() {
        long start = System.currentTimeMillis();
        // Business Logic
        long executionTime = System.currentTimeMillis() -
                            start;
        System.out.println("Time taken: " + executionTime + "
                            ms");
    }
}
```

This code captures the time taken by itself. If you have hundreds of methods in your application, then you need to add a time-capturing piece of code in each one of them to monitor time. Moreover, what if you want to modify the code? It means you have to modify the code in all those places. You don't want to do that. This is where AOP helps you. It makes your cross-cutting code modular.

Let's create an AOP example for capturing the time a method takes to execute. Here, logging monitoring time will be our aspect that will capture the time taken by a method for its execution.

As the first step, you'll define an annotation (`TimeMonitor`) to target the method. Methods annotated with `@TimeMonitor` will log the time taken by that method. This will help us to identify the pointcut. The pointcut is defined in the `Aspect` class's code explanation, highlighted with bold text:

```
@Target(ElementType.METHOD)
@Retention(RetentionPolicy.RUNTIME)
public @interface TimeMonitor {}
```

Next, we need to define the `Aspect`. Aspects insert additional logic during the execution of the program at a certain point. This point is known as the **Join point**. The **Join point** could be a field being modified, a method being called, or an exception being thrown.

The @Aspect annotation is used to mark the class as an Aspect. Aspect-monitoring time has been defined using the following code:

```
@Aspect
@Component
public class TimeMonitorAspect {
    @Around("@annotation(com.packt.modern.api.TimeMonitor)")
    public Object logTime(ProceedingJoinPoint joinPoint) throws
            Throwable {
        long start = System.currentTimeMillis();
        Object proceed = joinPoint.proceed();
        long executionTime = System.currentTimeMillis() -
                        start;
        System.out.println(joinPoint.getSignature() + "
                        takes:   " + executionTime + " ms");
        return proceed;
    }
}
```

@Around is a method annotation that defines the Advice. The **Advice** is an action taken by the Aspect at a specific time (**Joinpoint**). This Advice could be any of the following:

- @Before: Advice executes before the JoinPoint.
- After: Advice executed after the JoinPoint. It has three subtypes:

  a. @After: Advice executes after the JoinPoint irrespective of the method's outcome – successful or failed.

  b. @AfterReturning: Advice executes after the JoinPoint executes successfully.

  c. @AfterThrowing: Advice executes after the JoinPoint throws an exception.
- @Around: Advice executes before and after the JoinPoint.

TimeMonitorAspect executes at the method level because the MonitorTime Advice target is a method.

@Around also takes an expression argument, @annotation(com.packt.modern. api.TimeMonitor). This predicate expression is known as the **Pointcut** that determines whether the Advice needs to be executed or not. The logTime method will be executed for all the methods, which are annotated with @TimeMonitor. Spring supports the AspectJ expression syntax. These expressions are dynamic in nature, allowing flexibility while defining the Pointcut.

JoinPoint is added as a method logTime() parameter. With the JoinPoint object, you can capture all the information of the target and proxy. You can capture method's full signature, class name, method name, arguments, and so on using the JoinPoint object.

That's all we need to implement TimeMonitorAspect. Now you can simply add the @ TimeMonitor annotation to log the computed time taken by the methods as shown:

```
class Test

    @TimeMonitor
    public void performSomeTask() {
        // Business Logic
    }
}
```

JoinPoint also allows you to capture the **target object** and **proxy**. You must be wondering what these are. These are created by a Spring AOP module and are important for AOP to work. An Advice is applied to the target object. Spring AOP creates a subclass of the target object and overrides the methods and advice is inserted. On the other hand, the proxy is an object that is created after an advice is applied to the target object using the CGLIB or JDK proxy lib.

Why use Spring Boot?

Nowadays, Spring Boot is the obvious choice for developing state-of-the-art, production-ready web applications specific to Spring. Its website (https://projects.spring. io/spring-boot/) also states its real advantages.

Spring Boot is an amazing Spring tool created by **Pivotal** that was released in April 2014 (GA). It was developed based on the request of SPR-9888 (https://jira. spring.io/browse/SPR-9888) with the title *Improved support for 'containerless' web application architectures.*

You must be wondering: Why containerless? Because today's cloud environment, or PaaS, provides most of the features offered by container-based web architectures, such as reliability, management, or scaling. Therefore, Spring Boot focuses on making itself an ultralight container.

Spring Boot has default configurations and supports auto-configuration to make production-ready web applications simple. **Spring Initializr** (http://start.spring. io) is a web page, where you can opt in for build tools, such as Maven or Gradle, along with project metadata, such as a group, artifacts, and dependencies. Once you fill in the required fields, you can just click on the **Generate Project** button, which will give you a Spring Boot project that you can use for your production application.

On this page, the default **Packaging** option is **Jar**, which we are going to use throughout this book. You can use WAR packaging if you want to deploy the application on any web server, such as WebLogic or Tomcat.

In simple words, it does all the configuration part for us; we can just focus on writing our biz logic and APIs.

Purpose of servlet dispatcher

In the previous chapter, you learned that RESTful web services are developed on top of the HTTP protocol. Java has a Servlets feature to work with HTTP protocol. Servlets allow you to have path mapping that can work at REST endpoints and provides the HTTP method for identification. It also allows you to form different types of response objects, including JSON and XML. However, it is a crude way of implementing REST endpoints. You have to handle the request URI, parse the parameters, convert JSON/XML, and handle the responses.

Spring MVC comes to your rescue. Spring MVC is based on the **Model-View-Controller (MVC)** pattern and has been part of the Spring Framework since its first release. MVC is a well-known design pattern:

- **Model**: Models are Java objects (POJOs) that contain the application data. They also represent the state of the application.

- **View**: The view is a presentation layer that consists of HTML/JSP/template files. The view renders the data from models and generates the HTML output.

- **Controller**: The controller processes the user requests and builds the model.

`DispatcherServlet` is part of the Spring MVC. It works as a front controller, that is, it handles all the incoming HTTP requests. Spring MVC is a web framework that allows you to develop traditional web applications where UI apps are also part of the backend. However, you'll develop RESTful web services and the UI will be based on the React JavaScript library, therefore we'll keep the Servlet Dispatcher role limited to implementing the REST endpoints using `@RestController`.

Let's have a look at the flow of a user request in Spring MVC for the REST controller:

1. The user sends the HTTP request, which is received by `DispatcherServlet`.

2. `DispatcherServlet` passes the baton to `HandlerMapping`. `HandlerMapping` does the job of finding the correct controller for the requested URI and passes it back to `DispatcherServlet`.

3. DispatcherServlet then makes use of HandlerAdaptor to handle Controller.

4. HandlerAdaptor calls the appropriate method inside Controller.

5. Controller then executes the associated business logic and forms the response.

6. Spring makes use of the marshaling/unmarshalling of request and response objects for JSON/XML conversion from Java and vice versa.

Let's see a visual representation of this process:

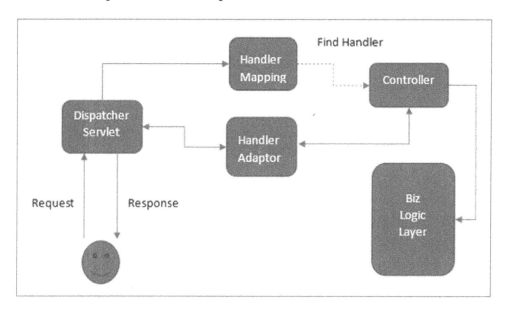

Figure 2.2 – DispatcherServlet

Now, you will have understood the importance of DispatcherServlet, which is key for REST API implementation.

Summary

This chapter helped you learn about Spring's key concepts: beans, dependency injection, and AOP. You also learned how to define the scope of beans, and create ApplicationContext programmatically and use it to get the beans. Now, you can define beans' configuration metadata using Java and annotations and have learned how to use different beans of the same type.

You also implemented a sample `Aspect` – a cross-cutting concern using a modular approach, and learned the key concepts of the AOP programming paradigm.

Since we are going to implement REST APIs in this book, it is important to understand the Servlet Dispatcher concept.

In the next chapter, we'll implement our first REST API application using the OpenAPI Specification and use the Spring controller to implement it.

Questions

1. How do you define a bean with the prototype scope?

2. What is the difference between prototype and singleton beans?

3. What is required for a session and request scope to work?

4. What is the relationship between `Advice` and `Pointcut` in terms of AOP?

5. Write an `Aspect` for logging that prints the method name and argument names before the method execution and prints the message with the return type, if any, after the method's successful execution.

Further reading

- *Inversion of Control Containers and the Dependency Injection pattern* (https://martinfowler.com/articles/injection.html)

- The Spring Framework documentation (5.2.9 was the latest at the time of writing this book) (https://docs.spring.io/spring-framework/docs/current/spring-framework-reference/)

- *Spring Boot 2 Fundamentals* (https://www.packtpub.com/product/spring-boot-2-fundamentals/9781838821975)

- *Developing Java Applications with Spring and Spring Boot* (https://www.packtpub.com/in/application-development/developing-java-applications-spring-and-spring-boot)

3
API Specifications and Implementation

In previous chapters, we learned about the design aspects of REST APIs and the Spring fundamentals required for developing RESTful web services. In this chapter, you'll make use of these two areas to implement REST APIs. We have chosen a design-first approach for implementation. You will make use of the **OpenAPI Specification** (**OAS**) for first designing an API and later implementing it. You will also learn how to handle errors that occur while serving the request. Here, an API of a sample e-commerce app will be designed and implemented for reference.

We'll cover the following topics as part of this chapter:

- Designing APIs with OAS
- Converting OAS to Spring code
- Implementing the OAS code interfaces
- Adding the Global Exception Handler

Technical requirements

You need the following to execute the instructions in this chapter:

- Any Java IDE, such as NetBeans, IntelliJ, or Eclipse

- **Java Development Kit (JDK)** 14

- An internet connection to download the dependencies and Gradle

You can find the code files for this chapter on GitHub at `https://github.com/PacktPublishing/Modern-API-Development-with-Spring-and-Spring-Boot/tree/main/Chapter03`.

Designing APIs with OAS

You could directly start coding the API; however, this approach leads to many issues, such as frequent modifications, difficulty in API management, and difficulty in reviews specifically lead by non-technical domain teams. Therefore, you should use the **design-first** approach.

The first question that comes to mind is, how can we design REST APIs? You learned in *Chapter 1*, *RESTful Web Service Fundamentals*, that there is no existing standard to govern the REST API implementation. OAS was introduced to solve at least the aspects of the REST API's specification and description. It allows you to write REST APIs in the **YAML Ain't Markup Language** (**YAML**) or **JavaScript Object Notation** (**JSON**) markup languages.

We'll use version 3.0 of OAS (`https://github.com/OAI/OpenAPI-Specification/blob/master/versions/3.0.3.md`) for implementing the e-commerce app REST API. We'll use YAML (pronounce as *yamel*, rhymes with *camel*), which is cleaner and easier to read. YAML is space-sensitive. It uses space for indentation; for example, it represents the `key: value` pair (pay attention to the space after colon `:`). You can read more about YAML at `https://yaml.org/spec/`.

OAS was earlier known as the Swagger Specification. However, OAS-supporting tools are still known as **Swagger tools**. Swagger tools are open source projects that help the overall development life cycle of REST APIs. We'll make use of the following Swagger tools in this chapter:

- **Swagger Editor** (`https://editor.swagger.io/`) for designing and describing the e-commerce app REST APIs. It allows you to write and preview, at the same time, your REST APIs' design and description. Please make sure that you use OAS 3.0. At the time of writing this book, the default is OpenAPI version 2.0. You can change that from **Edit | Convert to OpenAPI 3**. An embedded message will appear with the **Cancel** and **Convert** options. Click on the **Convert** button to convert OAS from version 2.0 to 3.0.

- **Swagger Codegen** (`https://github.com/swagger-api/swagger-codegen`) for generating the Spring-based API interface. You'll use the Gradle plugin (`https://github.com/int128/gradle-swagger-generator-plugin`) for generating code that works on top of Swagger Codegen. There is also an OpenAPI tool Gradle plugin – **OpenAPI Generator** (`https://github.com/OpenAPITools/openapi-generator/tree/master/modules/openapi-generator-gradle-plugin`). However, we'll prefer the former one because of the open issues count, which is 1.7k (multiple for Java/Spring as well) at the time of writing.

- **Swagger UI** (`https://swagger.io/swagger-ui/`) for generating the REST API documentation. The same Gradle plugin will be used to generate the API documentation.

Next, let's discuss an OAS overview.

Understanding the basic structure of OAS

The OpenAPI definition structure can be divided into the following sections (all are keyword- and case-sensitive):

- `openapi` (version)
- `info`
- `externalDocs`
- `servers`
- `tags`
- `paths`
- `components`

These all are part of root. The first three sections (`openapi`, `info`, and `externalDocs`) are used for defining the metadata of the API.

You can have an API's definition either in a single file or divided into multiple files. OAS supports both. We'll use a single file for defining the sample e-commerce API.

Instead of discussing all these sections theoretically and then writing the e-commerce API definitions, we'll discuss both together. First, we'll cover each section definition of the e-commerce API, and then we'll discuss why we have used it and what it implies.

The metadata sections of OAS

Let's have a look at the metadata sections of the e-commerce API definitions:

```yaml
openapi: 3.0.3
info:
  title: Sample Ecommerce App
  description: >
    'This is a ***sample ecommerce app API***.  You can find
    out more about Swagger at [swagger.io](http://swagger.io).
    Description supports markdown markup. For example, you can
    use the `inline code` using back ticks.'
  termsOfService: https://github.com/PacktPublishing/Modern-
                  API-Development-with-Spring-and-Spring-
                  Boot/blob/master/LICENSE
  contact:
    email: support@packtpub.com
  license:
    name: MIT
    url: https://github.com/PacktPublishing/Modern-API-
         Development-with-Spring-and-Spring-Boot/blob/master/
         LICENSE
  version: 1.0.0
externalDocs:
  description: Document link you want to generate along with
               API.
  url: http://swagger.io
```

https://github.com/PacktPublishing/Modern-API-Development-with-Spring-and-Spring-Boot/tree/main/Chapter03/src/main/resources/api/openapi.yaml

Now, we have written the metadata definitions of our API. Let's discuss each in detail.

openapi

The `openapi` section tells us which OAS is being used for writing the API's definition. OpenAPI uses semantic versioning (`https://semver.org/`), which means the version will be in `major:minor:patch` form. If you look at the `openapi` metadata value, we are using `3.0.3`. This reveals we are using major version 3 with patch 3 (the minor version is 0).

info

The `info` section contains the metadata about the API. This information is used for generating the documentation and may be used by the client. It contains the following fields, out of which only `title` and `version` are mandatory fields, while the others are optional fields:

- `title`: Title of the API.
- `description`: This is used for describing the API details. As you can see, we can use Markdown (`https://spec.commonmark.org/`) here. An > (angular bracket) symbol is used for adding multi-line values.
- `termsOfService`: A URL that links to the terms of services. Make sure it follows the proper URL format.
- `contact`: Contact information of the API provider. The `email` attribute should be the email address of the contact person/organization. Other attributes that we have not used are `name` and `URL`. The `name` attribute represents the name of the contact person or organization. The `URL` attribute provides the link to the contact page. This is an optional field and all attributes are also optional.
- `license`: License information. The `name` attribute is a required field that represents the correct license name, such as MIT. `url` is optional and provides a link to the license document.
- `version`: Exposes the API version in string format.

externalDocs

`externalDocs` is an optional field that points to extended documentation of the exposed API. It has two attributes: `description` and `url`. `description` is an *optional* field that defines a summary of the external documentation. You can use the Markdown syntax for the description. The `url` attribute is *mandatory* and links to external documentation.

Let's continue building our API definition. We are done with the metadata section. Let's discuss the `servers` and `tags` sections.

The servers and tags sections of OAS

After the metadata section, we can now describe the `servers` and `tags` sections. Let's have a look at the following code:

```
servers:
  - url: https://ecommerce.swagger.io/v2
tags:
  - name: cart
    description: Everything about cart
    externalDocs:
      description: Find out more (extra document link)
      url: http://swagger.io
  - name: order
    description: Operation about orders
  - name: user
    description: Operations about users
```

https://github.com/PacktPublishing/Modern-API-Development-with-Spring-and-Spring-Boot/tree/main/Chapter03/src/main/resources/api/openapi.yaml

servers

`servers` is an optional section that contains a list of servers that host the API. If the hosted API document is interactive, then it can be used by Swagger UI to directly call the API and show the response. If it is not provided, then it points to the root (/) of the hosted document server. Server URLs are shown using the `url` attribute.

tags

The `tags` section, defined at the root level, contains the collection of tags and their metadata. Tags are used for grouping the operations performed on the resources. The `tags` metadata contains `name`, which is a *mandatory* field, and two additional *optional* attributes: `description` and `externalDocs`.

The `name` attribute contains the tag name. We have already discussed the `description` and `externalDocs` fields in the previous section on metadata.

Let's discuss the last two sections of OAS.

The components section of OAS

If we were going through it sequentially, we would have discussed `path` first. However, conceptually, we would like to write our models first before we use them in the `path` section. Therefore, we'll discuss the `components` section first.

Here is a code snippet from the `components` section of the sample e-commerce app:

```yaml
components:
  schemas:
    Cart:
      description: Shopping Cart of the user
      type: object
      properties:
        customerId:
          description: Id of the customer who possesses the
cart
          type: string
        items:
          description: Collection of items in cart.
          type: array
          items:
            $ref: '#/components/schemas/Item'
```

https://github.com/PacktPublishing/Modern-API-Development-with-Spring-and-Spring-Boot/tree/main/Chapter03/src/main/resources/api/openapi.yaml

If you are working with YAML for the first time, you may find it a bit complex. However, once you go through this section, you'll feel more comfortable with YAML.

Here, we have defined a model called `Cart`. The `Cart` model is of the `object` type and contains two fields, namely `Id` (a string) and `items` (an array).

The object data type

You can define any model or field as an object. Once you mark a type as an object, the next attribute is `properties`, which consists of all the object's fields. For example, the `Cart` model in the previous code will have the following syntax:

```
type: object
properties:
    <field name>:
        type: <data type>
```

OAS supports six basic data types, which are as follows (all are in lowercase):

- `string`
- `number`
- `integer`
- `boolean`
- `object`
- `array`

Let's discuss the `Cart` model, in which we have used the `string`, `object`, and `array` data types. Other data types are `number`, `integer`, and `boolean`. Now, you must be wondering how to define the date, time, and float types, and so on. You can do that with the `format` attribute, which you can use along with the `object` type. For example, have a look at the following code:

```
orderDate:
    type: string
    format: date-time
```

In the previous code, `orderDate` is defined with `string`, but `format` determines what string value it will contain. Since `format` is marked with `date-time`, the `orderDate` field would contain the date and time in the format defined in *RFC 3339, section 5.6* (`https://tools.ietf.org/html/rfc3339#section-5.6`), for example, `2020-10-22T19:31:58Z`.

There are some other common formats you can use along with types, as follows:

- `type: number` with `format: float`: This would contain the floating-point number.

- `type: number` with `format: double`: This would contain the floating-point number with double precision.

- `type: integer` with `format: int32`: This would contain the `int` type (signed 32-bit integer).

- `type: integer` with `format: int64`: This would contain the `long` type (signed 64-bit integer).

- `type: string` with `format: date`: This would contain the date as per *RFC 3339*, for example, `2020-10-22`.

- `type: string` with `format: byte`: This would contain the Base64-encoded values.

- `type: string` with `format: binary`: This would contain the binary data (can be used for files).

Our `Cart` model's `items` field is an array of the user-defined `Item` type. Here, `Item` is another model and referenced using `$ref`. **In fact, all user-defined types are reference using $ref**. The `Item` model is also part of the `components/schema` section. Therefore, the value of `$ref` contains an anchor to user-defined types with `#/component/schemas/{type}`.

`$ref` represents the reference object. It is based on JSON reference (`https://tools.ietf.org/html/draft-pbryan-zyp-json-ref-03`) and follows the same semantics in YAML. It can refer to an object in the same document or external documents. Therefore, it is used when you have API definitions divided into multiple files. You have already seen one way of its usage in the previous code. Let's see one more example:

```
# Relative Schema Document
$ref: Cart.yaml


# Relative Document with embedded Schema
$ref: definitions.yaml#/Cart
```

There is another caveat to the previous code. If you look closely, you will find two *items* – one is a property of the `Cart` **object type** and another one is an attribute of the **array type**. The former one is simple, a field of the `Cart` object. However, the latter belongs to `array` and is a part of the array syntax.

Array syntax

```
type: array
items:
    type: <type of object>
```

i. You could have a nested array, if you place `type of object` as `array`.

ii. You can also refer to the user-defined type using `$ref` as shown in the code (Then, the `type` attribute is not required for `items`.)

Let's see what the `Item` model looks like:

```
Item:
  description: Items in shopping cart
  type: object
  properties:
    id:
      description: Item Identifier
      type: string
    quantity:
      description: The item quantity
      type: integer
      format: int32
    unitPrice:
      description: The item's price per unit
      type: number
      format: double
```

https://github.com/PacktPublishing/Modern-API-Development-with-Spring-and-Spring-Boot/tree/main/Chapter03/src/main/resources/api/openapi.yaml

The `Item` model is also part of the `components/schema` section. We have defined several models used by the e-commerce app API. You can find them in the GitHub code at `https://github.com/PacktPublishing/Modern-API-Development-with-Spring-and-Spring-Boot/tree/main/Chapter03/src/main/resources/api/openapi.yaml`.

Now, you have learned how we can define models under the `components/schema` section. We'll next discuss how to define an API's endpoints in the `path` section of OAS.

> **Important note**
>
> Similar to **schemas**, you can also define **requestBodies** (request payload) and **responses** in the `components` section. This is useful when you have common request bodies and responses.

The path section of OAS

`path` is the last section of OAS (sequence-wise, it is second-to-last, but we have already discussed `components` in the previous subsection), where we define the endpoints. This is the place where we form the URI and attach the HTTP methods.

Let's write the definition for `POST /api/v1/carts/{customerId}/items`. This API adds an item to the cart associated with a given customer identifier:

```
paths:
  /api/v1/carts/{customerId}/items:
    post:
      tags:
        - cart
      summary: Adds an item in shopping cart
      description: Adds an item to the shopping cart
      operationId: addCartItemsByCustomerId
      parameters:
        - name: customerId
          in: path
          description: Customer Identifier
          required: true
          schema:
            type: string
      requestBody:
```

```
        description: Item object
        content:
          application/xml:
            schema:
              $ref: '#/components/schemas/Item'
          application/json:
            schema:
              $ref: '#/components/schemas/Item'
      responses:
        201:
          description: Item added successfully
          content:
            application/json:
              schema:
                type: array
                items:
                  $ref: '#/components/schemas/Item'
        404:
          description: Given customer ID doesn't exist
          content: {}
```

https://github.com/PacktPublishing/Modern-API-Development-with-Spring-and-Spring-Boot/tree/main/Chapter03/src/main/resources/api/openapi.yaml

If you just go through it, you know what the endpoint is, what HTTP method and parameter this API is using, and most importantly, what response you can expect. Let's discuss it in more detail. Here, `v1` represents the version of the API. Each endpoint path (such as `/api/v1/carts/{customerId}/items`) has an HTTP method (such as `post`) associated with it. The endpoint path always starts with `/`.

Each method can then have seven fields: `tags`, `summary`, `description`, `operationId`, `parameters`, `responses`, and `requestBody`. We will talk about each in the following subsections.

tags

Tags are used for grouping APIs, as shown in the following screenshot for APIs tagged with `cart`. Tags can also be used by **Swagger Codegen** to generate the code:

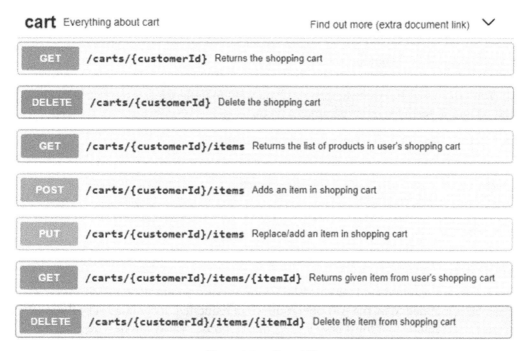

Figure 3.1 — Cart APIs

As we can see, for each tag, it creates a separate API interface.

summary and description

The summary and description sections are the same as we discussed earlier in the *metadata sections of OAS* section. They contain the given API's operation summary and detailed description, respectively. As usual, you can use Markdown in the description field as it refers to the same schema.

operationId

This represents the name of the operation. As you can see in the previous code, we have assigned the addCartItemsByCustomerId value to it. This same operation name would be used by Swagger Codegen as a method name in the generated API interface.

parameters

If you look closely, you'll find a - (hyphen) in front of the name field. This is used for declaring it as an array element. The parameters field can contain multiple parameters, in fact, a combination of the path and query parameters, therefore it is declared as an array.

For path parameters, you need to make sure that the value of name, under parameters, is the same as given in path inside curly braces.

The parameters field contains the API query, path, header, and cookie parameters. In the previous code, we were using the path parameter (the value of the **in** field). You can change the value to query if you want to declare it as a query parameter and so on and so forth for other parameter types.

description, as usual, describes the defined parameter.

You can mark a field as required or optional using the **required** field inside the parameters section, which is a Boolean parameter.

At last, you have to declare the data type of the parameter, which is where the schema field is used.

responses

responses is a required field for all API operations. This defines the type of responses that can be responded to by the API operation when requested. It contains HTTP status codes as the default field. It must have at least one response, which can be a default response or any successful HTTP status code, such as 200. As the name suggests, the default response will be used when no other response is defined or available in the API operation.

The response type (such as 200 or default) field contains three types of fields: description, content, and headers:

- description is used for describing the response.

- headers is used for defining the header and its value. A headers example is shown as follows:

```
responses:
  200:
    description: operation successful
    headers:
      X-RateLimit-Limit:
        schema:
          type: integer
```

- `content`, as we have in previous code, defines the type of content that denotes the different media types. We have used `application/json`. Similarly, you can define other media types, such as `application/xml`. The `content` type field contains the actual response object that can be defined using the `schema` field, as we have defined an array of the `Item` model inside it.

As mentioned earlier, you can create a reusable response under the `components` section and can directly use it using `$ref`.

requestBody

`requestBody` is used for defining the request payload object. Like the `responses` object, `requestBody` also contains the `description` and `content` fields. Content can be defined in a similar fashion to the way it is defined for the `responses` object. You can refer to the previous code of `POST /carts/{customerId}/items` for an example. As a response, you can also create reusable request bodies under the `components` section and can directly use them using `$ref`.

That's great, now you have learned how to define the API specification using OAS. Here, we have just described part of a sample e-commerce app's API. Similarly, you can describe other APIs. You can refer to `openapi.yaml` (`https://github.com/PacktPublishing/Modern-API-Development-with-Spring-and-Spring-Boot/tree/main/Chapter03/src/main/resources/api/openapi.yaml`) for the complete code of our e-commerce API definitions.

I would suggest you copy the code from `openapi.yaml` and paste it into the editor at `https://editor.swagger.io` to view the API in a nice user interface and play around with it. Make sure to convert the API to OpenAPI version 3 using the **Edit** menu if the default version is not set to 3.0.

We are done with designing our APIs, so now let's generate the code using `openapi.yaml` and enjoy the fruits of our hard work.

Converting OAS to Spring code

I am sure you are as excited as I am to start implementing the API. So far, we have learned about the RESTful web service theory and concepts and Spring fundamentals, as well as designing our first API specs for a sample e-commerce application.

Either you can clone the Git repository (`https://github.com/PacktPublishing/Modern-API-Development-with-Spring-and-Spring-Boot`) or you can start to create a Spring project from scratch using Spring Initializr (`https://start.spring.io/`) with the following options:

- **Project**: `Gradle`

- **Language**: `Java`

- **Spring Boot**: `2.3.5.RELEASE`

- Project metadata with your preferred values

- **Packaging**: `JarAdded`

- **Java**: `15` (You can change it to 14 in the `build.gradle` file later.)

- **Dependencies**: `'org.springframework.boot:spring-boot-starter-web'` (Spring Web in Spring Initializer)

Once you open the project in your favorite IDE (IntelliJ, Eclipse, or NetBeans), you can add the following extra dependencies required for OpenAPI support under `dependencies` in the `build.gradle` file:

```
swaggerCodegen 'org.openapitools:openapi-generator-cli:4.3.1'
compileOnly 'io.swagger:swagger-annotations:1.6.2'
compileOnly 'org.springframework.boot:spring-boot-starter-
            validation'
compileOnly 'org.openapitools:jackson-databind-nullable:0.2.1'
implementation 'com.fasterxml.jackson.dataformat:jackson-
               dataformat-xml'
implementation 'org.springframework.boot:spring-boot-starter-
               hateoas'
```

https://github.com/PacktPublishing/Modern-API-Development-with-Spring-and-Spring-Boot/tree/main/Chapter03/build.gradle

As mentioned earlier, we are going to use a Swagger plugin for code generation from the API definitions we have just written. You can follow the next seven steps to generate the code.

Step 1 – adding the Gradle plugin

To make use of the OpenAPI Generator CLI tool, you can add the Swagger Gradle plugin under `plugins {}` in `build.gradle` as shown:

```
plugins {
    ...

    ...

    id 'org.hidetake.swagger.generator' version '2.18.2'
}
```

Step 2 – defining the OpenAPI config for code generation

You need certain configurations, such as what model and API package names OpenAPI Generator's CLI should use, or the library it should use for generating the REST interfaces or date/time-related objects. All these and other configurations can be defined in `config.json` (`/src/main/resources/api/config.json`):

```
{
    "library": "spring-mvc",
    "dateLibrary": "java8",
    "hideGenerationTimestamp": true,
    "modelPackage": "com.packt.modern.api.model",
    "apiPackage": "com.packt.modern.api",
    "invokerPackage": "com.packt.modern.api",
    "serializableModel": true,
    "useTags": true,
    "useGzipFeature" : true,
    "hateoas": true,
    "withXml": true,
    "importMappings": {
        "Link": "org.springframework.hateoas.Link"
    }
}
```

https://github.com/PacktPublishing/Modern-API-Development-with-Spring-and-Spring-Boot/tree/main/Chapter03/src/main/resources/api/config.json

All the properties are self-explanatory except `importMappings`. It contains the mapping of type from YAML file to Java or a type existing in the external library. Therefore, once code is generated for the `importMapping` object, it uses the mapped class in the generated code. If we are using `Link` in any of the models, then the generated models would use the mapped `'org.springframework.hateoas.Link'` class instead of the model defined in the YAML file.

`hateoas` allows us to use the Spring HATEOAS library and add HATEOAS links.

`withXML` allows us to generate the models with XML annotations and support the `application/xml` content type.

You can find more information about the configuration at `https://github.com/swagger-api/swagger-codegen#customizing-the-generator`.

Step 3 – defining the OpenAPI Generator ignore file

You can also add a `.gitignore`-like file to ignore certain code you don't want to generate. Add the following line of code to the file (`/src/main/resources/api/.openapi-generator-ignore`):

```
**/*Controller.java
```

We don't want to generate controllers. After its addition, only API interfaces and models will be generated. We'll add controllers manually.

Step 4 – defining a swaggerSources task in the Gradle build file

Now, let's add the logic to the `swaggerSources` task in the `build.gradle` file:

```
swaggerSources {
  def typeMappings = 'URI=URI'
  def importMappings = 'URI=java.net.URI'
  eStore {
    def apiYaml = "${rootDir}/src/main/resources/api
                   /openapi.yaml"
    def configJson = "${rootDir}/src/main/resources/api
                      /config.json"
    inputFile = file(apiYaml)
    def ignoreFile = file("${rootDir}/src/main/resources/api
                           /.openapi-generator-ignore")
```

```
code {
  language = 'spring'
  configFile = file(configJson)
  rawOptions = ['--ignore-file-override', ignoreFile,
                '--type-mappings',
       typeMappings, '--import-mappings', importMappings] as
                  List<String>
  components = [models: true, apis: true, supportingFiles:
                'ApiUtil.java']
  //depends On validation // Should be uncommented once
  //plugin starts supporting OA 3 validation
  }
 }
}
```

Here, we have defined eStore (user-defined name) that contains inputFile pointing to the location of the openapi.yaml file. After defining the input, the generator needs to produce the output, which is configured in code.

We have defined language (it supports various languages), configFile pointing to config.json, rawOptions (contains the type and import mappings), and components in the code block. Aside from language, all are optional.

We just want to generate models and APIs. You can generate other files too, such as clients or test files. ApiUtil.java is required in the generated API interface else it will give a compilation error during build time; therefore, it is added in components.

Step 5 – adding swaggerSources to the compileJava task dependency

Next, we need to add swaggerSources to the compileJava task as a dependent task. It points to the code block defined under eStore:

```
compileJava.dependsOn swaggerSources.eStore.code
```

Step 6 – adding the generated source code to Gradle sourceSets

We also need to add the generated source code and resources to `sourceSets`. This makes the generated source code and resources available for development and build:

```
sourceSets.main.java.srcDir "${swaggerSources.eStore.code.
outputDir}/src/main/java"
```

```
sourceSets.main.resources.srcDir "${swaggerSources.eStore.code.
outputDir}/src/main/resources"
```

The source code will be generated in the `/build` directory of the project, such as `modern-api-with-spring-and-sprint-boot\Chapter03\build\swagger-code-eStore`. This will append the generated source code and resources to Gradle `sourceSets`.

Step 7 – running the build for generating, compiling, and building the code

The last step is to execute the build. Make sure you have an executable Java code in the build path. The Java version should match the version defined in the property of `build.gradle(sourceCompatibility = '1.14')` or in the IDE settings:

```
$ gradlew clean build
```

Once the build is executed successfully, you can find the generated code in the build directory, as shown in the following screenshot:

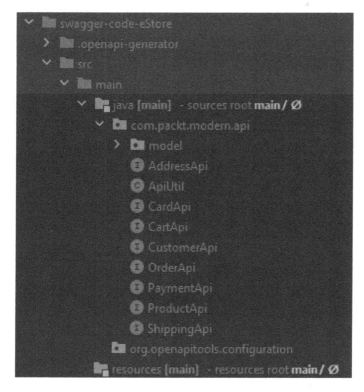

Figure 3.2 — OpenAPI generated code

In the next section, you'll implement the API interfaces generated by **OpenAPI Codegen**.

Implementing the OAS code interfaces

So far, we have generated code that consists of e-commerce app models and API interfaces. These generated interfaces contain all the annotations as per the YAML description provided by us. For example, in CartApi.java, @RequestMapping, @PathVariable, and @RequestBody contain the endpoint path (/api/v1/carts/{customerId}/items), the value of the path variable (such as {customerId} in path), and the request payload (such as Item), respectively. Similarly, generated models contain all the mapping required for supporting JSON and XML content types.

Swagger Codegen writes the Spring code for us. We just need to implement the interface and write the business logic inside it. Swagger Codegen generates the API interfaces for each of the provided tags. For example, it generates the CartApi and PaymentAPI Java interfaces for the cart and payment tags, respectively. All the paths are clubbed together into a single Java interface based on the given tag. For example, all the APIs with the cart tag will be clubbed together into a single Java interface, CartApi.

Now, we just need to create a class for each of the interfaces and implement it. We'll create `CartController.java` in the `com.packt.modern.api.controllers` package and implement `CartApi`:

```java
@RestController
public class CartsController implements CartApi {

  private static final Logger log = LoggerFactory.
    getLogger(CartsController.class);

  @Override
  public ResponseEntity<List<Item>> addCartItemsByCustomerId
      (String customerId, @Valid Item item) {
    log.info("Request for customer ID: {}\nItem: {}",
            customerId, item);
    return ok(Collections.EMPTY_LIST);
  }

  @Override
  public ResponseEntity<List<Cart>> getCartByCustomerId(String
      customerId) {
    throw new RuntimeException("Manual Exception thrown");
  }

  // Other method implementations (omitted)

}
```

https://github.com/PacktPublishing/Modern-API-Development-with-Spring-and-Spring-Boot/blob/main/Chapter03/src/main/java/com/packt/modern/api/controllers/CartsController.java

Here, we have just implemented the two methods for demonstration purposes. We'll implement the actual business logic in the next chapter.

To add an item (`POST /api/v1/carts/{customerId}/items`) request, we are just logging the incoming request payload and customer ID inside the `addCartItemsByCustomerId` method. Another method, `getCartByCustomerId`, simply throws an exception. This will allow us to demonstrate the Global Exception Handler in the next section.

Adding a Global Exception Handler

We'll have multiple controllers that consist of multiple methods. Each method may have checked exceptions or throw runtime exceptions. We should have a centralized place to handle all these errors for better maintainability and modularity and clean code.

Spring provides an AOP feature for this. We just need to write a single class annotated with `@ControllerAdvice`. Then, we just need to add `@ExceptionHandler` for each of the exceptions. This exception handler method will generate user-friendly error messages with other related information.

You can make use of the Lombok library if approved by your respective organization for third-party library usage. This will remove the verbosity of the code for getters, setters, constructors, and so on.

Let's first write the `Error` class in the `exceptions` package that contains all the error information:

```
public class Error {

  private static final long serialVersionUID = 1L;
  /**
   * App error code, which is different from HTTP error code.
   */
  private String errorCode;


  /**
   * Short, human-readable summary of the problem.
   */
  private String message;


  /**
   * HTTP status code.
   */
```

```
    private Integer status;

    /**
     * Url of request that produced the error.
     */
    private String url = "Not available";

    /**
     * Method of request that produced the error.
     */
    private String reqMethod = "Not available";

    // getters and setters (omitted)

}
```

https://github.com/PacktPublishing/Modern-API-Development-with-Spring-and-Spring-Boot/blob/main/Chapter03/src/main/java/com/packt/modern/api/exceptions/Error.java

You can add other fields here if required. The `exceptions` package will contain all the code for user-defined exceptions and global exception handling.

After that, we'll write an enum called `ErrorCode` that will contain all the exception keys, including user-defined errors and their respective error codes:

```
public enum ErrorCode {
    // Internal Errors: 1 to 0999
    GENERIC_ERROR("PACKT-0001", "The system is unable to complete
                    the request. Contact system support."),
    HTTP_MEDIATYPE_NOT_SUPPORTED("PACKT-0002", "Requested
    media type is not supported. Please use application/json or
    application/xml as 'Content-Type' header value"),
    HTTP_MESSAGE_NOT_WRITABLE("PACKT-0003", "Missing 'Accept'
    header. Please add 'Accept' header."),
    HTTP_MEDIA_TYPE_NOT_ACCEPTABLE("PACKT-0004", "Requested
    'Accept' header value is not supported. Please use
    application/json or application/xml as 'Accept' value"),
    JSON_PARSE_ERROR("PACKT-0005", "Make sure request payload
```

```
    should be a valid JSON object."),
    HTTP_MESSAGE_NOT_READABLE("PACKT-0006", "Make sure request
    payload should be a valid JSON or XML object according to
    'Content-Type'.");


    private String errCode;
    private String errMsgKey;

    ErrorCode(final String errCode, final String errMsgKey) {
        this.errCode = errCode;
        this.errMsgKey = errMsgKey;
    }


    /**
     * @return the errCode
     */
    public String getErrCode() {
        return errCode;
    }


    /**
     * @return the errMsgKey
     */
    public String getErrMsgKey() {
        return errMsgKey;
    }
}
```

https://github.com/PacktPublishing/Modern-API-Development-with-Spring-and-Spring-Boot/blob/main/Chapter03/src/main/java/com/packt/modern/api/exceptions/ErrorCode.java

Here, we have just added actual error messages instead of message keys. You can add message keys and add the resource file to `src/main/resources` for internationalization.

Next, we'll add a utility to create the `Error` object, as shown:

```
public class ErrorUtils {

  private ErrorUtils() {}

  /**
   * Creates and return an error object
   *
   * @param errMsgKey
   * @param errorCode
   * @param httpStatusCode
   * @param url
   * @return error
   */
  public static Error createError(final String errMsgKey, final
      String errorCode, final Integer httpStatusCode) {
    Error error = new Error();
    error.setMessage(errMsgKey);
    error.setErrorCode(errorCode);
    error.setStatus(httpStatusCode);
    return error;
  }
}
```

https://github.com/PacktPublishing/Modern-API-Development-with-Spring-and-Spring-Boot/blob/main/Chapter03/src/main/java/com/packt/modern/api/exceptions/ErrorUtils.java

Finally, we'll create a class for implementing the Global Exception Handler, as shown:

```
@ControllerAdvice
public class RestApiErrorHandler {
  private static final Logger log = LoggerFactory.
    getLogger(RestApiErrorHandler.class);
  private final MessageSource messageSource;

  @Autowired
```

```java
  public RestApiErrorHandler(MessageSource messageSource) {
    this.messageSource = messageSource;
  }


  @ExceptionHandler(Exception.class)
  public ResponseEntity<Error> handleException
      (HttpServletRequest request, Exception ex, Locale locale)
{

    Error error = ErrorUtils
        .createError(ErrorCode.GENERIC_ERROR.getErrMsgKey(),
          ErrorCode.GENERIC_ERROR.getErrCode(),
            HttpStatus.INTERNAL_SERVER_ERROR.value()).
              setUrl(request.getRequestURL().toString())
        .setReqMethod(request.getMethod());
    return new ResponseEntity<>(error, HttpStatus.INTERNAL_
                                SERVER_ERROR);
  }


  @ExceptionHandler(HttpMediaTypeNotSupportedException.class)
  public ResponseEntity<Error>
      handleHttpMediaTypeNotSupportedException(
      HttpServletRequest request,
      HttpMediaTypeNotSupportedException ex, Locale locale) {
    Error error = ErrorUtils
        .createError(ErrorCode.HTTP_MEDIATYPE_NOT_SUPPORTED.
                      getErrMsgKey(),
            ErrorCode.HTTP_MEDIATYPE_NOT_SUPPORTED.
                      getErrCode(),
            HttpStatus.UNSUPPORTED_MEDIA_TYPE.value()).
                      setUrl(request.getRequestURL().toString())
        .setReqMethod(request.getMethod());
    log.info("HttpMediaTypeNotSupportedException :: request.
              getMethod(): " + request.getMethod());
    return new ResponseEntity<>(error, HttpStatus. UNSUPPORTED_
                                MEDIA_TYPE);
  }
```

https://github.com/PacktPublishing/Modern-API-Development-with-Spring-and-Spring-Boot/blob/main/Chapter03/src/main/java/com/packt/modern/api/exceptions/RestApiErrorHandler.java

As you can see, we have marked the class with `@ControllerAdvice`, which enables this class to trace all the request and response processing by the REST controllers and allows handling exceptions using `@ExceptionHandler`.

In the previous code, we are handling two exceptions: a generic `internal server error` exception and `HttpMediaTypeNotSupportException`. The handling method just populates the `Error` object using `ErrorCode`, `HttpServletRequest`, and `HttpStatus`. At the end, it returns the error wrapped inside `ResponseEntity` with the appropriate HTTP status.

Here, you can add user-defined exceptions too. You can also make use of the `Locale` instance (a method parameter) and the `messageSource` class member for supporting internationalized messages.

Testing

Once the code is ready to run, you can compile and build the artifact using the following command from the root folder of the project:

```
gradlew clean build
```

The previous command removes the `build` folder and generates the artifact (compiled classes and JAR). After the successful build, you can run the application using the following command:

```
java -jar build\libs\Chapter03-0.0.1-SNAPSHOT.jar
```

Now, we can perform the tests using the `curl` command:

```
$ curl --request GET 'http://localhost:8080/api/v1/carts/1'
--header 'Accept: application/xml'
```

This command calls the GET request for `/carts` with ID 1. Here, we demand the XML response using the `Accept` header, and we get the following response:

```
<Error>
    <errorCode>PACKT-0001</errorCode>
    <message>The system is unable to complete the
request. Contact system support.</message>
    <status>500</status>
    <url>http://localhost:8080/api/v1/carts/1</url>
    <reqMethod>GET</reqMethod>
</Error>
```

If you changed the `Accept` header from `application/xml` to `application/json`, you would get the following JSON response:

```
{
    "errorCode": "PACKT-0001",
    "message": "The system is unable to complete
                the request. Contact system support.",
    "status": 500,
    "url": "http://localhost:8080/api/v1/carts/1",
    "reqMethod": "GET"
}
```

Similarly, we can also call the *add item to cart* call, as shown:

```
$ curl --request POST 'http://localhost:8080/api/v1/carts/1/
items' \
> --header 'Content-Type: application/json' \
> --header 'Accept: application/json' \
> --data-raw '{
>     "id": "1",
>     "quantity": 1,
>     "unitPrice": 2.5
> }'
```

Here, we get [] (empty array) as a response because in the implementation, we are just returning the empty collection. You need to provide the `Content-Type` header in this request because we are sending the payload (item object) along with the request. You can change `Content-Type` to `application/xml` if the payload is written in XML. If the `Accept` header value is `application/xml`, it will return the `<List/>` value. You can remove/change the `Content-Type` and `Accept` headers or use the malformed JSON or XML to test the other error response.

This way, we can generate the API description using OpenAPI and then use the generated models and API interfaces to implement the APIs.

Summary

In this chapter, we opted for the design-first approach for writing the RESTful web services. You learned how to write an API description using OAS and how to generate models and API interfaces using the Swagger Codegen tool (using the Gradle plugin). We also implemented a Global Exception Handler to centralize the handling of all the exceptions. Once you have the API Java interfaces, you can write their implementations for business logic. Now, you know how to use OAS and Swagger Codegen for writing RESTful APIs. You have also learned how to handle exceptions globally.

In the next chapter, we'll implement fully fledged API interfaces with business logic with database persistence.

Questions

1. What is OpenAPI and how does it help?

2. How can you define a nested array in a model in a YAML OAS-based file?

3. What annotations do we need to implement a Global Exception Handler?

4. How you can use models or classes written in Java code in your OpenAPI description?

5. Why do we only generate models and API interfaces using Swagger Codegen?

Further reading

*   The OpenAPI Specification 3.0: `https://github.com/OAI/OpenAPI-Specification/blob/master/versions/3.0.3.md`

*   The Gradle plugin for OpenAPI Codegen: `https://github.com/int128/gradle-swagger-generator-plugin`

*   OAS Code Generator configuration options for Spring: `https://openapi-generator.tech/docs/generators/spring/`

*   YAML specifications: `https://yaml.org/spec/`

*   Semantic versioning: `https://semver.org/`

4
Writing Business Logic for APIs

We discussed APIs using OpenAPI in the previous chapter. API interfaces and models were generated by the Swagger Codegen. In this chapter, you will implement the API's code in terms of both business logic and data persistence. You will write services and repositories for implementation and also add hypermedia and ETags to API responses. It is worth noting that the code provided only consists of the important lines and not the whole file in the interest of brevity. You can always access the links given below the code to view the complete file.

This chapter includes the following topics:

- Overview of the service design
- Adding repository components
- Adding service components
- Implementing hypermedia
- Enhancing the controller with a service and HATEOAS
- Adding ETags to API responses

Technical requirements

You need the following to execute instructions in this chapter:

- Any Java IDE such as NetBeans, IntelliJ, or Eclipse

- The **Java Development Kit (JDK)** 15+

- An internet connection to download the dependencies and Gradle

- The Postman tool (`https://learning.postman.com/docs/getting-started/sending-the-first-request/`)

You can find the code files for this chapter on GitHub at `https://github.com/PacktPublishing/Modern-API-Development-with-Spring-and-Spring-Boot/tree/main/Chapter04`.

Overview of the service design

We are going to implement a multilayered architecture that comprises four layers – the presentation layer, application layer, domain layer, and infrastructure layer. Multilayered architecture is a fundamental building block in the architecture style known as **Domain-Driven Design (DDD)**. Let's have a brief look at each of these layers:

- **Presentation layer**: This layer represents the **User Interface (UI)**. In the upcoming *Chapter 7, Designing a User Interface*, we'll develop the UI for our e-commerce app.

- **Application layer**: The application layer contains the application logic and maintains and coordinates the overall flow of the application. Just to remind you, it only contains the application logic and **not** the business logic. RESTful web services, async APIs, gRPC APIs, and GraphQL APIs are a part of this layer.

  We already covered the REST API interfaces and controllers (implementing REST API interfaces) in *Chapter 3, API Specifications and Implementation*, which are part of the application layer. We implemented the controllers for demonstration purposes in the previous chapter. In this chapter, we'll implement a controller extensively to serve real data.

- **Domain layer**: This layer contains the business logic and domain information. It contains the state of the business objects such as Order, Product, and so on. It is responsible for reading/persisting these objects to the infrastructure layer. The domain layer consists of services and repositories too. We'll also be covering these in this chapter.

- **Infrastructure layer**: The infrastructure layer provides support to all other layers. It is responsible for communication such as interaction with the database, message brokers, filesystems, and so on. Spring Boot works as an infrastructure layer and provides support for communication and interaction with both external and internal systems such as databases, message brokers, and so on.

We'll use the bottom-to-top approach. Let's start implementing the domain layer with the `@Repository` component.

Adding a Repository component

We'll use the bottom-to-top approach to add a `@Repository` component. Let's start implementing the domain layer with a `@Repository` component. We'll implement the service and enhance the `Controller` component in subsequent sections accordingly. We will code the `@Repository` component first, then use it in the `@Service` component using constructor injection. The `@Controller` component will be enhanced using the `@Service` component, which will also be injected into the `Controller` using constructor injection.

@Repository annotation

Repository components are Java classes marked with the `@Repository` annotation. This is a special Spring component that is used for interacting with databases.

`@Repository` is a general-purpose stereotype that represents both DDD's Repository and the Java **Enterprise Edition (EE)** pattern, the **Data Access Object (DAO)**. Developers and teams should handle Repository objects based on the underlying approach. In DDD, a Repository is a central object that carries references to all the objects and should return the reference of a requested object. We need to have all the required dependencies and configurations in place before we start writing classes marked with `@Repository`.

We'll use the following libraries as database dependencies:

- **H2 database for persisting data**: We are going to use H2's memory instance, however, you can also use a file-based instance.

- **Hibernate Object Relational Mapping (ORM)**: For database object mapping.

- **Flyway for database migration**: This helps maintain the database and maintains a database changes history that allows rollbacks, version upgrades, and so on.

Let's add these dependencies to the `build.gradle` file. `org.springframework.boot:spring-boot-starter-data-jpa` adds all the required JPA dependencies including Hibernate:

```
implementation 'org.springframework.boot:spring-boot-starter-
data-jpa'
implementation 'org.flywaydb:flyway-core'
runtimeOnly 'com.h2database:h2'
```

https://github.com/PacktPublishing/Modern-API-Development-with-Spring-and-Spring-Boot/blob/main/Chapter04/build.gradle

After adding the dependencies, we can add the configuration related to the database.

Database and JPA configuration

We also need to modify the `application.properties` file with the following configuration:

1. **Data source configuration**

    The following is the Spring data source configuration:

    ```
    spring.datasource.name=ecomm
    spring.datasource.url=jdbc:h2:mem:ecomm;DB_CLOSE_DELAY=-
    1;IGNORECASE=TRUE;DATABASE_TO_UPPER=false

    spring.datasource.driverClassName=org.h2.Driver
    spring.datasource.username=sa
    spring.datasource.password=
    ```

    We need to add H2-specific properties to the data source. The URL value suggests that a memory-based H2 database instance will be used.

2. **H2 database configuration**

    The following are the two H2 database configurations:

    ```
    spring.h2.console.enabled=true
    spring.h2.console.settings.web-allow-others=false
    ```

    The H2 console is enabled for local access only; it means you can access the H2 console only on localhost. Also, remote access is disabled by setting `web-allow-others` to `false`.

3. **JPA configuration**

The following are the JPA/Hibernate configurations:

```
spring.jpa.properties.hibernate.default_schema=ecomm
spring.jpa.database-platform=org.hibernate.dialect.
H2Dialect
spring.jpa.show-sql=true
spring.jpa.format_sql=true
spring.jpa.generate-ddl=false
spring.jpa.hibernate.ddl-auto=none
```

We don't want to generate the DDL or to process the SQL file, because we want to use Flyway for database migrations. Therefore, `generate-ddl` is marked with `false` and `ddl-auto` is set to `none`.

4. **Flyway configuration**

The following are the Flyway configurations:

```
spring.flyway.url=jdbc:h2:mem:ecomm
spring.flyway.schemas=ecomm
spring.flyway.user=sa
spring.flyway.password=
```

https://github.com/PacktPublishing/Modern-API-Development-with-Spring-and-Spring-Boot/blob/main/Chapter04/src/main/resources/application.properties

---

Accessing the H2 database

You can access the H2 database console using `/h2-console`. For example, if your server is running on localhost and on port `8080` then you can access it using `http://localhost:8080/h2-console/`.

The database and seed data script

Now, we are done configuring the `build.gradle` and `application.properties` files and we can now start writing the code. First, we'll add the Flyway database migration script. This script can be written in SQL only. You can place this file in the `db/migration` directory inside the `src/main/resources` directory. We'll follow the Flyway naming convention (`V<version>.<name>.sql`) and create the `V1.0.0.Init.sql` file inside the `db/migration` directory. You can then add the following script in this file:

```sql
create schema if not exists ecomm;


-- Other script tags


create TABLE IF NOT EXISTS ecomm.cart (
  id uuid NOT NULL,
 user_id uuid NOT NULL,
 FOREIGN KEY (user_id)
  REFERENCES ecomm.user(id),
 PRIMARY KEY(id)
);


create TABLE IF NOT EXISTS ecomm.cart_item (
  cart_id uuid NOT NULL,
  item_id uuid NOT NULL,
  FOREIGN KEY (cart_id)
   REFERENCES ecomm.cart(id),
  FOREIGN KEY(item_id)
   REFERENCES ecomm.item(id)
);


-- other SQL scripts
```

https://github.com/PacktPublishing/Modern-API-Development-with-Spring-and-Spring-Boot/blob/main/Chapter04/src/main/resources/db/migration/V1.0.0__Init.sql

This script creates the `ecomm` schema and adds all the tables required for our sample e-commerce app. It also adds `insert` statements for the seed data.

Adding entities

Now, we can add the entities. An entity is a special object marked with the `@Entity` annotation that maps directly to the database table using an ORM implementation such as *Hibernate*. Another popular ORM is *EclipseLink*. You can place all entity objects in the `com.packt.modern.api.entity` package. Let's create the `CartEntity.java` file:

```
@Entity
@Table(name = "cart")
public class CartEntity {

  @Id
  @GeneratedValue
  @Column(name = "ID", updatable = false, nullable = false)
  private UUID id;

  @OneToOne
  @JoinColumn(name = "USER_ID", referencedColumnName = "ID")
  private UserEntity user;

  @ManyToMany(
    cascade = CascadeType.ALL
  )
  @JoinTable(
    name = "CART_ITEM",
    joinColumns = @JoinColumn(name = "CART_ID"),
    inverseJoinColumns = @JoinColumn(name = "ITEM_ID")
  )
  private List<ItemEntity> items = Collections.emptyList();

// Getters/Setter and other codes are removed for brevity
```

https://github.com/PacktPublishing/Modern-API-Development-with-Spring-and-Spring-Boot/blob/main/Chapter04/src/main/java/com/packt/modern/api/entity/CartEntity.java

Here, the @Entity annotation is part of the javax.persistence package that denotes that it is an Entity and should be mapped to the database table. By default, it takes the entity name, however we are using the @Table annotation to map to the database table.

We are also using one-to-one and many-to-many annotations for mapping the Cart entity to the User Entity and Item Entity respectively. The ItemEntity list is also associated with @JoinTable, because you are using the CART_ITEM join table to map the cart and product items based on the CART_ID and ITEM_ID columns in their respective tables.

In UserEntity, the Cart entity has also been added to maintain the relationship as shown in the next code block. FetchType is marked as LAZY, which means the user's cart will be loaded only when asked for explicitly. Also, you want to remove the cart if it is not referenced by the user, which can be done by configuring orphanRemoval to true:

```java
@Entity
@Table(name = "user")
public class UserEntity {

    // other code

    @OneToOne(mappedBy = "user", fetch = FetchType.LAZY,
                                orphanRemoval = true)
    private CartEntity cart;

    // other code...
```

https://github.com/PacktPublishing/Modern-API-Development-with-Spring-and-Spring-Boot/blob/main/Chapter04/src/main/java/com/packt/modern/api/entity/UserEntity.java

All other entities are being added to the entity package located at https://github.com/PacktPublishing/Modern-API-Development-with-Spring-and-Spring-Boot/blob/main/Chapter04/src/main/java/com/packt/modern/api/entity.

Now, we can add the repository.

Adding repositories

All the repository have been added to `https://github.com/PacktPublishing/Modern-API-Development-with-Spring-and-Spring-Boot/blob/main/Chapter04/src/main/java/com/packt/modern/api/repository`.

Repositories are simplest to add for CRUD operations, thanks to Spring Data JPA. You just have to extend the interfaces with default implementations, such as `CrudRepository,` which provides all the CRUD operation implementation such as `save, saveAll, findById, findAll, findAllById, delete,` and `deleteById`. The `Save(Entity e)` method is used for both create and update entity operations.

Let's create `CartRepository`:

```
public interface CartRepository extends
    CrudRepository<CartEntity, UUID> {


  @Query("select c from CartEntity c join c.user u where u.id =
      :customerId")
  public Optional<CartEntity> findByCustomerId(@
      Param("customerId") UUID customerId);


}
```

https://github.com/PacktPublishing/Modern-API-Development-with-Spring-and-Spring-Boot/blob/main/Chapter04/src/main/java/com/packt/modern/api/repository/CartRepository.java

The `CartRepository` interface extends the `CrudRepository` part of the `org.springframework.data.repository` package. You can also add methods supported by the JPA Query Language marked with the `@Query` annotation (part of the `org.springframework.data.jpa.repository` package). The query inside the `@Query` annotation is written in **Java Persistence Query Language (JPQL)**. JPQL is very similar to SQL, however, here you used the Java class name mapped to a database table instead of using the actual table name. Therefore, we have used `CartEntity` as the table name instead of `Cart`.

> **Note**
>
> Similarly, for attributes, you should use the variable names given in the class for the fields, instead of using the database table fields. In any case, if you use the database table name or field name and it does not match with the class and class members mapped to the actual table, you will get an error.

You must be wondering, *"What if I want to add my own custom method with JPQL or native SQL?"* Well let me tell you, you can do that too. For orders, we have added a custom interface for this very purpose. First, let's have a look at `OrderRepository`, which is very similar to `CartRepository`:

```
@Repository
public interface OrderRepository extends
        CrudRepository<OrderEntity, UUID>, OrderRepositoryExt {

    @Query("select o from OrderEntity o join o.userEntity u where
        u.id = :customerId")
    public Iterable<OrderEntity> findByCustomerId(@
        Param("customerId") UUID customerId);
}
```

https://github.com/PacktPublishing/Modern-API-Development-with-Spring-and-Spring-Boot/blob/main/Chapter04/src/main/java/com/packt/modern/api/repository/OrderRepository.java

If you look closely, we have extended an extra interface – `OrderRepositoryExt`. This is our extra interface for the `Order` repository and consists of the following code:

```
public interface OrderRepositoryExt {
    Optional<OrderEntity> insert(NewOrder m);
}
```

https://github.com/PacktPublishing/Modern-API-Development-with-Spring-and-Spring-Boot/blob/main/Chapter04/src/main/java/com/packt/modern/api/repository/OrderRepositoryExt.java

We already have a `save()` method for this purpose in `CrudRepository`, however, we want to use a different implementation. For this purpose, and to demonstrate how you can create your own repository method implementation, we are adding this extra repository interface.

Now, let's create the `OrderRepositoryExt` interface implementation as shown here:

```
@Repository
@Transactional
public class OrderRepositoryImpl implements OrderRepositoryExt
{
```

```
@PersistenceContext
private EntityManager em;

private ItemRepository itemRepo;

private ItemService itemService;


public OrderRepositoryImpl(EntityManager em, ItemRepository
itemRepo, ItemService itemService) {
    this.em = em;
    this.itemRepo = itemRepo;
    this.itemService = itemService;
}


// other code
```

https://github.com/PacktPublishing/Modern-API-Development-with-Spring-and-Spring-Boot/blob/main/Chapter04/src/main/java/com/packt/modern/api/repository/OrderRepositoryImpl.java

This way we can also have our own implementation in JPQL/**Hibernate Query Language** (**HQL**), or in native SQL. Here, the @Repository annotation tells the Spring container that this special component is a Repository and should be used for interacting with the database using the underlying JPA.

It is also marked as @Transactional, which is a special annotation that means that transactions performed by methods in this class will be managed by Spring. It removes all the manual work of adding commits and rollbacks. You can also add this annotation to a specific method inside the class.

We are also using @PersistenceContext for the EntityManager class, which allows us to create and execute the query manually as shown in the following code:

```
@Override
 public Optional<OrderEntity> insert(NewOrder m) {
 // Items are already in cart and saved in db when user places
 // order
 // Here you can also populate other Order details like address
 // etc.
   Iterable<ItemEntity> dbItems =
                  itemRepo.findByCustomerId(m.
```

```
getCustomerId());
  List<ItemEntity> items =
                 StreamSupport.stream(dbItems.spliterator(),
                                      false)
              .collect(toList());
  if (items.size() < 1) {
    throw new ResourceNotFoundException(String.format(
        "There is no item found in customer's (ID: %s)
        cart.", m.getCustomerId()));
  }
  BigDecimal total = BigDecimal.ZERO;
  for (ItemEntity i : items) {
    total = (BigDecimal.valueOf(i.getQuantity()).multiply(
                              i.getPrice())).add(total);
  }
  Timestamp orderDate = Timestamp.from(Instant.now());
  em.createNativeQuery("""
    INSERT INTO ecomm.orders (address_id, card_id, customer_id
    order_date, total, status) VALUES(?, ?, ?, ?, ?, ?)
    """)
    .setParameter(1, m.getAddress().getId())
    .setParameter(2, m.getCard().getId())
    .setParameter(3, m.getCustomerId())
    .setParameter(4, orderDate)
    .setParameter(5, total)
    .setParameter(6, StatusEnum.CREATED.getValue())
    .executeUpdate();
  Optional<CartEntity> oCart =
      cRepo.findByCustomerId(UUID.fromString(m.
                              getCustomerId()));
  CartEntity cart = oCart.orElseThrow(() -> new
          ResourceNotFoundException(String.format("Cart not
          found for given customer (ID: %s)",
          m.getCustomerId())));
  itemRepo.deleteCartItemJoinById(cart.getItems().stream()
            .map(i -> i.getId()).collect(toList()), cart.
                getId());
  OrderEntity entity = (OrderEntity) em.createNativeQuery("""
```

```
    SELECT o.* FROM ecomm.orders o WHERE o.customer_id = ? AND
    o.order_date >= ?
    """, OrderEntity.class)
    .setParameter(1, m.getCustomerId())
    .setParameter(2, OffsetDateTime.ofInstant(orderDate.
               toInstant(),
       ZoneId.of("Z")).truncatedTo(ChronoUnit.MICROS))
    .getSingleResult();
  oiRepo.saveAll(cart.getItems().stream()
      .map(i -> new OrderItemEntity().setOrderId(entity.
          getId())
        .setItemId(i.getId())).collect(toList()));
  return Optional.of(entity);
}
```

This method basically first fetches the items in the customer's cart. Then, it calculates the order total, creates a new order, and saves it in the database. Next, it removes the items from the cart by removing the mapping because cart items are now part of the order. Next, it saves the mapping of the order and cart items.

Order creation is done using the native SQL query with the prepared statement.

If you look closely, you'll also find that we have used the official *Java 15* feature, **text blocks** (https://docs.oracle.com/en/java/javase/15/text-blocks/index.html), in it.

Similarly, you can create a repository for all other entities. All entities are available at https://github.com/PacktPublishing/Modern-API-Development-with-Spring-and-Spring-Boot/tree/main/Chapter04/src/main/java/com/packt/modern/api/repository.

Now that we have created the repositories, we can move on to adding services.

Adding a Service component

The Service component is an interface that works between controllers and repositories and is where we'll add the business logic. Though you can directly call repositories from controllers, it is not a good practice as repositories should only be part of the data retrieval and persistence functionalities. Service components also help in sourcing data from various sources, such as databases and other external applications.

Service components are marked with the @Service annotation, which is a specialized Spring @Component that allows implemented classes to be auto-detected using class-path scanning. Service classes are used for adding business logic. Like Repository, the Service object also represents both DDD's Service and Java EE's Business Service Façade pattern. Like Repository, it is also a general-purpose stereotype and can be used according to the underlying approach.

First we'll create the service interface, which is a normal Java interface with all the desired method signatures. This interface will expose all the operations that can be performed by CartService:

```java
public interface CartService {

    public List<Item> addCartItemsByCustomerId(String customerId,
        @Valid Item item);

    public List<Item> addOrReplaceItemsByCustomerId(String
        customerId, @Valid Item item);

    public void deleteCart(String customerId);

    public void deleteItemFromCart(String customerId, String
        itemId);

    public CartEntity getCartByCustomerId(String customerId);

    public List<Item> getCartItemsByCustomerId(String
        customerId);

    public Item getCartItemsByItemId(String customerId, String
        itemId);
}
```

https://github.com/PacktPublishing/Modern-API-Development-with-Spring-and-Spring-Boot/blob/main/Chapter04/src/main/java/com/packt/modern/api/service/CartService.java

The methods added to CartService are directly mapped to serve each of the APIs defined in the CartController class. Now, we can implement each of the methods in the CartServiceImpl class, which is an implementation of the CartService interface. Each method in CartServiceImpl makes use of a specific Repository object to carry out the operation:

```
@Service
public class CartServiceImpl implements CartService {
  private CartRepository repository;
  private UserRepository userRepo;
  private ItemService itemService;

  public CartServiceImpl(CartRepository repository,
      UserRepository userRepo, ItemService itemService) {
    this.repository = repository;
    this.userRepo = userRepo;
    this.itemService = itemService;
  }

  @Override
  public List<Item> addCartItemsByCustomerId(
                          String customerId, @Valid Item item)
{
    CartEntity entity = getCartByCustomerId(customerId);
    long count = entity.getItems().stream().filter(i ->
                    i.getProduct().getId().equals(
                    UUID.fromString(item.getId()))).count();
    if (count > 0) {
      throw new GenericAlreadyExistsException(
          String.format("Item with Id (%s) already exists.
            You can update it.", item.getId())));
    }
    entity.getItems().add(itemService.toEntity(item));
    return itemService.toModelList(
          repository.save(entity).getItems());
  }
  // rest of the code trimmed for brevity
```

https://github.com/PacktPublishing/Modern-API-Development-with-Spring-and-Spring-Boot/blob/main/Chapter04/src/main/java/com/packt/modern/api/service/CartServiceImpl.java

The `CartServiceImpl` class is annotated with `@Service`, therefore would be auto-detected and available for injection. The `CartRepository`, `UserRepository`, and `ItemService` class dependencies are injected using constructor injection.

Let's have a look at one more method implementation of the `CartService` interface. Check the following code. It adds an item, or updates the price and quantity if the item already exists:

```java
@Override
public List<Item> addOrReplaceItemsByCustomerId(
                    String customerId, @Valid Item item) {
  // 1
  CartEntity entity = getCartByCustomerId(customerId);
  List<ItemEntity> items =Objects.nonNull(entity.getItems()) ?
                    entity.getItems() : Collections.
emptyList();
  AtomicBoolean itemExists = new AtomicBoolean(false);

  // 2
  items.forEach(i -> {
    if (i.getProduct().getId()
          .equals(UUID.fromString(item.getId()))) {
      i.setQuantity(item.getQuantity()).setPrice(i.getPrice());
      itemExists.set(true);
    }
  });
  if (!itemExists.get()) {
      items.add(itemService.toEntity(item));
  }
  // 3
  return itemService.
          toModelList(repository.save(entity).getItems());
}
```

In the preceding code, we are not managing the application state, but are instead writing the sort of business logic that queries the database, sets the entity object, persists the object, and then returns the model class. Let's have a look at the statements one by one:

1. The method only has `customerId` as a parameter and there is no `Cart` parameter. Therefore, first we get `CartEntity` from the database based on the given `customerId`.

2. The program control iterates through the items retrieved from the `CartEntity` object. If the given item already exists then the quantity and price are changed. Else, it creates a new `Item` entity from the given `Item` model and then saves it to the `CartEntity` object. The `itemExists` flag is used to find out whether we need to update the existing `Item` or add a new one.

3. Finally, the updated `CartEntity` object is saved in the database. The latest `Item` entity is retrieved from the database, and then gets converted to a model collection and returned back to the calling program.

Similarly, you can write `Service` components for others the way you have implemented it for `Cart`. Before we start enhancing the Controller classes, we need to add a final frontier to our overall feature.

Implementing hypermedia

We have learned about hypermedia and **Hypermedia As The Engine Of Application State** (**HATEOAS**) in *Chapter 1, RESTful Web Service Fundamentals*. Spring provides state-of-the-art support to HATEOAS using the `org.springframework.boot:spring-boot-starter-hateoas` dependency.

First of all, we need to make sure that all models returned as part of the API response contain the link field. There are different ways to associate links (that is, the `org.springframework.hateoas.Link` class) with models, either manually or via auto-generation. Spring HATEOAS's links and its attributes are implemented according to *RFC 8288* (`https://tools.ietf.org/html/rfc8288`). For example, you can create a self-link manually as follows:

```
import static org.springframework.hateoas.server.mvc.
    WebMvcLinkBuilder.linkTo;
import static org.springframework.hateoas.server.mvc.
    WebMvcLinkBuilder.methodOn;
// other code blocks...
```

```
responseModel.setSelf(linkTo(methodOn(CartController.class)
    .getItemsByUserId(userId,item)).withSelfRel())
```

Here, `responseModel` is a model object that is returned by the API. It has a field called `_self` that is set using the `linkTo` and `methodOn` static methods. The `linkTo` and `methodOn` methods are provided by the Spring HATEOAS library and allow us to generate a self-link for a given controller method.

This can also be done automatically by using Spring HATEOAS's `RepresentationModelAssembler` interface. This interface mainly exposes two methods – `toModel(T model)` and `toCollectionModel(Iterable<? extends T> entities)` – that convert the given entity/entities to `Model` and `CollectionModel` respectively.

Spring HATEOAS provides the following classes to enrich the user-defined models with hypermedia. It basically provides a class that contains links and methods to add those to the model:

- `RepresentationModel`: Models/DTOs can extend this to collect the links.

- `EntityModel`: This extends `RepresentationModel` and wraps the domain object (that is, the model) inside it with the `content` private field. Therefore, it contains the domain model/DTO and the links.

- `CollectionModel`: `CollectionModel` also extends `RepresentationModel`. It wraps the collection of models and provides a way to maintain and store the links.

- `PageModel`: `PageModel` extends `CollectionModel` and provides ways to iterate through the pages, such as `getNextLink()` and `getPreviousLink()`, and through page metadata with `getTotalPages()`, among others.

The default way to work with Spring HATEOAS is to extend `RepresentationModel` with domain models as shown in the following snippet:

```java
public class Cart extends RepresentationModel<Cart>  implements
Serializable {
  private static final long serialVersionUID = 1L;


  @JsonProperty("customerId")
  @JacksonXmlProperty(localName = "customerId")
  private String customerId;
```

```
@JsonProperty("items")
@JacksonXmlProperty(localName = "items")
@Valid
private List<Item> items = null;
```

Extending `RepresentationModel` enhances the model with additional methods including `getLink()`, `hasLink()`, and `add()`.

You know that all these models are being generated by the Swagger Codegen, therefore we need to configure the Swagger Codegen to generate new models that support hypermedia. This can be done by configuring the Swagger Codegen using the following `config.json` file:

```
{
    // ...
    "apiPackage": "com.packt.modern.api",
    "invokerPackage": "com.packt.modern.api",
    "serializableModel": true,
    "useTags": true,
    "useGzipFeature" : true,
    "hateoas": true,
    "withXml": true,
    // ...
}
```

https://github.com/PacktPublishing/Modern-API-Development-with-Spring-and-Spring-Boot/blob/main/Chapter04/src/main/resources/api/config.json

Adding the `hateoas` property and setting it to `true` would automatically generate models that would extend the `RepresentationModel` class.

We are halfway there to implement the API business logic. Now, we need to make sure that links will be populated with the appropriate URL automatically. For that purpose, we'll extend the `RepresentationModelAssemblerSupport` abstract class that internally implements `RepresentationModelAssembler`. Let's write the assembler for `Cart` as shown in the following code block:

```java
@Component
public class CartRepresentationModelAssembler extends
    RepresentationModelAssemblerSupport<CartEntity, Cart> {

  private ItemService itemService;

  public CartRepresentationModelAssembler(ItemService
      itemService) {
    super(CartsController.class, Cart.class);
    this.itemService = itemService;
  }


  @Override
  public Cart toModel(CartEntity entity) {
    String uid = Objects.nonNull(entity.getUser()) ?
                      entity.getUser().getId().toString() :
                      null;
    String cid = Objects.nonNull(entity.getId()) ?
                      entity.getId().toString() : null;
    Cart resource = new Cart();
    BeanUtils.copyProperties(entity, resource);
    resource.id(cid).customerId(uid)
              .items(itemService.toModelList(entity.
                      getItems())));
    resource.add(linkTo(methodOn(CartsController.class)
              .getCartByCustomerId(uid)).withSelfRel());
    resource.add(linkTo(methodOn(CartsController.class)
              .getCartItemsByCustomerId(uid.toString()))
          .withRel("cart-items"));
    return resource;
  }
```

```
public List<Cart> toListModel(Iterable<CartEntity>
    entities) {
  if (Objects.isNull(entities)) return Collections.
    emptyList();
  return StreamSupport.stream(entities.spliterator(), false)
            .map(e -> toModel(e)).collect(toList());
}


}
```

The important part in the `Cart` assembler is extending `RepresentationModelAssemblerSupport` and overriding the `toModel()` method. If you look closely, you'll see that `CartController.class` along with the `Cart` model is also passed to `Rep` using the `super()` call. This allows the assembler to generate the links appropriately as is required for the `methodOn` method shared earlier. This way, you can generate the link automatically.

You may also need to add additional links to other resource controllers. This you can achieve by writing a bean that implements `RepresentationModelProcessor` and then override the `process()` method as shown here:

```
@Override
public Order process(Order model) { model.add(Link.of("/
payments/{orderId}").withRel(LinkRelation.of("payments"))
    .expand(model.getOrderId()));


  return model;
}
```

You can always refer to `https://docs.spring.io/spring-hateoas/docs/current/reference/html/` for more information.

Enhancing the controller with a service and HATEOAS

In *Chapter 3, API Specifications and Implementation*, we created the `Controller` class for the Cart API – `CartController`, which just implements the Swagger Codegen-generated API specification interface – `CartApi`. It was just a mere block of code without any business logic or data persistence calls.

Now, since we have written the repositories, services, and HATEOAS assemblers, we can enhance the API controller class as shown here:

```
@RestController
public class CartsController implements CartApi {

  private static final Logger log = LoggerFactory
                        .getLogger(CartsController.class);
  private CartService service;
  private final CartRepresentationModelAssembler assembler;

  public CartsController(CartService service,
          CartRepresentationModelAssembler assembler) {
    this.service = service;
    this.assembler = assembler;
  }
```

https://github.com/PacktPublishing/Modern-API-Development-with-Spring-and-Spring-Boot/blob/main/Chapter04/src/main/java/com/packt/modern/api/controller/CartsController.java

You could see that `CartService` and `CartRepresentationModelAssembler` are injected using the constructor. The Spring container injects these dependencies at runtime. Then, these can be used as shown in the following code block:

```
@Override
public ResponseEntity<Cart> getCartByCustomerId(String
    customerId) {

  return ok(



      assembler.toModel(service.getCartByCustomerId
        (customerId)));
```

https://github.com/PacktPublishing/Modern-API-Development-with-Spring-and-Spring-Boot/blob/main/Chapter04/src/main/java/com/packt/modern/api/controller/CartsController.java

In the preceding code, you can see that the service retrieves the `Cart` entity based on `customerId` (which internally retrieves it from the repository). This `Cart` entity then gets converted into a model that also contains the hypermedia links made available by Spring HATEOAS's `RepresentationModelAssemblerSupport` class.

The `ok()` static method of `ResponseEntity` is used for wrapping the returned model that also contains the status `200 OK`.

This way you can also enhance and implement the other controllers. Now, we can also add an ETag to our API responses.

Adding ETags to API responses

An **Entity Tag** (**ETag**) is an HTTP response header that contains a computed hash or equivalent value of the response entity and a minor change in the entity must change its value. HTTP request objects can then contain the `If-None-Match` and `If-Match` headers for receiving the conditional responses.

Let's call an API for retrieving the response with an ETag as shown next:

```
$ curl -v --location --request GET 'http://localhost:8080/
api/v1/products/6d62d909-f957-430e-8689-b5129c0bb75e' –header
'Content-Type: application/json' --header 'Accept: application/
json'
Note: Unnecessary use of -X or --request, GET is already
inferred.
*    Trying ::1...
* TCP_NODELAY set
* Connected to localhost (::1) port 8080 (#0)
> GET /api/v1/products/6d62d909-f957-430e-8689-b5129c0bb75e
  HTTP/1.1
> Host: localhost:8080
> User-Agent: curl/7.55.1
> Content-Type: application/json
> Accept: application/json
>
< HTTP/1.1 200
< ETag: "098e97de3b61db55286f5f2812785116f"
< Content-Type: application/json
< Content-Length: 339
<
```

```json
{
    "_links": {
        "self": {
            "href": "http://localhost:8080/6d62d909-f957-430e-
                8689-b5129c0bb75e"
        }
    },
    "id": "6d62d909-f957-430e-8689-b5129c0bb75e",
    "name": "Antifragile",
    "description": "Antifragile - Things that gains from
                disorder. By Nassim Nicholas Taleb",
    "imageUrl": "/images/Antifragile.jpg",
    "price": 17.1500,
    "count": 33,
    "tag": [
        "psychology",
        "book"
    ]
}
```

Then, you can copy the value from the ETag header to the If-None-Match header and send the same request again with the If-None-Match header:

```
$ curl -v --location --request GET 'http://localhost:8080/
api/v1/products/6d62d909-f957-430e-8689-b5129c0bb75e'
--header 'Content-Type: application/json' --header
'Accept: application/json' --header 'If-None-Match:
"098e97de3b61db55286f5f2812785116f"'
Note: Unnecessary use of -X or --request, GET is already
inferred.
*   Trying ::1...
* TCP_NODELAY set
* Connected to localhost (::1) port 8080 (#0)
> GET /api/v1/products/6d62d909-f957-430e-8689-b5129c0bb75e
HTTP/1.1
> Host: localhost:8080
> User-Agent: curl/7.55.1
> Content-Type: application/json
> Accept: application/json
```

```
> If-None-Match: "098e97de3b61db55286f5f2812785116f"
>
< HTTP/1.1 304
< ETag: "098e97de3b61db55286f5f2812785116f"
```

You can see that since there is no change to the entity in the database, and it contains the same entity, it sends a 304 response instead of sending the proper response with 200 OK.

The easiest and simplest way to implement ETags is using Spring's ShallowEtagHeaderFilter as shown here:

```
@Bean
public ShallowEtagHeaderFilter shallowEtagHeaderFilter() {

  return new ShallowEtagHeaderFilter();
```

https://github.com/PacktPublishing/Modern-API-Development-with-Spring-and-Spring-Boot/blob/main/Chapter04/src/main/java/com/packt/modern/api/AppConfig.java

For this implementation, Spring calculates the MD5 hash from the cached content written to the response. Next time, when it receives a request with the If-None-Match header, it again creates the MD5 hash from the cached content written to the response and then compares these two hashes. If both are the same, it sends the 304 NOT MODIFIED response. This way it will save bandwidth but computation will be performed there using the same CPU computation.

We can use the HTTP cache control (org.springframework.http.CacheControl) class and use the version or similar attribute that gets updated for each change, if available, to avoid unecessary CPU computation and for better ETag handling as shown next:

```
Return ResponseEntity.ok()
       .cacheControl(CacheControl.maxAge(5, TimeUnit.DAYS))
       .eTag(prodcut.getModifiedDateInEpoch())
       .body(product);
```

Adding an ETag to the response also allows UI apps to determine whether a page/object refresh is required, or an event needs to be triggered, especially where data changes frequently in applications such as providing live scores or stock quotes.

Testing the APIs

Now, you must be looking forward to testing. You can find the Postman (API client) collection at the following location, which is based on Postman Collection version 2.1. You can import it and then test the APIs:

https://github.com/PacktPublishing/Modern-API-Development-with-Spring-and-Spring-Boot/blob/main/Chapter04/Chapter04.postman_collection.json

---

Building and running the service

You can build the code by running `gradlew clean build` from the root of the project, and run the service using `java -jar build/libs/Chapter04-0.0.1-SNAPSHOT.jar`. Make sure to use Java 15 in the path.

Summary

In this chapter, we have learned about database migration using Flyway, maintaining and persisting data using repositories, and writing business logic to services. You have also learned how hypermedia can automatically be added to API responses using Spring HATEOAS assemblers. You have now learned about the complete RESTful API development practices, which allows you to use these skill in your day-to-day work involving RESTful API development.

So far we have written synchronous APIs. In the next chapter, you will learn about async APIs and how to implement them using Spring.

Questions

1. Why is the `@Repository` class used?

2. Is it possible to add extra imports or annotations to Swagger-generated classes or models?

3. How is ETag useful?

Further reading

- Spring HATEOAS: `https://docs.spring.io/spring-hateoas/docs/current/reference/html/`

- *RFC-8288*: `https://tools.ietf.org/html/rfc8288`

- A video on Spring HATEOAS: `https://subscription.packtpub.com/video/programming/9781788993241/p3/video3_6/using-spring-hateoas`

- The Postman tool: `https://learning.postman.com/docs/getting-started/sending-the-first-request/`

5
Asynchronous API Design

So far, we have developed RESTful web services based on the traditional model, where calls are synchronous. What if you want to make code async and non-blocking? This is what we are going to do in this chapter. You'll learn about asynchronous API design in this chapter, where calls are asynchronous and non-blocking. We'll develop these APIs using Spring WebFlux, which is itself based on Project Reactor (`https://projectreactor.io`).

First, we'll walk through the Reactive programming fundamentals, and then we'll migrate the existing e-commerce REST API (which we learned about in *Chapter 4, Writing Business Logic for APIs*) to an asynchronous (Reactive) API to make things easier by co-relating and comparing the existing (imperative) way and Reactive way of programming.

We'll discuss the following topics in this chapter:

- Understanding Reactive Streams
- Exploring Spring WebFlux
- Understanding DispatcherHandler
- Controllers
- Functional endpoints
- Implementing Reactive APIs for our e-commerce app

At the end of this chapter, you will learn how to develop and implement the reactive APIs and explore the async API development.

Technical requirements

You need the following to execute the code in this chapter:

- Any Java IDE, such as NetBeans, IntelliJ, or Eclipse
- **Java Development Kit (JDK)** 15
- An internet connection to clone the code and download the dependencies and Gradle
- Postman/cURL (for API testing)

The code present in this chapter is found at `https://github.com/ PacktPublishing/Modern-API-Development-with-Spring-and-Spring-Boot/tree/main/Chapter05`.

Understanding Reactive Streams

Normal Java code achieves asynchronicity by using thread pools. Your web server uses a thread pool for serving requests – it assigns a thread to each incoming request. The application uses the thread pool for database connections. Each database call uses a separate thread and waits for the result. Therefore, each web request and database call uses its own thread. However, there is a wait associated with this and therefore, these are blocking calls. The thread waits and utilizes the resources until a response is received back from the database or a response object is written. This is kind of a limitation when you scale as you can only use the resources available to JVM. You overcome this limitation by using a load balancer with other instances of the service, which is a type of horizontal scaling.

In the last decade, there has been a rise in client-server architecture. Lots of IoT-enabled devices, smartphones that have native apps, first-class web apps, and traditional web applications have emerged. Applications not only have third-party services but also have various sources of data, which leads to higher-scale applications. On top of that, microservice-based architecture has increased the communication among services themselves. You need lots of resources to serve this higher-network communication demand. This makes scaling a necessity. Threads are expensive and not infinite. You don't want to block them for effective utilization. This is where asynchronicity helps. In asynchronous calls, threads become free as soon as a call is done and use a callback utility such as JavaScript. When data is available at the source, it pushes the data. Reactive Streams uses the **publisher-subscriber model**, where the source of data, the publisher, pushes the data to the subscriber.

You might be aware that, on the other hand, Node.js uses a single thread to make use of most resources. It is based on an asynchronous non-blocking design, known as an **event loop**.

Reactive APIs are also based on an event loop design and use push-style notifications. If you look closely, Reactive Streams also supports Java Streams operations such as `map`, `flatMap`, and `filter`. Internally, Reactive Streams uses a push style, whereas a Java Stream works on a pull model; that is, items are pulled from the source, such as a `Collection`. In Reactive, the source (publisher) pushes the data.

In **Reactive Streams**, streams of data are asynchronous and non-blocking and support back-pressure. (Refer to the *Subscriber* section of this chapter for an explanation of back-pressure.)

There are four basic types as per the Reactive Streams specification:

- Publisher
- Subscriber
- Subscription
- Processor

Let's have a look at each.

Publisher

The publisher provides a stream of data to one or more subscribers. A subscriber uses the `subscriber()` method to subscribe to a publisher. Each subscriber should only subscribe once to a publisher. Most importantly, the publisher pushes data according to the demand received from subscribers. Reactive streams are lazy; therefore, the publisher will only push an element if there is a subscriber.

A publisher is defined as follows:

```
package org.reactivestreams;

// T - type of element Publisher sends
public interface Publisher<T> {
    public void subscribe(Subscriber<? super T> s);
}
```

Subscriber

The subscriber consumes the data pushed by the publisher. Publisher-subscriber communication works as follows:

1. When a `Subscriber` instance is passed to the `Publisher.subscribe()` method, it triggers the `onSubscribe()` method. It contains a `Subscription` parameter that controls the back-pressure, that is, how much data a subscriber demands from the publisher.

2. After the first step, `Publisher` waits for the `Subscription.request(long)` call. It only pushes data to `Subscriber` after the `Subscription.request()` call is made. This method demands the number of elements from `Publisher`.

    Normally, the publisher pushes the data to the subscriber, irrespective of whether the subscriber can handle it safely or not. However, the subscriber knows best how much data it can handle safely; therefore, in Reactive streams, `Subscriber` uses the `Subscription` instance to communicate the demand for the number of elements to `Publisher`. This is known as **back-pressure** or **flow control**.

    You must be wondering, what if `Publisher` asks `Subscriber` to slow down but it can't? In that case, `Publisher` has to decide whether to fail, drop, or buffer.

3. Once the demand is made using *step 2*, Publisher sends the data notifications and the onNext() method is used to consume it. It will be triggered until the data notifications are pushed by Publisher according to the demand communicated by Subscription.request().

4. At the end, either onError() or onCompletion() will be triggered as the terminal state. No notification will be sent after one of these invocations has been triggered even if you call Subscription.request(). The following are the terminal methods:

    a. onError() would be invoked the moment any error occurs.

    b. onCompletion() would be invoked when all elements are pushed.

The Subscriber interface is defined as follows:

```
package org.reactivestreams;


// T - type of element Publisher sends
public interface Subscriber<T> {

    public void onSubscribe(Subscription s);
    public void onNext(T t);
    public void onError(Throwable t);
    public void onComplete();
}
```

Subscription

A subscription is a mediator between the publisher and subscriber. It is the subscriber's responsibility to invoke the Subscription.subscriber() method and let the publisher know of the demand. It can be invoked as and when required by the subscriber. The cancel() method asks the publisher to stop sending data notifications and to clean up the resources.

A subscription is defined as follows:

```
package org.reactivestreams;

public interface Subscription {
    public void request(long n);
    public void cancel();
}
```

Processor

The processor is a bridge between the publisher and subscriber and represents the processing stage. It works as both a publisher and subscriber and obeys the contract defined by both. It is defined as follows:

```
package org.reactivestreams;

public interface Processor<T, R> extends Subscriber<T>,
Publisher<R> {
}
```

Let's have a look at the following example. Here, we are creating `Flux` by using the `Flux.just()` static factory method. `Flux` is a type of publisher in Project Reactor. This publisher contains four integer elements. Then, we use the `reduce` operator (similar to Java Streams) to perform a `sum` operation on it:

```
Flux<Integer> fluxInt = Flux.just(1, 10, 100, 1000).log();
fluxInt.reduce(Integer::sum)
    .subscribe(sum -> System.out.printf("Sum is: %d", sum));
```

When you run this code, it prints the following output:

```
11:00:38.074 [main] INFO reactor.Flux.Array.1 - |
onSubscribe([Synchronous Fuseable] FluxArray.ArraySubscription)
11:00:38.074 [main] INFO reactor.Flux.Array.1 - |
request(unbounded)
11:00:38.084 [main] INFO reactor.Flux.Array.1 - | onNext(1)
11:00:38.084 [main] INFO reactor.Flux.Array.1 - | onNext(10)
11:00:38.084 [main] INFO reactor.Flux.Array.1 - | onNext(100)
11:00:38.084 [main] INFO reactor.Flux.Array.1 - | onNext(1000)
```

```
11:00:38.084 [main] INFO reactor.Flux.Array.1 - | onComplete()
Sum is: 1111
Process finished with exit code 0
```

Looking at the output, when `Publisher` is subscribed, `Subscriber` sends an unbounded `Subscription.request()`. When the first element is notified, `onNext()` is called, and so on. At the end, when the publisher is done with the push elements, the `onComplete()` event is called. This is how Reactive streams work.

Now that you have an idea of how Reactive streams work, let's see how and why Spring makes use of these Reactive streams in the Spring WebFlux module.

Exploring Spring WebFlux

Existing Servlet APIs are blocking APIs. They use input and output streams, which block APIs. Servlet 3.0 containers evolve and use the underlying event loop. Async requests are processed asynchronously but read and write operations still use the input/output streams that are blocking. The Servlet 3.1 container evolves further and supports asynchronicity, and has the non-blocking I/O stream APIs. However, there are certain Servlet APIs, such as `request.getParameters()`, that parse the request body that is blocking and provide synchronous contracts such as `Filter`. The **Spring MVC** framework is based on the Servlet API and Servlet containers.

Therefore, Spring provides **Spring WebFlux**, which is fully non-blocking and provides back-pressure functionality. It provides concurrency with a small number of threads and scales with fewer hardware resources. WebFlux provides fluent, functional, and continuation-style APIs to support the declarative composition of asynchronous logic. Writing asynchronous functional code is more complex than writing imperative-style code. However, once you get hands-on with it, you will love it because it allows you to write precise and readable code.

Both Spring WebFlux and Spring MVC can co-exist; however, you should never mix a Reactive flow with blocking calls for the effective use of the Reactive programming model.

Spring WebFlux supports the following features and archetypes:

- The event loop concurrency model
- Both annotated controllers and functional endpoints
- Reactive clients
- Netty and Servlet 3.1 container-based web servers, such as Tomcat, Undertow, and Jetty

Let's dig deep to understand how WebFlux works by understanding the Reactive APIs and Reactor Core.

Reactive APIs

Spring WebFlux APIs are Reactive APIs and accept `Publisher` as plain input. WebFlux then adapts it to a type supported by a Reactive library such as Reactor Core or RxJava. It then processes the input and returns the output based on the supported Reactive library's type. This allows WebFlux APIs to be interoperable with other Reactive libraries.

By default, Spring WebFlux uses Reactor (`https://projectreactor.io`) as a core dependency. Project Reactor provides Reactive Streams library. As stated in the previous paragraph, WebFlux accepts the input as `Publisher`, then adapts it to a Reactor type, and then returns it as `Mono` or `Flux` output.

You know that `Publisher` in Reactive Streams pushes the data to its subscribers based on demand. It can push one or more (possibly infinite) elements. Project Reactor takes it further and provides two `Publisher` implementations, namely `Mono` and `Flux`. `Mono` can return either 0 or 1 to `Subscriber`, whereas `Flux` returns 0 to N elements. Both are abstract classes that implement the `CorePublisher` interface. The `CorePublisher` interface extends the publisher.

Normally, we have the following methods in the repository:

```
public Product findById(UUID id);
public List<Product> getAll();
```

These can be replaced with `Mono` and `Flux`:

```
Public Mono<Product> findById(UUID id);
public Flux<Product> getAll();
```

There is the concept of hot and cold streams. The source is restarted if there are multiple subscribers for cold streams and the same source is used for multiple subscribers in hot streams. Project Reactor streams are, by default, cold. Therefore, once you consume a stream, you can't reuse it until restarted. However, Project Reactor allows you to turn a cold stream into a hot one by using `cache()` methods. These two methods are available in both `Mono` and `Flux` abstract classes.

Let's understand the cold and hot stream concepts with some examples:

```
Flux<Integer> fluxInt = Flux.just(1, 10, 100).log();
fluxInt.reduce(Integer::sum)
    .subscribe(sum -> System.out.printf("Sum is: %d\n", sum));
fluxInt.reduce(Integer::max)
    .subscribe(max -> System.out.printf("Maximum is: %d",
                                        max));
```

Here, we are creating `Flux` of three numbers. Then, we are performing two operations separately – `sum` and `max`. You can see that there are two subscribers. By default, Project Reactor streams are cold; therefore, when a second subscriber registers, it restarts, as shown in the following output:

```
11:23:35.060 [main] INFO reactor.Flux.Array.1 - |
onSubscribe([Synchronous Fuseable] FluxArray.ArraySubscription)
11:23:35.060 [main] INFO reactor.Flux.Array.1 - |
request(unbounded)
11:23:35.060 [main] INFO reactor.Flux.Array.1 - | onNext(1)
11:23:35.060 [main] INFO reactor.Flux.Array.1 - | onNext(10)
11:23:35.060 [main] INFO reactor.Flux.Array.1 - | onNext(100)
11:23:35.060 [main] INFO reactor.Flux.Array.1 - | onComplete()
Sum is: 111
11:23:35.076 [main] INFO reactor.Flux.Array.1 - |
onSubscribe([Synchronous Fuseable] FluxArray.ArraySubscription)
11:23:35.076 [main] INFO reactor.Flux.Array.1 - |
request(unbounded)
11:23:35.076 [main] INFO reactor.Flux.Array.1 - | onNext(1)
11:23:35.076 [main] INFO reactor.Flux.Array.1 - | onNext(10)
11:23:35.076 [main] INFO reactor.Flux.Array.1 - | onNext(100)
11:23:35.076 [main] INFO reactor.Flux.Array.1 - | onComplete()
Maximum is: 100
```

The source is created in the same program, but what if the source is somewhere else, such as in an HTTP request, or you don't want to restart the source? In these cases, you can turn the cold stream into a hot stream by using `cache()`, as shown in the next code block. The only difference between this one and the previous code is that we have added a `cache()` call to `Flux.just()`:

```
Flux<Integer> fluxInt = Flux.just(1, 10, 100).log().cache();
fluxInt.reduce(Integer::sum)
    .subscribe(sum -> System.out.printf("Sum is: %d\n", sum));
fluxInt.reduce(Integer::max)
    .subscribe(max -> System.out.printf("Maximum is: %d",
                                                max));
```

Now, look at the output. The source has not restarted; instead, the same source is used again:

```
11:29:25.665 [main] INFO reactor.Flux.Array.1 - |
onSubscribe([Synchronous Fuseable] FluxArray.ArraySubscription)
11:29:25.665 [main] INFO reactor.Flux.Array.1 - |
request(unbounded)
11:29:25.665 [main] INFO reactor.Flux.Array.1 - | onNext(1)
11:29:25.665 [main] INFO reactor.Flux.Array.1 - | onNext(10)
11:29:25.665 [main] INFO reactor.Flux.Array.1 - | onNext(100)
11:29:25.665 [main] INFO reactor.Flux.Array.1 - | onComplete()
Sum is: 111
Maximum is: 100
```

Now we have got to the crux of Reactive APIs, let's see what Spring WebFlux's Reactive Core consists of.

Reactive Core

This provides a foundation for developing a Reactive web application with Spring. A web application needs three levels of support for serving HTTP web requests:

- Handling of web requests by the server:

  a. `HttpHandler`: An interface that is an abstraction of a request/response handler over different HTTP server APIs, such as Netty or Tomcat:

  ```
  public interface HttpHandler {
    Mono<Void> handle(ServerHttpRequest request,
  ```

```
    ServerHttpResponse response);
    }
```

b. `WebHandler`: Provides support for user sessions, request and session attributes, a locale and principal for the request, form data, and so on

- Handling of a web request call by the client using `WebClient`

- Codecs (`Encoder`, `Decoder`, `HttpMessageWriter`, `HttpMessageReader`, and `DataBuffer`) for the serialization and deserialization of content at both the server and client level for the request and response

These components are at the core of Spring WebFlux. WebFlux application configuration also contains the following beans – `webHandler` (`DispatcherHandler`), `WebFilter`, `WebExceptionHandler`, `HandlerMapping`, `HandlerAdapter`, and `HandlerResultHandler`.

For REST service implementation, there are specific `HandlerAdapter` instances for the following web servers – Tomcat, Jetty, Netty, and Undertow. A web server such as Netty, which supports Reactive Streams, handles the subscriber's demands. However, if the server handler does not support Reactive Streams, then the `org.springframework.http.server.reactive.ServletHttpHandlerAdapter` HTTP `HandlerAdapter` is used. It handles the adaptation between Reactive Streams and Servlet 3.1 container async I/O and implements a `Subscriber` class. This uses the OS TCP buffers. OS TCP uses its own back-pressure (control flow); that is, when the buffer is full, the OS uses the TCP back-pressure to stop incoming elements.

The browser, or any HTTP client, consumes REST APIs using the HTTP protocol. When a request is received by the web server, it forwards it to the Spring WebFlux application. Then, WebFlux builds the Reactive pipeline that goes to the controller. `HttpHandler` is an interface between WebFlux and the web server that communicates using the HTTP protocol. If the underlying server supports Reactive Streams, such as Netty, then the subscription is done by the server natively. Else, WebFlux uses `ServletHttpHandlerAdapter` for Servlet 3.1 container-based servers. `ServletHttpHandlerAdapter` then adapts the streams to async I/O Servlet APIs and vice versa. Then, the subscription of Reactive Streams happens with `ServletHttpHandlerAdapter`.

Therefore, in summary, `Mono`/`Flux` streams are subscribed by WebFlux internal classes, and when the controller sends a `Mono`/`Flux` stream, these classes convert it into HTTP packets. The HTTP protocol does support event streams. However, for other media types, such as JSON, Spring WebFlux subscribes the `Mono`/`Flux` streams and waits until `onComplete()` or `onError()` is triggered. Then, it serializes the whole list of elements, or a single element in the case of `Mono`, in one HTTP response.

Spring WebFlux needs a component similar to `DispatcherServlet` in Spring MVC –
a front controller. Let's discuss this in the next section.

Understanding DispatcherHandler

`DispatcherHandler`, a front controller in Spring WebFlux, is what
`DispatcherServlet` is in the Spring MVC framework. `DispatcherHandler`
contains an algorithm that makes use of special components – `HandlerMapping`
(maps requests to the handler), `HandlerAdapter` (a `DispatcherHandler` helper to
invoke a handler mapped to a request), and `HandlerResultHandler` (a palindrome
of words, for processing the result and forming results) – for processing requests. The
`DispatcherHandler` component is identified by a bean named `webHandler`.

It processes requests in the following way:

1. A web request is received by `DispatcherHandler`.

2. `DispatcherHandler` uses `HandlerMapping` to find a matching handler for
   the request and uses the first match.

3. It then uses the respective `HandlerAdapter` to process the request, which
   exposes the `HandlerResult` (return value after processing). The return value
   could be one of the following – `ResponseEntity`, `ServerResponse`, values
   returned from `@RestController`, or values (`CharSequence`, `view`, `map`, and
   so on) returned by a view resolver.

4. Then, it makes use of the respective `HandlerResultHandler` to
   write the response or render a view based on the `HandlerResult` type
   received from *step 2*. `ResponseEntityResultHandler` is used for
   `ResponseEntity`, `ServerResponseResultHandler` is used for
   `ServerResponse`, `ResponseBodyResultHandler` is used for values
   returned by `@RestController` or `@ResponseBody`-annotated methods, and
   `ViewResolutionResultHandler` is used for values returned by the view
   resolver.

5. The request is completed.

You can create REST endpoints in Spring WebFlux using either an annotated controller
such as Spring MVC or functional endpoints. Let's explore these in the next sections.

Controllers

The Spring team has kept the same annotations for both Spring MVC and Spring WebFlux as these annotations are non-blocking. Therefore, you can use the same annotations we have used in previous chapters for creating REST controllers. There, the annotation runs on Reactive Core and provides a non-blocking flow. However, you, as the developer, have the responsibility of maintaining a fully non-blocking flow and maintaining the Reactive chain (pipeline). Any blocking calls in a Reactive chain would convert the Reactive chain into a blocking call.

Let's create a simple REST controller that supports non-blocking and Reactive calls:

```java
@RestController
public class OrderController {


  @RequestMapping(value = "/api/v1/orders",
                  method = RequestMethod.POST)
  public ResponseEntity<Order> addOrder(@RequestBody NewOrder
                                                   newOrder){
     // ...
  }


  @RequestMapping(value = "/api/v1/orders/{id}",
                  method = RequestMethod.GET)
  public ResponseEntity<Order> getOrderById(@PathVariable("id")
                                                   String id){
     // ...
  }


}
```

You can see that it uses all the annotations that we have used in Spring MVC:

- `@RestController` is used for marking a class as a REST controller. Without this, the endpoint won't register and the request will be returned as NOT FOUND 404.

- `@RequestMapping` is used for defining the path and HTTP method. Here, you can also use `@PostMapping` with just the path. Similarly, for each of the HTTP methods, a respective mapping is there, such as `@GetMapping`.

- The @RequestBody annotation marks a parameter as a request body, and an appropriate codec would be used for conversion. Similarly, there is @PathVariable and @RequestParam for the path parameter and query parameter, respectively.

We are going to use an annotation-based model for writing the REST endpoints. You'll get a closer look when we implement the e-commerce app controllers using WebFlux. Spring WebFlux also provides a way to write a REST endpoint using a functional programming style that you'll explore in the next section.

Functional endpoints

The REST controllers we coded using Spring MVC were written in imperative-style programming. Reactive programming, on the other hand, is functional-style programming. Therefore, Spring WebFlux also allows an alternative way to define REST endpoints, using functional endpoints. These also use the same Reactive Core foundation.

Let's see how we can write the same order REST endpoint using a functional endpoint:

```java
import static org.springframework.http.MediaType.APPLICATION_
    JSON;
import static org.springframework.web.reactive.function.server.
    RequestPredicates.*;
import static org.springframework.web.reactive.function.server.
    RouterFunctions.route;

OrderRepository repository = ...
OrderHandler handler = new OrderHandler(repository);

RouterFunction<ServerResponse> route = route()
    .GET("/v1/api/orders/{id}", accept(APPLICATION_JSON),
            handler::getOrderById)
    .POST("/v1/api/orders", handler::addOrder)
    .build();
public class OrderHandler {
    public Mono<ServerResponse> addOrder(ServerRequest req){
        // ...
    }

    public Mono<ServerResponse> getOrderById(ServerRequest req)
    {
```

```
        // ...
    }
}
```

You can see that the `RouterFunctions.route()` builder allows you to write all the REST routes in a single statement using the functional programming style. Then, it uses the method reference of the handler class to process the request, which is exactly the same as the `@RequestMapping` body of an annotation-based model.

Let's add the following code in the `OrderHandler` methods:

```java
public class OrderHandler {
    public Mono<ServerResponse> addOrder(ServerRequest req){
        Mono<NewOrder> order = req.bodyToMono(NewOrder.class);
        return ok().build(repository.save(toEntity(order)));
    }
    public Mono<ServerResponse> getOrderById(ServerRequest req)
{
        String orderId = req.pathVariable("id");
        return repository.getOrderById(UUID.fromString(orderId))
           .flatMap(order -> ok()
              .contentType(APPLICATION_JSON).
                       bodyValue(toModel(order)))
           .switchIfEmpty(ServerResponse.notFound().build());
    }
}
```

Unlike the `@RequestMapping()` mapping methods in the REST controller, handler methods don't have multiple parameters such as body, path, or query parameters. They just have a `ServerRequest` parameter, which can be used to extract the body, path, and query parameters. In the `addOrder` method, the `Order` object is extracted using `request.bodyToMono()`, which parses the request body and then converts it into an `Order` object. Similarly, the ID is extract from a request using `request.pathVariable()` in the `getOrderById()` handler method.

Now, let's discuss the response. The handler method uses the `ServerResponse` object in comparison to `ResponseEntity` in Spring MVC. Therefore, the `ok()` static method looks like it's from `ResponseEntity`, but it is from `org.springframework.web.reactive.function.server.ServerResponse.ok`. The Spring team has tried to keep the API as similar as possible to Spring MVC; however, the underlying implementation differs and provides a non-blocking Reactive interface.

The last point about these handler methods is the way a response is written. It uses a functional style instead of an imperative style and makes sure that the Reactive chain does not break. The repository returns the `Mono` object (a publisher) in both cases and returns it as a response wrapped inside `ServerResponse`.

You can find interesting code in the `getOrderById()` handler method. It performs a `flatMap` operation on the received `Mono` object from the repository. It converts it from an entity into a model, then wraps it in a `ServerResponse` object and returns the response. You must be wondering what happens if the repository returns null. The repository returns `Mono` as per the contract, which is similar in nature to the Java `Optional` class. Therefore, the `Mono` object can be empty but not null, as per the contract. If the repository returns an empty `Mono`, then the `switchIfEmpty()` operator will be used and a `NOT FOUND 404` response will be sent.

In the case of an error, there are different error operators that can be used, such as `doOnError()` or `onErrorReturn()`.

We have discussed the logic flow using the `Mono` type; the same explanation will apply if you use the `Flux` type in place of the `Mono` type.

We have discussed a lot of theory relating to Reactive, asynchronous, and non-blocking programming in a Spring context. Let's jump into coding and migrate the e-commerce API developed in *Chapter 4, Writing Business Logic for APIs*, to a Reactive API.

Implementing Reactive APIs for our e-commerce app

Now that you have an idea of how Reactive streams work, we can go ahead and implement REST APIs that are asynchronous and non-blocking.

You'll recall that we are following the design-first approach, so we need the API design specification first. However, we can reuse the e-commerce API specification we created previously in *Chapter 3, API Specifications and Implementation*.

OpenAPI Codegen is used for generating the API interface/contract that generates the Spring MVC-compliant API Java interfaces. Let's see what changes we need to do to generate the Reactive API interfaces.

Changing OpenAPI Codegen for Reactive APIs

You need to tweak few OpenAPI Codegen configurations to generate Spring WebFlux-compliant Java interfaces, as shown next:

```
{
    "library": "spring-boot",
    "dateLibrary": "java8",
    "hideGenerationTimestamp": true,
    "modelPackage": "com.packt.modern.api.model",
    "apiPackage": "com.packt.modern.api",
    "invokerPackage": "com.packt.modern.api",
    "serializableModel": true,
    "useTags": true,
    "useGzipFeature" : true,
    "reactive": true,
    "interfaceOnly": true,
        ...
        ...
}
```

https://github.com/PacktPublishing/Modern-API-Development-with-Spring-and-Spring-Boot/blob/main/Chapter05/src/main/resources/api/config.json

Reactive API support is only there if you opt for `spring-boot` as `library`. Also, you need to set the `reactive` flag to `true`. By default, the `reactive` flag is `false`.

Now, you can run the following command:

```
$ gradlew clean generateSwaggerCode
```

This will generate Reactive Streams-compliant Java interfaces, which are annotation-based REST controller interfaces. When you open any API interface, you'll find `Mono/Flux` reactor types in it, as shown in the following code block for the `OrderAPI` interface:

```
@ApiOperation(value = "Creates a new order for the given order
request", nickname = "addOrder", notes = "Creates a new order
for the given order request.", response = Order.class, tags = {
"order",})
@ApiResponses(value = {
    @ApiResponse(code = 201, message = "Order added
```

```
                   successfully", response = Order.class),
       @ApiResponse(code = 406, message = "If payment is not
                   authorized.")})
   @RequestMapping(value = "/api/v1/orders",
       produces = {"application/xml", "application/json"},
       consumes = {"application/xml", "application/json"},
       method = RequestMethod.POST)
   Mono<ResponseEntity<Order>> addOrder(
       @ApiParam(value = "New Order Request object") @Valid @
          RequestBody(required = false) Mono<NewOrder> newOrder,
          ServerWebExchange exchange);
```

You would have observed another change: an additional parameter, ServerWebExchange, is also required for Reactive controllers.

Now, when you compile your code, you may find compilation errors, because we haven't yet added the dependencies required for Reactive support. Let's learn how to add them in the next section.

Adding Reactive dependencies in build.xml

First, we'll remove spring-boot-starter-web as we don't need Spring MVC now. Second, we'll add spring-boot-starter-webflux and reactor-test for Spring WebFlux and Reactor support tests, respectively. Once these dependencies are added successfully, you should not see any compilation errors in the OpenAPI-generated code.

You can add the required Reactive dependencies in build.gradle as shown next:

```
implementation 'org.springframework.boot:spring-boot-starter-
webflux'
// implementation 'org.springframework.boot:spring-boot-
// starter-web'

testImplementation('org.springframework.boot:spring-boot-
                   starter-test')
testImplementation 'io.projectreactor:reactor-test'
```

https://github.com/PacktPublishing/Modern-API-Development-with-Spring-and-Spring-Boot/blob/main/Chapter05/build.gradle

We need to have a complete Reactive pipeline from the REST controller to the database. However, existing JDBC and Hibernate dependencies only support blocking calls. JDBC is a fully blocking API. Hibernate is also blocking. Therefore, we need to have Reactive dependencies for the database.

Hibernate Reactive (`https://github.com/hibernate/hibernate-reactive`) is in beta version at the time of writing this chapter. Hibernate Reactive beta only supports PostgresSQL, MySQL, and Db2. Hibernate Reactive does not support H2 at the time of writing. Therefore, we would simply use Spring Data, a Spring framework that provides the `spring-data-r2dbc` library for working with Reactive Streams.

Many NoSQL databases, such as MongoDB, already provide a Reactive database driver. An R2DBC-based driver should be used for relational databases in place of JDBC for fully non-blocking/Reactive API calls. **R2DBC** stands for **Reactive Relational Database Connectivity**. R2DBC is a Reactive API open specification that establishes a **Service Provider Interface** (**SPI**) for database drivers. Almost all the popular relation databases support R2DBC drivers – H2, MySQL, MariaDB, SQL Server, PostgresSQL, and Proxy. Oracle DB, at the time of writing, provides flow-based Reactive JDBC extensions (DB 20c) that integrate with Reactor, RxJava, and Akka Streams. However, an Oracle R2DBC driver is soon to be launched (it hasn't yet been launched as of 2020).

Let's add the R2DBC dependencies for Spring Data and H2 in the `build.gradle` file:

```
// DB Starts
implementation 'org.springframework.boot:spring-boot-starter-
data-r2dbc'
implementation 'com.h2database:h2'
runtimeOnly 'io.r2dbc:r2dbc-h2'
// DB Ends
```

Now, we can write end-to-end (from the controller to the repository) code without any compilation errors. Let's add global exception handling before we jump into writing an implementation for API interfaces.

Handling exceptions

We'll add the global exception handler the way it was added in Spring MVC in *Chapter 3, API Specifications and Implementation*. Before that, you must be wondering how to handle exceptions in a Reactive pipeline. Reactive pipelines are a flow of streams and you can't add exception handling the way you do in imperative code. You need to raise the error in a pipeline flow only.

Check out the following code:

```
.flatMap(card -> {
  if (Objects.isNull(card.getId())) {
    return service.registerCard(mono)
        .map(ce -> status(HttpStatus.CREATED)
            .body(assembler.entityToModel(ce, exchange)));
  } else {
    return Mono.error(
        () -> new CardAlreadyExistsException(" for user with ID
        - " + d.getId()));
  }
})
```

Here, a flatMap operation is performed. An error should be thrown if card is not valid, that is, if card does not have the requested ID. Here, Mono.error() is used because the pipeline expects Mono as a returned object. Similarly, you can use Flux.error() if Flux is expected as the returned type.

Let's assume you are expecting an object from a service or repository call, but instead you receive an empty object. Then you can use the switchIfEmpty() operator as shown in the next code:

```
Mono<List<String>> monoIds = itemRepo.findByCustomerId(
    customerId)
  .switchIfEmpty(Mono.error(new ResourceNotFoundException(
      ". No items found in Cart of customer with Id - " +
      customerId)))
  .map(i -> i.getId().toString())
  .collectList().cache();
```

Here, the code expects a Mono object of List from the item repository. However, if the returned object is empty, then it simply throws ResourceNotFoundException. switchIfEmpty() accepts the alternate Mono instance.

By now, you might have a question about the type of exception. It throws a runtime exception. See the ResourceNotFoundException class declaration here:

```
public class ResourceNotFoundException extends RuntimeException
```

Similarly, you can also use `onErrorReturn()`, `onErrorResume()`, or similar error operators from Reactive Streams. Look at the use of `onErrorReturn()` in the next code block:

```
return service.getCartByCustomerId(customerId)
    .map(cart ->
        assembler.itemfromEntities(cart.getItems().stream()
            .filter(i -> i.getProductId().toString().
                    equals(itemId.trim())).collect(toList()))
            .get(0)).map(ResponseEntity::ok)
    .onErrorReturn(notFound().build())
```

All exceptions should be handled and error response should be sent to the user. This is why we'll have a look at the global exception handler first.

Handling global exceptions for controllers

You created a global exception handler using `@ControllerAdvice` in Spring MVC. We'll take a slightly different route for handling errors in Spring WebFlux. First, we'll create the `ApiErrorAttributes` class, which can also be used in Spring MVC. This class extends `DefaultErrorAttributes`, which is a default implementation of the `ErrorAttributes` interface. The `ErrorAttributes` interface provides a way to handle maps, a map of error fields, and their values. These error attributes can then be used for displaying an error to the user or for logging.

The following attributes are provided by the `DefaultErrorAttributes` class:

- `timestamp`: The time that the error was captured
- `status`: The status code
- `error`: Error description
- `exception`: The class name of the root exception (if configured)
- `message`: The exception message (if configured)
- `errors`: Any `ObjectErrors` from a `BindingResult` exception (if configured)
- `trace`: The exception stack trace (if configured)
- `path`: The URL path when the exception was raised
- `requestId`: The unique ID associated with the current request

We have added two default values to the status and message – an internal server error and a generic error message (The system is unable to complete the request. Contact system support.), respectively – in ApiErrorAttributes, as shown next:

```
@Component
public class ApiErrorAttributes extends DefaultErrorAttributes
{

    private HttpStatus status = HttpStatus.INTERNAL_SERVER_ERROR;
    private String message = ErrorCode.GENERIC_ERROR.
        getErrMsgKey();

    @Override
    public Map<String, Object> getErrorAttributes(ServerRequest
        request, ErrorAttributeOptions options) {
      var attributes = super.getErrorAttributes(
                        request, options);
      attributes.put("status", status);
      attributes.put("message", message);
      attributes.put("code", ErrorCode.GENERIC_ERROR.
                    getErrCode());
      return attributes;
    }
 // Getter/Setters
```

https://github.com/PacktPublishing/Modern-API-Development-with-Spring-and-Spring-Boot/blob/main/Chapter05/src/main/java/com/packt/modern/api/exception/ApiErrorAttributes.java

Now, we can use this ApiErrorAttributes class in a custom global exception handler class. We'll create the ApiErrorWebExceptionHandler class, which extends the AbstractErrorWebExceptionHandler abstract class.

The AbstractErrorWebExceptionHandler class implements the ErrorWebExceptionHandler and InitializingBean interfaces. ErrorWebExceptionHandler is a functional interface that extends the WebExceptionHandler interface, which indicates that WebExceptionHandler is used for rendering exceptions. WebExceptionHandler is a contract for handling exceptions when server exchange processing takes place.

The `InitializingBean` interface is a part of the Spring core framework. It is used by components that react when all properties are populated. It can also be used to check whether all the mandatory properties are set.

Now that we have studied the basics, let's jump into writing the `ApiErrorAttributes` class:

```java
@Component
@Order(-2)
public class ApiErrorWebExceptionHandler extends
    AbstractErrorWebExceptionHandler {


  public ApiErrorWebExceptionHandler(ApiErrorAttributes
      errorAttributes,
      ApplicationContext applicationContext,
      ServerCodecConfigurer serverCodecConfigurer) {
    super(errorAttributes, new WebProperties().getResources(),
        applicationContext);
    super.setMessageWriters(
        serverCodecConfigurer.getWriters());
    super.setMessageReaders(
        serverCodecConfigurer.getReaders());
  }


  @Override
  protected RouterFunction<ServerResponse>
    getRoutingFunction(ErrorAttributes errorAttributes) {
    return RouterFunctions.route(
        RequestPredicates.all(), this::renderErrorResponse);
  }
```

https://github.com/PacktPublishing/Modern-API-Development-with-Spring-and-Spring-Boot/blob/main/Chapter05/src/main/java/com/packt/modern/api/exception/ApiErrorWebExceptionHandler.java

The first important observation about this code is that we have added the `@Order` annotation, which tells us the preference of execution. `ResponseStatusExceptionHandler` is ordered at 0 by the Spring Framework and `DefaultErrorWebExceptionHandler` is ordered at -1. Both are exception handlers like the one we have created. If you don't give a precedence order to `ApiErrorWebExceptionHandler` over both of these, then it won't ever execute. Therefore, the order is set at -2.

Next, this class overrides the `getRoutingFunction()` method, which calls the privately defined `renderErrorResponse()` method, where we have our own custom implementation for error handling, as shown next:

```java
private Mono<ServerResponse> renderErrorResponse(
    ServerRequest request) {
  Map<String, Object> errorPropertiesMap =
      getErrorAttributes(request,
      ErrorAttributeOptions.defaults());
  Throwable throwable = (Throwable) request
      .attribute("org.springframework.boot.web.reactive.error.
                DefaultErrorAttributes.ERROR")
      .orElseThrow(
          () -> new IllegalStateException("Missing exception
          attribute in ServerWebExchange"));

  ErrorCode errorCode = ErrorCode.GENERIC_ERROR;
  if (throwable instanceof IllegalArgumentException
      || throwable instanceof DataIntegrityViolationException
      || throwable instanceof ServerWebInputException) {
    errorCode = ILLEGAL_ARGUMENT_EXCEPTION;
  } else if (throwable instanceof CustomerNotFoundException) {
    errorCode = CUSTOMER_NOT_FOUND;
  } else if (throwable instanceof ResourceNotFoundException) {
    errorCode = RESOURCE_NOT_FOUND;
  } // other else-if
    ...
    ...
```

https://github.com/PacktPublishing/Modern-API-Development-with-Spring-and-Spring-Boot/blob/main/Chapter05/src/main/java/com/packt/modern/api/exception/ApiErrorWebExceptionHandler.java

Here, first we extract the error attributes in `errorPropertiesMap`. This will be used when we form the error response. Next, we capture the occurred exception using `throwable`. Then, we check the type of exception and assign appropriate code to it. We keep the default as `GenericError`, which is nothing more than `InternalServerError`.

Next, we use a `switch` statement to form an error response based on the raised exception, as shown here:

```
switch (errorCode) {
    case ILLEGAL_ARGUMENT_EXCEPTION -> {
      errorPropertiesMap.put(
        "status", HttpStatus.BAD_REQUEST);
      errorPropertiesMap.put("code", ILLEGAL_ARGUMENT_
        EXCEPTION.getErrCode());
      errorPropertiesMap.put("error",
        ILLEGAL_ARGUMENT_EXCEPTION);
      errorPropertiesMap.put("message", String
          .format("%s %s", ILLEGAL_ARGUMENT_EXCEPTION.
                  getErrMsgKey(), throwable.getMessage()));
      return ServerResponse.status(HttpStatus.BAD_REQUEST)
          .contentType(MediaType.APPLICATION_JSON)
          .body(BodyInserters.fromValue(errorPropertiesMap));
    }
    case CUSTOMER_NOT_FOUND -> {
      errorPropertiesMap.put("status", HttpStatus.NOT_FOUND);
      errorPropertiesMap.put("code",
        CUSTOMER_NOT_FOUND.getErrCode());
      errorPropertiesMap.put("error", CUSTOMER_NOT_FOUND);
      errorPropertiesMap.put("message", String
          .format("%s %s", CUSTOMER_NOT_FOUND.getErrMsgKey(),
              throwable.getMessage()));
      return ServerResponse.status(HttpStatus.NOT_FOUND)
          .contentType(MediaType.APPLICATION_JSON)
          .body(BodyInserters.fromValue(errorPropertiesMap));
    }
    case RESOURCE_NOT_FOUND -> {
                    // Rest of the code
        ...
```

```
    ...
}
```

https://github.com/PacktPublishing/Modern-API-Development-with-Spring-and-Spring-Boot/blob/main/Chapter05/src/main/java/com/packt/modern/api/exception/ApiErrorWebExceptionHandler.java

Probably in the next version of Java, we will be able to combine the `if-else` and `switch` blocks to make the code more concise. You can also create a separate method that takes `errorPropertiesMap` as an argument and returns the formed server response based on it. Then, you can use `switch`.

Custom application exception classes, such as `CustomerNotFoundException`, and other exception handling-supported classes, such as `ErrorCode` and `Error`, are being used from the existing code (from *Chapter 4, Writing Business Logic for APIs*).

Now that we have studied exception handling, we can concentrate on HATEOAS.

Adding hypermedia links to an API response

HATEOAS support for Reactive APIs is there and is a bit similar to what we did in the previous chapter using Spring MVC. We create these assemblers again for HATEOAS support. We also use the HATEOAS assembler classes for conversion from a model to an entity and vice versa.

Spring WebFlux provides the `ReactiveRepresentationModelAssembler` interface for forming hypermedia links. We would override its `toModel()` method for adding the links to response models.

Here, we will do some groundwork for populating the links. We will create an `HateoasSupport` interface with a single default method as shown next:

```
public interface HateoasSupport {
    default UriComponentsBuilder getUriComponentBuilder(
      @Nullable ServerWebExchange exchange) {
      if (exchange == null) {
        return UriComponentsBuilder.fromPath("/");
      }

        ServerHttpRequest request = exchange.getRequest();
        PathContainer contextPath = request.getPath().
                                    contextPath();
```

```
    return UriComponentsBuilder.fromHttpRequest(request)
        .replacePath(contextPath.toString())
        .replaceQuery("");
    }
}
```

This class contains a single default method, `getUriCompononentBuilder()`, which accepts `ServerWebExchange` as an argument and returns the `UriComponentsBuilder` instance. This instance can then be used to extract the server URI that would be used for adding the links with a protocol, host, and port. If you remember, the `ServerWebExchange` argument was added to controller methods. This interface is used for getting the HTTP request, response, and other attributes.

Now, we can use these two interfaces – `HateoasSupport` and `ReactiveRepresentationModelAssembler` – for defining the representation model assemblers.

Let's define the address's representational model assembler as shown next:

```
@Component
public class AddressRepresentationModelAssembler implements
    ReactiveRepresentationModelAssembler<AddressEntity,
        Address>, HateoasSupport {


  private static String serverUri = null;


  private String getServerUri(@Nullable ServerWebExchange
        exchange) {
    if (Strings.isBlank(serverUri)) {
      serverUri = getUriComponentBuilder(exchange).
        toUriString();
    }
    return serverUri;
  }
```

https://github.com/PacktPublishing/Modern-API-Development-with-Spring-and-Spring-Boot/blob/main/Chapter05/src/main/java/com/packt/modern/api/hateoas/AddressRepresentationModelAssembler.java

Here, we have defined another private method, `getServerUri()`, which extracts the server URI from `UriComponentBuilder`, which itself is returned from the default `getUriComponentBuilder()` method of the `HateoasSupport` interface.

Now, we can override the `toModel()` method as shown in the following code block:

```java
@Override
public Mono<Address> toModel(AddressEntity entity,
        ServerWebExchange exchange) {
  return Mono.just(entityToModel(entity, exchange));
}


public Address entityToModel(AddressEntity entity,
        ServerWebExchange exchange) {
  Address resource = new Address();
  if(Objects.isNull(entity)) {
    return resource;
  }
  BeanUtils.copyProperties(entity, resource);
  resource.setId(entity.getId().toString());
  String serverUri = getServerUri(exchange);

  resource.add(Link.of(String.format("%s/api/v1/addresses",
            serverUri)).withRel("addresses"));

  resource.add(
      Link.of(String.format("%s/api/v1/addresses/%s",
            serverUri, entity.getId())).withSelfRel());

  return resource;
}
```

https://github.com/PacktPublishing/Modern-API-Development-with-Spring-and-Spring-Boot/blob/main/Chapter05/src/main/java/com/packt/modern/api/hateoas/AddressRepresentationModelAssembler.java

The `toModel()` method returns the `Mono<Address>` object with hypermedia links formed from the `AddressEntity` instance using the `entityToModel()` method.

entityToModel() copies the properties from the entity instance to the model instance. Most importantly, it adds hypermedia links to the model using the resource.add() method. The add() method takes the org.springframework.hateoas.Link instance as an argument. Then, we use the Link class's of() static factory method to form the link. You can see that a server URI is used here to add it to the link. You can form as many links as you want and add these to the resource using the add() method.

The ReactiveRepresentationModelAssembler interface provides the toCollectionModel() method with a default implementation that returns the Mono<CollectionModel<D>> collection model. However, we can also add the toListModel() method as shown here, which returns the Flux instance of addresses:

```
public Flux<Address> toListModel(Flux<AddressEntity> entities,
        ServerWebExchange exchange) {
  if (Objects.isNull(entities)) {
    return Flux.empty();
  }
  return Flux.from(entities.map(e -> entityToModel(
                    e, exchange)));
}
```

https://github.com/PacktPublishing/Modern-API-Development-with-Spring-and-Spring-Boot/blob/main/Chapter05/src/main/java/com/packt/modern/api/hateoas/AddressRepresentationModelAssembler.java

This method internally uses the entityToModel() method. Similarly, you can create a representation model assembler for other API models. You can find all these models at https://github.com/PacktPublishing/Modern-API-Development-with-Spring-and-Spring-Boot/blob/main/Chapter05/src/main/java/com/packt/modern/api/hateoas.

Now that we are done with the basic code infrastructure, we can develop the API implementation based on the interfaces generated by OpenAPI Codegen. Here, we'll first develop the repositories that will be consumed by the services. At the end, we'll write the controller implementation. Let's start with the repositories.

Defining an entity

Entities are defined in more or less the same way as we defined and used them in *Chapter 4, Writing Business Logic for APIs*. However, instead of using Hibernate mappings and JPA, we'll use Spring Data annotations as shown here:

```java
import org.springframework.data.annotation.Id;
import org.springframework.data.relational.core.mapping.Column;
import org.springframework.data.relational.core.mapping.Table;
// other imports

@Table("ecomm.orders")
public class OrderEntity {

    @Id
    @Column("id")
    private UUID id;

    @Column("customer_id")
    private UUID customerId;

    @Column("address_id")
    private UUID addressId;

    @Column("card_id")
    private UUID cardId;

    @Column("order_date")
    private Timestamp orderDate;

    // other fields mapped to table columns

    private UUID cartId;

    private UserEntity userEntity;

    private AddressEntity addressEntity;
```

```
    private PaymentEntity paymentEntity;

    private List<ShipmentEntity> shipments = Collections.
        emptyList();

    // other entities fields and getter/setters
```

https://github.com/PacktPublishing/Modern-API-Development-with-Spring-and-Spring-Boot/blob/main/Chapter05/src/main/java/com/packt/modern/api/entity/OrderEntity.java

We use `@Table` to associate an entity class to a table name and `@Column` for mapping a field to a column of the table. As is obvious, `@Id` is used as the identifier column. Similarly, you can define the other entities.

Adding repositories

A repository is an interface between our application code and database. It is the same as what was there in Spring MVC. However, we are writing the code using the Reactive paradigm. Therefore, it is necessary to have repositories that use an R2DBC-/Reactive-based driver and return instances of Reactive types on top of Reactive streams. This is the reason why we can't use JDBC.

Spring Data R2DBC provides different repositories for Reactor and RxJava, such as `ReactiveCrudRepository`, `ReactiveSortingRepository`, `RxJava2CrudRepository`, and `RxJava3CrudRepository`. Also, you can write your own custom implementation.

We are going to use `ReactiveCrudRepository` and write a custom implementation also.

We'll write repositories for the `Order` entity. For other entities, you can find the repositories at `https://github.com/PacktPublishing/Modern-API-Development-with-Spring-and-Spring-Boot/tree/main/Chapter05/src/main/java/com/packt/modern/api/repository`.

First, let's write the **CRUD (Create, Read, Update,** and **Delete)** repository for the `Order` entity as shown next:

```
import org.springframework.data.r2dbc.repository.Query;
import org.springframework.data.repository.reactive.
    ReactiveCrudRepository;


@Repository
public interface OrderRepository extends
    ReactiveCrudRepository<OrderEntity, UUID>,
    OrderRepositoryExt {


    @Query("select o.* from ecomm.orders o join ecomm.user u on
        o.customer_id = u.id where u.id = :custId")
    Flux<OrderEntity> findByCustomerId(String custId);
}
```

https://github.com/PacktPublishing/Modern-API-Development-with-Spring-and-Spring-Boot/blob/main/Chapter05/src/main/java/com/packt/modern/api/repository/OrderRepository.java

This is as simple as shown. The `OrderRepository` interface extends `ReactiveCrudRepository` and our own custom repository interface, `OrderRepositoryExt`.

We'll discuss `OrderRepositoryExt` a bit later; let's discuss `OrderRepository` first. We have added one extra method, `findByCustomerId()`, in the `OrderRepository` interface, which finds the order by the given customer ID. The `ReactiveCrudRepository` interface and the `Query()` annotation are part of the Spring Data R2DBC library. `Query()` consumes native SQL queries unlike in Spring MVC.

> **Caution**
>
> The Spring Data R2DBC library, at the time of writing, does not support nested entities.

We can also write our own custom repository. Let's write a simple contract for it as shown next:

```
public interface OrderRepositoryExt {

  Mono<OrderEntity> insert(Mono<NewOrder> m);

  Mono<OrderEntity> updateMapping(OrderEntity orderEntity);
}
```

https://github.com/PacktPublishing/Modern-API-Development-with-Spring-and-Spring-Boot/blob/main/Chapter05/src/main/java/com/packt/modern/api/repository/OrderRepositoryExt.java

Here, we have written two method signatures – the first one inserts a new order record in the database and the second one updates the order item and cart item mapping. The idea is that once an order is placed, items should be removed from the cart and added to the order. If you want, you can also combine both operations together.

Let's first define the `OrderRepositoryExtImpl` class that extends the `OrderRepositoryExt` interface as shown in the following code block:

```
@Repository
public class OrderRepositoryExtImpl implements
    OrderRepositoryExt {

  private ConnectionFactory connectionFactory;
  private DatabaseClient dbClient;
  private ItemRepository itemRepo;
  private CartRepository cartRepo;
  private OrderItemRepository oiRepo;

  public OrderRepositoryExtImpl(ConnectionFactory
      connectionFactory, ItemRepository itemRepo,
      OrderItemRepository oiRepo, CartRepository cartRepo,
      DatabaseClient dbClient) {
```

```
    this.itemRepo = itemRepo;
    this.connectionFactory = connectionFactory;
    this.oiRepo = oiRepo;
    this.cartRepo = cartRepo;
    this.dbClient = dbClient;   }
```

https://github.com/PacktPublishing/Modern-API-Development-with-Spring-and-Spring-Boot/blob/main/Chapter05/src/main/java/com/packt/modern/api/repository/OrderRepositoryExtImpl.java

We have just defined a few class properties and added these properties in the constructor as an argument for constructor-based dependency injection.

As per the contract, it receives `Mono<NewOrder>`. Therefore, we need to add a method that converts a model to an entity in the `OrderRepositoryExtImpl` class. We also need an extra argument as `CartEntity` contains the cart items. Here it is:

```
private OrderEntity toEntity(NewOrder order, CartEntity c) {
    OrderEntity orderEntity = new OrderEntity();
    BeanUtils.copyProperties(order, orderEntity);
    orderEntity.setUserEntity(c.getUser());
    orderEntity.setCartId(c.getId());
    orderEntity.setItems(c.getItems())
        .setCustomerId(UUID.fromString(order.getCustomerId()))
        .setAddressId(UUID.fromString(
            order.getAddress().getId()))
        .setOrderDate(Timestamp.from(Instant.now()))
        .setTotal(c.getItems().stream().collect(Collectors.
            toMap(k -> k.getProductId(),
            v -> BigDecimal.valueOf(v.getQuantity()).multiply(v.
            getPrice())))
            .values().stream().reduce(BigDecimal::add).
            orElse(BigDecimal.ZERO));
    return orderEntity;
}
```

https://github.com/PacktPublishing/Modern-API-Development-with-Spring-and-Spring-Boot/blob/main/Chapter05/src/main/java/com/packt/modern/api/repository/OrderRepositoryExtImpl.java

This method is straightforward except for the code where the total is set. The total is calculated using the stream. Let's break it down as shown next to understand it:

1. First, it takes the items from the `cart` entity.

2. Then, it creates streams from items.

3. It creates a map with the key as the product ID and the value as the product of the quantity and price.

4. It takes the value from the map and converts it into a stream.

5. It performs the reduce operation by adding a method to `BigDecimal`. It then gives the total amount.

6. If values are not present, then it simply returns `0`.

After the `toEntity()` method, we also need another mapper that reads rows from the database and converts them to `OrderEntity`. For this purpose, we'll write `BiFunction`, which is a part of the `java.util.function` package:

```java
class OrderMapper implements BiFunction<Row, Object,
    OrderEntity> {

  @Override
  public OrderEntity apply(Row row, Object o) {
    OrderEntity oe = new OrderEntity();
    return oe.setId(row.get("id", UUID.class))
        .setCustomerId(row.get("customer_id", UUID.class))
        .setAddressId(row.get("address_id", UUID.class))
        .setCardId(row.get("card_id", UUID.class))
        .setOrderDate(Timestamp.from(
          ZonedDateTime.of((LocalDateTime)
          row.get("order_date"), ZoneId.of("Z")).toInstant()))
        .setTotal(row.get("total", BigDecimal.class))
        .setPaymentId(row.get("payment_id", UUID.class))
        .setShipmentId(row.get("shipment_id", UUID.class))
        .setStatus(StatusEnum.fromValue(row.get("status",
          String.class)));
  }
}
```

https://github.com/PacktPublishing/Modern-API-Development-with-Spring-and-Spring-Boot/blob/main/Chapter05/src/main/java/com/packt/modern/api/repository/OrderRepositoryExtImpl.java

We have overridden the `apply()` method, which returns `OrderEntity`, by mapping properties from the row to `OrderEntity`. The second parameter of the `apply()` method is not used because it contains metadata that we don't need.

Let's first implement the `updateMapping()` method from the `OrderRepositoryExt` interface:

```
@Override
public Mono<OrderEntity> updateMapping(OrderEntity orderEntity)
{
  return oiRepo.saveAll(orderEntity.getItems().stream()
      .map(i -> new OrderItemEntity()
        .setOrderId(orderEntity.getId()).setItemId(i.getId()))
      .collect(toList()))
      .then(
        itemRepo.deleteCartItemJoinById(orderEntity.getItems()
          .stream().map(i -> i.getId().toString())
          .collect(toList()),
        orderEntity.getCartId().toString())
          .then(Mono.just(orderEntity))
      );
}
```

https://github.com/PacktPublishing/Modern-API-Development-with-Spring-and-Spring-Boot/blob/main/Chapter05/src/main/java/com/packt/modern/api/repository/OrderRepositoryExtImpl.java

Here, we have created a pipeline of Reactive streams and performed two back-to-back database operations. First, it creates the order item mapping using `OrderItemRepository`, and then it removes the cart item mapping using `ItemRepository`.

Java Streams is used for creating an input list of `OrderItemEntity` instances in the first operation, and a list of item IDs in the second operation.

So far, we have made use of `ReactiveCrudRepository` methods. Let's implement a custom method using an entity template as shown next:

```
@Override
public Mono<OrderEntity> insert(Mono<NewOrder> mdl) {
  AtomicReference<UUID> orderId = new AtomicReference<>();

  Mono<List<ItemEntity>> itemEntities = mdl
      .flatMap(m -> itemRepo.findByCustomerId(m.
          getCustomerId())
      .collectList().cache());

  Mono<CartEntity> cartEntity = mdl
      .flatMap(m -> cartRepo.findByCustomerId(m.
          getCustomerId()))
      .cache();

  cartEntity = Mono.zip(cartEntity, itemEntities, (c, i) -> {
    if (i.size() < 1) {
      throw new ResourceNotFoundException(String
        .format("No item found in customer's (ID:%s) cart.",
            c.getUser().getId())));
    }
    return c.setItems(i);
  }).cache();
```

https://github.com/PacktPublishing/Modern-API-Development-with-Spring-and-Spring-Boot/blob/main/Chapter05/src/main/java/com/packt/modern/api/repository/OrderRepositoryExtImpl.java

Here, we override the `insert()` method from the `OrderRepositoryExt` interface. The `insert()` method is filled with fluent, functional, and Reactive APIs. The `insert()` method receives a `NewOrder` model `Mono` instance as an argument that contains the payload for creating a new order. Spring Data R2DBC does not allow fetching nested entities. However, you can write a custom repository for `Cart` similar to `Order` that can fetch `Cart` and its items together.

We are using `ReactiveCrudRepository` for `Cart` and `Item` entities. Therefore, we are fetching them one by one. First, we use the item repository to fetch the cart items based on the given customer ID. `Customer` has a one-to-one mapping with `Cart`. Then, we fetch the `Cart` entity using the cart repository by using the customer ID.

We get the two separate `Mono` objects – `Mono<List<ItemEntity>>` and `Mono<CartEntity>`. Now, we need to combine them. `Mono` has a `zip()` operator that allows you to take two `Mono` objects and then use the Java `BiFunction` to merge them. `zip()` returns a new `Mono` object only when both the given `Mono` objects produce the item. `zip()` is polymorphic and therefore other forms are also available.

We have the cart and its items, plus the `NewOrder` payload. Let's use these items to insert them into a database as shown in the next code block:

```
R2dbcEntityTemplate template = new
            R2dbcEntityTemplate(connectionFactory);

Mono<OrderEntity> orderEntity = Mono.zip(mdl, cartEntity,
                    (m, c) -> toEntity(m, c)).cache();

return orderEntity.flatMap(oe -> dbClient.sql("""
    INSERT INTO ecomm.orders (address_id, card_id, customer_
    id, order_date, total, status)
    VALUES($1, $2, $3, $4, $5, $6)""")
    .bind("$1", Parameter.fromOrEmpty(oe.getAddressId(),
                                    UUID.class))
    .bind("$2", Parameter.fromOrEmpty(oe.getCardId(), UUID.
                                    class))
    .bind("$3", Parameter.fromOrEmpty(oe.getCustomerId(),
                                    UUID.class))
    .bind("$4", OffsetDateTime.ofInstant(oe.getOrderDate().
                toInstant(),
            ZoneId.of("Z")).truncatedTo(ChronoUnit.MICROS))
    .bind("$5", oe.getTotal())
    .bind("$6", StatusEnum.CREATED.getValue())
                    .map(new OrderMapper()::apply)
    .one())
    .then(orderEntity.flatMap(x -> template.selectOne(
        query(where("customer_id").is(x.getCustomerId())
            .and("order_date")
```

```
                .greaterThanOrEquals(OffsetDateTime
                    .ofInstant(x.getOrderDate().toInstant(),
                                ZoneId.of("Z"))
                    .truncatedTo(ChronoUnit.MICROS))),
            OrderEntity.class).map(t -> x.setId(t.getId())
                .setStatus(t.getStatus())))
        ));
    }
```

https://github.com/PacktPublishing/Modern-API-Development-with-Spring-and-Spring-Boot/blob/main/Chapter05/src/main/java/com/packt/modern/api/repository/OrderRepositoryExtImpl.java

We again use `Mono.zip()` to create an `OrderEntity` instance. Now, we can use values from this instance to insert into the orders table.

There are two ways to interact with the database for running SQL queries – by using either `DatabaseClient` or `R2dbcEntityTemplate`. Now, `DatabaseClient` is a lightweight implementation that uses the `sql()` method to deal with SQL directly, whereas `R2dbcEntityTemplate` provides a fluent API for CRUD operations. We have used both classes to demonstrate their usage.

First, we use `DatabaseClient.sql()` to insert the new order in the orders table. We use `OrderMapper` to map the row returned from the database to the entity. Then, we use the `then()` reactive operator to select the newly inserted record and then map it back to `orderEntity` using the `R2dbcEntityTemplate.selectOne()` method.

Similarly, you can create repositories for other entities. Now, we can use these repositories in services. Let's define them in the next sub-section.

Adding services

Let's add a service for `Order`. There is no change in the server interface, as shown:

```
public interface OrderService {

  Mono<OrderEntity> addOrder(@Valid Mono<NewOrder>
      newOrder);

  Mono<OrderEntity> updateMapping(@Valid OrderEntity
      orderEntity);
```

```
Flux<OrderEntity> getOrdersByCustomerId(@NotNull @Valid
    String customerId);

Mono<OrderEntity> getByOrderId(String id);
}
```

https://github.com/PacktPublishing/Modern-API-Development-with-Spring-and-Spring-Boot/blob/main/Chapter05/src/main/java/com/packt/modern/api/service/OrderService.java

You just need to make sure that interface method signatures have Reactive types as returned types to keep the non-blocking flow in place.

Now, we can implement it in the following way:

```
@Override
public Mono<OrderEntity> addOrder(@Valid Mono<NewOrder>
    newOrder) {
  return repository.insert(newOrder);
}

@Override
public Mono<OrderEntity> updateMapping(@Valid OrderEntity
    orderEntity) {
  return repository.updateMapping(orderEntity);
}
```

https://github.com/PacktPublishing/Modern-API-Development-with-Spring-and-Spring-Boot/blob/main/Chapter05/src/main/java/com/packt/modern/api/service/OrderServiceImpl.java

The first two are straightforward; we just use the OrderRepository instance to call the respective methods. The third one is a bit tricky, as shown next:

```
private BiFunction<OrderEntity, List<ItemEntity>, OrderEntity>
        biOrderItems = (o, fi) -> o.setItems(fi);

@Override
public Flux<OrderEntity> getOrdersByCustomerId(String
```

```
        customerId) {
    // will use the dummy Card Id that doesn't exist
    // if it is null
    return repository.findByCustomerId(customerId).flatMap(order
    ->
        Mono.just(order)
            .zipWith(userRepo.findById(order.getCustomerId()))
            .map(t -> t.getT1().setUserEntity(t.getT2()))
            .zipWith(addRepo.findById(order.getAddressId()))
            .map(t -> t.getT1().setAddressEntity(t.getT2()))
            .zipWith(cardRepo.findById(
                order.getCardId() != null ? order.getCardId()
                : UUID.fromString(
                    "0a59ba9f-629e-4445-8129-b9bce1985d6a"))
                .defaultIfEmpty(new CardEntity()))
            .map(t -> t.getT1().setCardEntity(t.getT2()))
            .zipWith(itemRepo.findByCustomerId(
                order.getCustomerId().toString()).collectList(),
                biOrderItems)
    );
}
```

https://github.com/PacktPublishing/Modern-API-Development-with-Spring-and-Spring-Boot/blob/main/Chapter05/src/main/java/com/packt/modern/api/service/OrderServiceImpl.java

This method looks complicated but it's not. What you are doing here is basically fetching data from multiple repositories and then populating the nested entities inside `OrderEntity`. This is done by using the `zipWith()` operator by using either the `map()` operator alongside it or `BiFunction` as a separate argument.

It first fetches the orders by using the customer ID, then flat maps the orders to populate its nested entities. Therefore, we are using `zipWith()` inside the `flatMap()` operator. If you observer the first `zipWith()`, it fetches the user entity and then sets the nested user entity's property using the `map()` operator. Similarly, other nested entities are populated.

In the last `zipWith()` operator, we are using the `BiFunction biOrderItems` to set the item entities in the `OrderEntity` instance.

The same algorithm is used for implementing the last method of the `OrderService` interface, as shown in the following code:

```
@Override
public Mono<OrderEntity> getByOrderId(String id) {
  return repository.findById(UUID.fromString(id)).flatMap(order
->
      Mono.just(order)
          .zipWith(userRepo.findById(order.getCustomerId()))
          .map(t -> t.getT1().setUserEntity(t.getT2()))
          .zipWith(addRepo.findById(order.getAddressId()))
          .map(t -> t.getT1().setAddressEntity(t.getT2()))
          .zipWith(cardRepo.findById(order.getCardId()))
          .map(t -> t.getT1().setCardEntity(t.getT2()))
          .zipWith(itemRepo.findByCustomerId(
              order.getCustomerId().toString()).collectList(),
              biOrderItems)
  );
}
```

https://github.com/PacktPublishing/Modern-API-Development-with-Spring-and-Spring-Boot/blob/main/Chapter05/src/main/java/com/packt/modern/api/service/OrderServiceImpl.java

There is another way to merge two `Mono` instances using the `Mono.zip()` operator, as shown next:

```
private BiFunction<CartEntity, List<ItemEntity>, CartEntity>
        cartItemBiFun = (c, i) -> c
      .setItems(i);
  @Override
public Mono<CartEntity> getCartByCustomerId(String customerId)
{
  Mono<CartEntity> cart = repository.
        findByCustomerId(customerId)
      .subscribeOn(Schedulers.boundedElastic());
  Flux<ItemEntity> items = itemRepo.
        findByCustomerId(customerId)
      .subscribeOn(Schedulers.boundedElastic());
```

```
    return Mono.zip(cart, items.collectList(), cartItemBiFun);
}
```

https://github.com/PacktPublishing/Modern-API-Development-with-Spring-and-Spring-Boot/blob/main/Chapter05/src/main/java/com/packt/modern/api/service/CartServiceImpl.java

This example is taken from the `CartServiceImpl` class. Here, we make two separate calls – one using the cart repository and another one from the item repository. As a result, these two calls produce two `Mono` instances, and merge them using the `Mono.zip()` operator. This we call directly using `Mono`; the previous example was used on `Mono/Flux` instances with the `zipWith()` operator.

Using similar techniques, the remaining services have been created. Those you can find at `https://github.com/PacktPublishing/Modern-API-Development-with-Spring-and-Spring-Boot/tree/main/Chapter05/src/main/java/com/packt/modern/api/service`.

Let's move our focus on to the last sub-section of our Reactive API implementation.

Adding controller implementations

REST controller interfaces are already generated by the OpenAPI Codegen tool. We can now create an implementation of those interfaces. The only different thing this time is having the Reactive pipelines to call the services and assemblers. You should also only return `ResponseEntity` objects wrapped in either `Mono` or `Flux` based on the generated contract.

Let's implement `OrderApi`, which is the controller interface for the `Orders` REST API:

```
@RestController
public class OrderController implements OrderApi {

  private final OrderRepresentationModelAssembler assembler;
  private OrderService service;

  public OrderController(OrderService service,
            OrderRepresentationModelAssembler assembler) {
    this.service = service;
    this.assembler = assembler;
  }
```

https://github.com/PacktPublishing/Modern-API-Development-with-Spring-and-Spring-Boot/blob/main/Chapter05/src/main/java/com/packt/modern/api/controller/OrderController.java

Here, `@RestController` is a trick that combines `@Controller` and `@ResponseBody`. These are the same annotations we used in *Chapter 4*, *Writing Business Logic for APIs*, for creating the REST controller. However, the methods have different signatures now to apply the Reactive pipelines. Make sure you don't break the Reactive chain of calls or add any blocking calls. If you do, either the REST call will not be fully non-blocking or you may see undesired results.

We use constructor-based dependency injection to inject the order service and assembler. Let's add the method implementations:

```java
@Override
public Mono<ResponseEntity<Order>> addOrder(@Valid
Mono<NewOrder> newOrder,
    ServerWebExchange exchange) {
  return service.addOrder(newOrder.cache())
      .zipWhen(x -> service.updateMapping(x))
      .map(t -> status(HttpStatus.CREATED)
        .body(assembler.entityToModel(t.getT2(), exchange)))
      .defaultIfEmpty(notFound().build());
}
```

https://github.com/PacktPublishing/Modern-API-Development-with-Spring-and-Spring-Boot/blob/main/Chapter05/src/main/java/com/packt/modern/api/controller/OrderController.java

Both the method argument and return type are Reactive types (`Mono`), used as a wrapper. Reactive controllers also have an extra parameter, `ServerWebExchange`, which we discussed earlier.

In this method, we simply pass the `newOrder` instance to the service. We have used `cache()` because we need to subscribe to it more than once. We get the newly created `EntityOrder` by the `addOrder()` call. Then, we use the `zipWhen()` operator, which performs the `updateMapping` operation using the newly created order entity. At the end, we send it by wrapping it inside `ResponseEntity`. Also, it returns `NOT FOUND 404` when an empty instance is returned.

Let's have a look at other API interface implementations:

```
@Override
public Mono<ResponseEntity<Flux<Order>>>
    getOrdersByCustomerId(@NotNull @Valid String customerId,
    ServerWebExchange exchange) {
  return Mono.just(ok(assembler.toListModel(
    service.getOrdersByCustomerId(customerId), exchange)));
}


@Override
public Mono<ResponseEntity<Order>> getByOrderId(String id,
    ServerWebExchange exchange) {
  return service.getByOrderId(id).map(o ->
                            assembler.entityToModel(o, exchange))
      .map(ResponseEntity::ok)
      .defaultIfEmpty(notFound().build());
}
```

https://github.com/PacktPublishing/Modern-API-Development-with-Spring-and-Spring-Boot/blob/main/Chapter05/src/main/java/com/packt/modern/api/controller/OrderController.java

Both are kind of similar in nature; the service returns `OrderEntity` based on the given customer ID and order ID. It then gets converted into a model and is wrapped inside `ResponseEntity` and `Mono`.

Similarly, other REST controllers are implemented using the same approach. You can find the rest of them at `https://github.com/PacktPublishing/Modern-API-Development-with-Spring-and-Spring-Boot/tree/main/Chapter05/src/main/java/com/packt/modern/api/controller`.

We are almost done with Reactive API implementation. Let's look into some other minor changes.

Adding H2 Console to an application

The H2 Console app is not available by default in Spring WebFlux the way it is available in Spring MVC. However, you can add it by defining the bean on your own, as shown:

```
@Component
public class H2ConsoleComponent {

    private final static Logger log = LoggerFactory.
        getLogger(H2ConsoleComponent.class);
    private Server webServer;

    @Value("${modern.api.h2.console.port:8081}")
    Integer h2ConsolePort;

    @EventListener(ContextRefreshedEvent.class)
    public void start() throws java.sql.SQLException {
      log.info("starting h2 console at port
          "+h2ConsolePort);
      this.webServer = org.h2.tools.Server.createWebServer(
          "-webPort", h2ConsolePort.toString(),
          "-tcpAllowOthers").start();
    }

    @EventListener(ContextClosedEvent.class)
    public void stop() {
      log.info("stopping h2 console at port
          "+h2ConsolePort);
      this.webServer.stop();
    }
}
```

https://github.com/PacktPublishing/Modern-API-Development-with-Spring-and-Spring-Boot/blob/main/Chapter05/src/main/java/com/packt/modern/api/H2ConsoleComponent.java

This is straightforward; we have added the `start()` and `stop()` methods, which are executed on `ContextRefreshEvent` and `ContextStopEvent`, respectively. `ContextRefreshEvent` is an application event that gets fired when `ApplicationContext` is refreshed or initialized. `ContextStopEvent` is also an application event that gets fired when `ApplicationContext` is closed.

The `start()` method creates the web server using the H2 library and starts it on a given port. The `stop()` method stops the H2 web server, that is, the H2 Console app.

You need a different port to execute H2 Console, which can be configured by adding the `modern.api.h2.console.port=8081` property in the `application.properties` file. The `h2ConsolePort` property is annotated with `@Value("${modern.api.h2.console.port:8081}")`, therefore the value configured in `application.properties` will be picked and assigned to `h2ConsolePort` when the `H2ConsoleComponent` bean is initialized by the Spring Framework. The value `8081` will be assigned if the property is not defined in the `application.properties` file.

Since we are discussing `application.properties`, let's have a look at some of the other changes.

Adding application configuration

We are going to use Flyway for database migration. Let's add the configuration required for it:

```
spring.flyway.url=jdbc:h2:file:./data/ecomm;AUTO_
SERVER=TRUE;DB_CLOSE_DELAY=-1;IGNORECASE=TRUE;DATABASE_TO_
UPPER=FALSE;DB_CLOSE_ON_EXIT=FALSE
spring.flyway.schemas=ecomm
spring.flyway.user=
spring.flyway.password=
```

https://github.com/PacktPublishing/Modern-API-Development-with-Spring-and-Spring-Boot/blob/main/Chapter05/src/main/resources/application.properties

You must be wondering why we are using JDBC here, instead of R2DBC. Because Flyway hasn't yet started supporting R2DBC (at the time of writing). You can change it to R2DBC once support is added.

We have specified the `ecomm` schema and set a blank username and password.

Let's see the Spring Data configuration:

```
spring.r2dbc.url=r2dbc:h2:file://././data/ecomm?options=AUTO_
SERVER=TRUE;DB_CLOSE_DELAY=-1;IGNORECASE=TRUE;DATABASE_TO_
UPPER=FALSE;DB_CLOSE_ON_EXIT=FALSE
spring.r2dbc.driver=io.r2dbc:r2dbc-h2
spring.r2dbc.name=
spring.r2dbc.password=
```

https://github.com/PacktPublishing/Modern-API-Development-with-Spring-and-Spring-Boot/blob/main/Chapter05/src/main/resources/application.properties

Spring Data supports R2DBC, therefore we are using an R2DBC-based URL. We have set io.r2dbc:r2dbc-h2 for the driver to H2 and set a blank username and password.

Similarly, we have added the following logging properties in logback-spring.xml for adding debug statements in the console for Spring R2DBC and H2, as shown next:

```
<logger name="org.springframework.r2dbc" level="debug"
    additivity="false">
  <appender-ref ref="STDOUT"/>
</logger>
<logger name="reactor.core" level="debug" additivity="false">
  <appender-ref ref="STDOUT"/>
</logger>
<logger name="io.r2dbc.h2" level="debug" additivity="false">
  <appender-ref ref="STDOUT"/>
</logger>
```

This concludes our implementation of Reactive RESTful APIs. Now, you can test them.

Testing Reactive APIs

Now, you must be looking forward to testing. You can find the Postman (API client) collection at the following location, which is based on Postman Collection version 2.1. You can import it and then test the APIs:

https://github.com/PacktPublishing/Modern-API-Development-with-Spring-and-Spring-Boot/blob/main/Chapter05/Chapter05.postman_collection.json

Summary

I hope you enjoyed learning about Reactive API development with an asynchronous, non-blocking, and functional paradigm. At first glance, you may find it complicated if you are not very familiar with the fluent and functional paradigm, but with practice, you'll start writing only functional-style code. Definitely, familiarity with Java Streams and functions gives you an edge to grasp the concepts easily.

Now that you have reached the end of the chapter, you have the skills to write functional and Reactive code. Now you can write Reactive, asynchronous, and non-blocking code and REST APIs. You also learned about R2DBC, which will become more solid and enhanced in the future as long as Reactive programming is there.

In the next chapter, we'll explore the security aspect of RESTful service development.

Questions

1. Do you really need the Reactive paradigm for application development?

2. Are there any disadvantages to using the Reactive paradigm?

3. Who plays the role of the subscriber in the case of an HTTP request in Spring WebFlux?

Further reading

- Project Reactor:

  ```
  https://projectreactor.io
  ```

- Spring Reactive documentation:

  ```
  https://docs.spring.io/spring-framework/docs/current/
  reference/html/web-reactive.html
  ```

- Spring Data R2DBC – reference documentation

  ```
  https://docs.spring.io/spring-data/r2dbc/docs/current/
  reference/html/#preface
  ```

- *Hands-On Reactive Programming in Spring 5* (book)

  https://www.packtpub.com/product/hands-on-reactive-programming-in-spring-5/9781787284951

- *Hands-On Reactive Programming with Java 12* (video)

  https://www.packtpub.com/product/hands-on-reactive-programming-with-java-12-video/9781789808773

Section 2: Security, UI, Testing, and Deployment

In this section, you will learn how to secure REST APIs with JWT using Spring Security. After completing this section, you will also be able to secure RESTful web services using Spring Security. You will learn how developed APIs are being consumed by the UI app, and you will learn how to automate the unit testing and integration testing of APIs. By the end, you will be able to containerize the built app and then deploy it in a Kubernetes cluster.

This section comprises the following chapters:

- *Chapter 6, Security (Authorization and Authentication)*
- *Chapter 7, Designing a User Interface*
- *Chapter 8, Testing APIs*
- *Chapter 9, Deployment of Web Services*

6
Security (Authorization and Authentication)

In previous chapters, we have developed RESTful web services (where **REST** stands for **REpresentational State Transfer**) using imperative and reactive coding styles. Now, you'll learn how you can secure these REST endpoints using Spring Security. You'll implement token-based authentication and authorization for REST endpoints. A successful authentication provides two types of tokens—a **JavaScript Object Notation (JSON) Web Token (JWT)** as an access token, and a refresh token in response. This JWT-based access token is then used to access the secured **Uniform Resource Locators** (**URLs**). A refresh token is used to request a new JWT if the existing JWT has expired, and a valid request token provides a new JWT to use.

You'll associate users with roles such as **Admin**, **User**, and so on. These roles will be used as authorization to make sure that REST endpoints can only be accessed if a user holds certain roles. We'll also briefly discuss briefly **cross-site request forgery** (**CSRF**) and **cross-origin resource sharing** (**CORS**). The topics of this chapter are divided into the following sections:

- Implementing authentication using Spring Security and JWT
- Securing REST **application programming interfaces** (**APIs**) with JWT
- Configuring CORS and CSRF
- Understanding authorization
- Testing security

By the end of the chapter, you will have learned how to implement authentication and authorization by using Spring Security.

Technical requirements

You need the following to develop and execute the code featured in this chapter:

- Any Java **integrated development environment** (**IDE**) such as NetBeans, IntelliJ IDEA, or Eclipse
- **Java Development Kit** (**JDK**) 15
- An internet connection to clone the code and download the dependencies from Gradle
- Postman/cURL (for API testing)

Please visit the following link to download the code files:
`https://github.com/PacktPublishing/Modern-API-Development-with-Spring-and-Spring-Boot/tree/main/Chapter06`

Implementing authentication using Spring Security and JWT

Spring Security is a framework consisting of a collection of libraries that allow you to implement enterprise application security without worrying about writing the boilerplate code. In this chapter, we will use the Spring Security framework to implement token-based (JWT) authentication and authorization. Throughout the course of this chapter, we will also learn about CORS and CSRF configuration.

It's useful to know that Spring Security also provides support for opaque tokens similar to JWTs. The main difference between them is how information is read from the token. You can't read the information from an opaque token in the way you can with a JWT— only the issuer is aware of how to do this.

> **Note**
>
> A token is a string of characters such as
> `5rm1tc1obfshrm23541u9dlt5reqm1ddjchqh81`
> `7rbk37q95b768bib0jf44df6suk1638sf78cef7`
> `hfolg4ap3bkighbnk7inr68ke780744fpej0gtd`
> `9qflm99908q`. It allows you to call secured **HyperText Transfer Protocol (HTTP)** endpoints or resources statelessly using various authorization flows.

You have learned about `DispatcherServlet` in *Chapter 2, Spring Concepts and REST APIs*. This is an interface between a client request and the REST controller. Therefore, if you want to place a logic for token-based authentication and authorization, you would have to do this before the request reaches `DispatcherServlet`. Spring Security libraries provide the servlet pre-filters (as a part of filter chain) that are processed before the request reaches `DispatcherServlet`. A pre-filter is a servlet filter that is processed before it reaches the actual servlet, which in Spring Security's case is `DispatcherServlet`. Similarly, post filters get processed after a request has been processed by the servlet/controller.

There are two ways you can implement token-based (JWT) authentication: by using either `spring-boot-starter-security` or `spring-boot-starter-oauth2-resource-server`. **We are going to use the latter.**

The former contains the following libraries:

- `spring-security-core`
- `spring-security-config`
- `spring-security-web`

`spring-boot-starter-oauth2-resource-server` provides the following, along with all three previously mentioned **Java ARchive** files (**JARs**):

- `spring-security-oauth2-core`
- `spring-security-oauth2-jose`

- `spring-security-oauth2-resource-server`
- When you start this chapter's code, you can find the following log. You can see that, by default, `DefaultSecurityFilterChain` is auto-configured. The `log` statement lists down the configured filters in the `DefaultSecurityFilterChain`, as shown in the following code block:

```
INFO [Chapter06,,,]        [null] [null] [null]
[null] 24052 --- [          main] o.s.s.web.
DefaultSecurityFilterChain      : Will secure any request
with [org.springframework.security.web.context.request.async.
WebAsyncManagerIntegrationFilter@413e8246, org.springframework.
security.web.context.SecurityContextPersistenceFilter@659565ed,
org.springframework.security.web.header.
HeaderWriterFilter@770c3ca2, org.springframework.web.
filter.CorsFilter@4c7b4a31, org.springframework.security.
web.csrf.CsrfFilter@1de6f29d, org.springframework.security.
web.authentication.logout.LogoutFilter@5bb90b89, org.
springframework.security.oauth2.server.resource.web.
BearerTokenAuthenticationFilter@732fa176, org.springframework.
security.web.savedrequest.RequestCacheAwareFilter@2ae0eb98,
org.springframework.security.web.servletapi.
SecurityContextHolderAwareRequestFilter@3f473daf,
org.springframework.security.web.authentication.
AnonymousAuthenticationFilter@2df7766b, org.
springframework.security.web.session.
SessionManagementFilter@711261c7, org.springframework.
security.web.access.ExceptionTranslationFilter@1a7f2d34,
org.springframework.security.web.access.intercept.
FilterSecurityInterceptor@3390621a]
```

Therefore, when a client fires an HTTP request, it will go through all the following security filters before reaching the REST controller (the order may vary based on the authentication outcome):

1. `WebAsyncManagerIntegrationFilter`
2. `SecurityContextPersistenceFilter`
3. `HeaderWriterFilter`
4. `CorsFilter`
5. `CsrfFilter`
6. `LogoutFilter`

7. `BearerTokenAuthenticationFilter`

8. `RequestCacheAwareFilter`

9. `SecurityContextHolderAwareRequestFilter`

10. `AnonymousAuthenticationFilter`

11. `SessionManagementFilter`

12. `ExceptionTranslationFilter`

13. `FilterSecurityInterceptor`

14. Finally reaches the controller

This filter chain could be changed in future releases. Also, the security filter chain would be different if you just used `spring-boot-starter-security` or changed the configuration. You can find all the filters available in `springSecurityFilterChain` at `https://docs.spring.io/spring-security/site/docs/current/reference/html5/#servlet-security-filters`. Let's see how we can use filters to perform authentication.

Learning how to authenticate using filters

You can skip this section and jump to the *Authentication using OAuth 2.0 Resource Server* section if you are already aware of filter-based authentication.

We are going to use the `spring-boot-starter-oauth2-resource-server` dependency for authentication implementation, which does not need a manual filter configuration. `oauth2-resource-server` makes use of `BearerTokenAuthenticationFilter` for authentication. However, an understanding of the filter-based authentication implementation and its configuration would simplify the Spring Security concepts. For filter-based authentication and authorization, you can simply add `spring-boot-starter-spring-security`.

You can add the authentication logic in the appropriate pre-filter. If a request fails to authenticate, then a response would be sent to the client with an access denied exception (`AccessDeniedException`) with a resulting `HTTP 401 Unauthorized error` status response code.

Username/password authentication flow using filters

Authentication with a username/password works as shown in the following diagram. If a user submits a valid username/password combination, the call succeeds and the user gets a token with a 200 OK status code (a successful response). If a call fails due to an invalid username/password combination, you'll receive a response with a 401 Unauthorized status code:

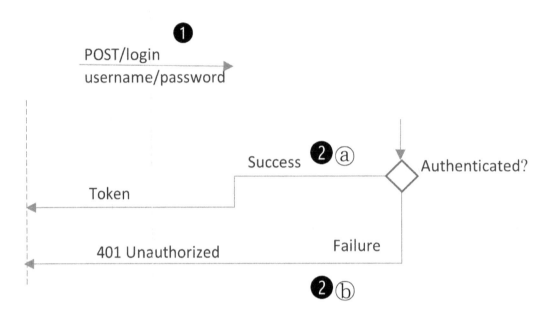

Figure 6.1 – Username/password authentication flow

Now, let's have a look at a token authorization flow using filters.

Token authorization flow using filters

Authorization with tokens works as shown in the following diagram. If a user submits a valid bearer token with an Authorization header, the call succeeds and invokes FilterChain.doFilter(request, response). Therefore, the call routes to the controller via DispatcherServlet. At the end, the client gets a response with the appropriate status code.

If the call fails due to an invalid token, AccessDeniedException is raised and a response is sent from AuthenticationEntryPoint with a 401 Unauthorized status code. You can override this behavior by implementing an AuthenticationEntryPoint interface and overriding its commence() method:

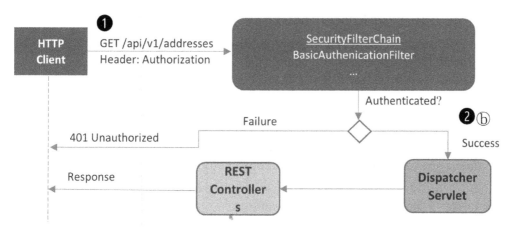

Figure 6.2 – Token authorization flow using BasicAuthenticationFilter

Let's first add the required dependencies in a Gradle build file.

Adding the required Gradle dependencies

Let's add the following dependencies into the `build.gradle` file, as shown next:

```
implementation 'org.springframework.boot:spring-boot-starter-
oauth2-resource-server'
implementation 'com.auth0:java-jwt:3.12.0'
```

**https://github.com/PacktPublishing/Modern-API-Development-with-Spring-and-
Spring-Boot/blob/main/Chapter06/build.gradle**

The Spring Boot Starter OAuth 2.0 resource server dependency would add the following JARs:

- `spring-security-core`
- `spring-security-config`
- `spring-security-web`
- `spring-security-oauth2-core`
- `spring-security-oauth2-jose`
- `spring-security-oauth2-resource-server`

For JWT implementation, we are going to use the `java-jwt` library from `auth0.com`.

You can now explore how to code these two filters—login- and token-based authentication.

Coding the filters for login functionality

The client will receive a JWT token after performing a successful login by providing a valid username/password combination. Spring Security provides `UsernamePasswordAuthenticationFilter`, which we can extend and then override its `attemptAuthentication()` and `successfulAuthentication()` methods. Let's first create a `LoginFilter` class, as follows:

```
public class LoginFilter extends
    UsernamePasswordAuthenticationFilter {

  private final AuthenticationManager
    authenticationManager;
  private final JwtManager tokenManager;
  private final ObjectMapper mapper;

  public LoginFilter(AuthenticationManager
      authenticationManager, JwtManager tokenManager,
      ObjectMapper mapper) {
    this.authenticationManager = authenticationManager;
    this.tokenManager = tokenManager;
    this.mapper = mapper;
    super.setFilterProcessesUrl("/api/v1/auth/token");
  }
```

https://github.com/PacktPublishing/Modern-API-Development-with-Spring-and-Spring-Boot/blob/main/Chapter06/src/main/java/com/packt/modern/api/security/UNUSED/LoginFilter.java

Here, `JwtManager` is a class that allows you to create a JWT based on the given `org.springframework.security.core.userdetails.User`. We will explore more about this in the *Adding a JWT to REST APIs* section. This constructor also allows you to configure the login URL using `setFilterProcessesUrl()`. If you don't see this, then by default it takes the `/login` URL.

Let's first override the `attemptAuthentication()` method, as follows:

```
@Override
public Authentication attemptAuthentication(
    HttpServletRequest req, HttpServletResponse res)
```

```java
    throws AuthenticationException {
  if (!req.getMethod().equals(HttpMethod.POST.name())) {
    throw new MethodNotAllowedException(req.getMethod(),
        List.of(HttpMethod.POST));
  }
  try (InputStream is = req.getInputStream()) {
    SignInReq user = new ObjectMapper().readValue(
        is, SignInReq.class);
    return authenticationManager.authenticate(
        new UsernamePasswordAuthenticationToken(
            user.getUsername(),
            user.getPassword(),
            Collections.emptyList())
    );
  } catch (IOException e) {
    throw new RuntimeException(e);
  }
 }
```

`org.springframework.security.authentication.`
`AuthenticationManager` is a component of Spring Security. We simply use it to authenticate by passing the username, password, and authorities. Authorities are passed as an empty list; however, you can pass the authorities too if they are either received from the request payload or by fetching them from the database/memory store. `SignInReq` is a **Plain Old Java Object** (**POJO**) that contains the `username` and `password` fields.

Once the login is successful, we need to return a JWT in response. The `successfulAuthentication()` method is overridden for the same purpose, as illustrated in the following code snippet:

```java
@Override
protected void successfulAuthentication(
    HttpServletRequest req,
    HttpServletResponse res,
    FilterChain chain,
    Authentication auth) throws IOException {
  User principal = (User) auth.getPrincipal();
  String token = tokenManager.create(principal);
  SignedInUser user = new SignedInUser()
```

```
                .username(principal.getUsername()).accessToken(
                        token);
    res.setContentType(MediaType.APPLICATION_JSON_VALUE);
    res.setCharacterEncoding("UTF-8");
    res.getWriter().print(mapper.writeValueAsString(user));
    res.getWriter().flush();
}
```

Here, a token is created using the `tokenManager.create()` method.
`SignedInUser` is a POJO that contains the `username` and `token` fields.

The client receives the username and token as a response after successful authentication and can then use this token in the `Authorization` header by prefixing the token value with `"Bearer"`. Let's see how you can configure and add these new filters in Spring Security filter chains.

Configuring Spring Security

There is a final piece missing here about how `AuthenticationManager.`
`authenticate()` works. `AuthenticationManager` uses the
`UserDetailsService` bean internally— this has a single method, as shown next:

```
UserDetails loadUserByUsername(String username) throws
                                UsernameNotFoundException
```

You just have to create an implementation of the `UserDetailsService` interface and expose it as a bean, as shown in the following code snippet:

```
@Bean
@Override
protected UserDetailsService userDetailsService() {
    return userService;
}
```

This bean is exposed in the `SecurityConfig` class, which extends
`WebSecurityConfigurerAdapter`, as shown in the following code snippet:

```
@EnableWebSecurity
public class SecurityConfig extends
WebSecurityConfigurerAdapter {
```

Here, `@EnableWebSecurity` annotation is applied to
configure `WebSecurityConfigurer` or the class that extends
`WebSecurityConfigurerAdapter` and customize the `WebSecurity` class.
This class does the auto-configuration for security, with some override methods. The
`configure()` method allows you to configure the HTTP security as a **domain-specific
language** (**DSL**), using the methods shown here:

```
@Override
protected void configure(HttpSecurity http) throws
    Exception {
  http.authorizeRequests()
        .antMatchers(HttpMethod.POST, SIGN_UP_URL).permitAll()
        .anyRequest().authenticated()
        .and()
        .addFilter(new LoginFilter(
            super.authenticationManager(), mapper))
        .addFilter(new JwtAuthenticationFilter(
            super.authenticationManager()))
        .sessionManagement().sessionCreationPolicy(
            SessionCreationPolicy.STATELESS);
}
```

The `configure()` method demonstrates how you can configure the HTTP security
using DSL and override its default implementation. Let's understand the code, as follows:

- You can see that the login URL is permitted without authentication by using
  the `antMatchers()` method. If you don't pass the HTTP method to
  `antMatchers()`, it will be applicable for all the HTTP methods. `permitAll()`
  removes the restriction on endpoints and their attached HTTP methods.

- You can add custom filters using `addFilter()`. All other URLs need
  authentication.

- Two extra filters have been added for sign-in operations and JWT token-based
  authentication.

- At the end, the session policy is set to `STATELESS` because we are going to use
  REST endpoints.

We have coded the login filter in this section. Now, let's add the
`JwtAuthenticationFilter` class for token verification.

Coding the filters for token verification

Let's explore how you can implement authentication using filters. If you use the `spring-boot-starter-security` dependency, then you can extend the `BasicAuthenticationFilter` class and override the `doFilterInternal` method for token verification, as shown in the code blocks that follow.

First, create a new class that extends the `BasicAuthenticationFilter` class, as follows:

```
public class JwtAuthenticationFilter extends
                              BasicAuthenticationFilter {
  public JwtAuthenticationFilter(AuthenticationManager
                                   authenticationManager) {
    super(authenticationManager);
  }
}
```

https://github.com/PacktPublishing/Modern-API-Development-with-Spring-and-Spring-Boot/blob/main/Chapter06/src/main/java/com/packt/modern/api/security/UNUSED/JwtAuthenticationFilter.java

Now, we can override the `doFilterInternal` method. Here, we check whether the request contains the `Authorization` header with a bearer token or not. If it has the `Authorization` header then it performs the authentication, adds the token to security context, and then passes the call to the next security filter. The code is illustrated in the following snippet:

```
@Override
protected void doFilterInternal(HttpServletRequest req,
        HttpServletResponse res, FilterChain chain) throws
                    IOException, ServletException {
  String header = req.getHeader("Authorization");
  if (Objects.isNull(header) || !header.startsWith("Bearer "))
{
    chain.doFilter(req, res);
    return;
  }

  Optional<UsernamePasswordAuthenticationToken>
    authentication = getAuthentication(req);
  authentication.ifPresentOrElse(e ->
```

```
        SecurityContextHolder.getContext().setAuthentication(
            e), SecurityContextHolder::clearContext);
    chain.doFilter(req, res);
}
```

The getAuthentication() method performs the token authentication logic, as shown in the following code snippet:

```
private Optional<UsernamePasswordAuthenticationToken>
        getAuthentication(HttpServletRequest request) {

    String token = request.getHeader("Authorization");
    if (Objects.nonNull(token)) {
        DecodedJWT jwt = JWT.require(Algorithm.HMAC512(
            SECRET_KEY.getBytes(StandardCharsets.UTF_8)))
            .build()
            .verify(token.replace(TOKEN_PREFIX, ""));
        String user = jwt.getSubject();

        @SuppressWarnings("unchecked")
        List<String> authorities = (List)
            jwt.getClaim("roles");
        if (Objects.nonNull(user)) {
            return Optional.of(
                    new UsernamePasswordAuthenticationToken(
                user, null, Objects.nonNull(authorities) ?
                authorities.stream().map(SimpleGrantedAuthority::
                    new)
                .collect(Collectors.toList()) :
                Collections.emptyList())));
        }
    }
    return Optional.empty();
}
```

Here, JWT and DecodedJWT are part of the com.auth0:java-jwt library. Calling verify() performs the verification of the given token and returns a DecodedJWT instance. If verification fails, it returns JWTVerificationException. Once verification is done, we simply create and return the UsernamePasswordAuthenticationToken token that takes the principal, credentials, and collection of GrantedAuthority objects. GrantedAuthority is an interface that represents the authority associated with an authentication object. OAuth2 Resource Server lets you add *scope* authority by default. However, you can add custom authority such as *roles*.

So far, we have learned about the authentication and token authorization flow using a Spring filter chain. Next, we are going to implement the authentication using the spring-boot-starter-oauth2-resource-server dependency. In the next section, we'll explore the authentication and authorization flow using OAuth 2.0 Resource Server.

Authentication using OAuth 2.0 Resource Server

Spring Security OAuth 2.0 Resource Server allows you to implement authentication and authorization using BearerTokenAuthenticationFilter. This contains bearer token authentication logic. However, you still need to write the REST endpoint for generating the token. Let's explore how the authentication flow works in OAuth2.0 Resource Server. Have a look at the following diagram:

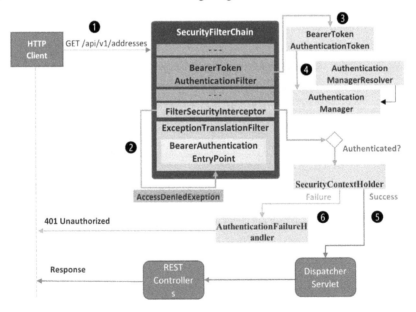

Figure 6.3 – Token authentication flow using OAuth 2.0 Resource Server

Let's understand the flow depicted in *Figure 6.3* , as follows:

1. The client sends a GET HTTP request to /api/v1/addresses.

2. BearerTokenAuthenciationFilter comes into action. If
    the request doesn't contain the Authorization header then
    BearerTokenAuthenticationFilter does not authenticate
    the request since it did not find the bearer token. It passes the call to
    FilterSecurityInterceptor, which does the authorization. It throws
    an AccessDeniedException exception (marked as **2** in *Figure 6.3*).
    ExceptionTranslationFilter springs into action. Control is moved to
    BearerTokenAuthenticationEntryPoint, which responds with a 401
    Unauthorized status and a WWW-Authenticate header with a Bearer value.
    If the client receives a WWW-Authenticate header with a Bearer value in
    response, it means it has to retry with the Authorization header that holds the
    valid bearer token. At this stage, the request cache is NullRequestCache (that is,
    empty) due to security reasons because the client can replay the request.

3. Let's assume the HTTP request contains an Authorization header.
    It extracts the Authorization header from the HTTP request and,
    apparently, the token from the Authorization header. It creates
    an instance of BearerTokenAuthenticationToken using the
    token value. BearerTokenAuthenticationToken is a type
    of AbstractAuthenticationToken class that implements an
    Authentication interface representing the token/principal for the authenticated
    request.

4. The HTTP request is passed to AuthenticationManagerResolver,
    which provides the AuthenticationManager based on
    the configuration. AuthenticationManager verifies the
    BearerTokenAuthenticationToken token.

5. If authentication is successful, then Authentication is set on
    the SecurityContext instance. This instance is then passed to
    SecurityContextHolder. setContext(). The request is passed to the
    remaining filters for processing and then routes to DispatcherServlet and
    then, finally, to AddressController.

6. If authentication fails, then `SecurityContextHolder.clearContext()` is called to clear the context value. `ExceptionTranslationFilter` springs into action. Control is moved to `BearerTokenAuthenticationEntryPoint`, which responds with a `401 Unauthorized` status and a `WWW-Authenticate` header with a value that contains the appropriate error message, such as `Bearer error="invalid_token", error_description="An error occurred while attempting to decode the Jwt: Jwt expired at 2020-12-14T17:23:30Z", error_uri="https://tools.ietf.org/html/rfc6750#section-3.1"`.

Exploring the fundamentals of JWT

You need an authority in the form of permissions or rights to carry out any activity or access any information. This authority is known as a claim. A claim is represented as a key-value pair. The key contains the claim name and the value contains the claim that can be a valid JSON value. A claim can also be metadata about the JWT.

How is JWT pronounced?

As per `https://tools.ietf.org/html/rfc7519`, the suggested pronunciation of JWT is the same as the English word *jot*.

A **JWT** is an encoded string that contains of set of claims. These claims are either digitally signed by a **JSON Web Signature** (**JWS**) or encrypted by **JSON Web Encryption** (**JWE**). *JWT is a self-contained way to transmit claims securely between parties.* The links for these **Request for Comments** (**RFC**) proposed standards are provided in the *Further reading* section of this chapter.

JWT structure

A JWT is an encoded string such as `aaa.bbb.ccc`, consisting of the following three parts separated by dots (`.`):

*   Header
*   Payload
*   Signature

A few websites such as `https://jwt.io/` or `https://www.jsonwebtoken.io/` allow you to view the content of a JWT and generate a JWT.

Let's have a look at the following sample JWT string. You can paste it into one of the previously mentioned websites to decode the content:

```
eyJ0eXAiOiJKV1QiLCJhbGciOiJIUzI1NiJ9.
eyJzdWIiOiJzY290dCIsInJvbGVzIjpbIlVTRVIiXSwiaX
NzIjoiTW9kZXJuIEFQSSBEZXZlbG9wbWVudCB3aXRoIFNw
cmluZyBhbmQgU3ByaW5nIEJvb3QiLCJleHAiOjE2MTA1Mj
A2MjksImlhdCI6MTYxMDE5ODIzNywianRpIjoiMjk3ZGY4
YTctNTE4Zi00ZWQ3LWJhNjYtOTJkYTQ5NGRkZDc2In0.MW-
QOgAcNwLoEYINzqnDSm73-N86yf29-RUJsrApDyg
```

This sample token demonstrates how a JWT is formed and divided into three parts using dots.

Header

A header consists of a Base64URL-encoded JSON string, normally containing two key-value pairs: a type of token (with a `typ` key) and a signing algorithm (with an `alg` key).

A sample JWT string contains the following header:

```
{
    "typ": "JWT",
    "alg": "HS256"
}
```

The preceding header contains `typ` and `alg` fields, representing type and algorithm respectively.

Payload

A payload is a second part of the JWT and contains the claims. This is also a Base64URL-encoded JSON string. There are three types of claims—registered, public, and private. These are outlined as follows:

- **Registered claims**: A few claims are registered in the **Internet Assigned Numbers Authority (IANA) JSON Web Token Claims** registry, therefore these claims are known as Registered claims. These are not mandatory but are recommended. Some Registered claims are listed here:

  a. Issuer claim (`iss` key): This claim identifies the principal who issued a token.

  b. Subject claim (`sub` key): This should be a unique value that represents the subject of the JWT.

c. Expiration Time claim (`exp` key): This is a numeric value representing the expiration time on or after which a JWT should be rejected.

d. Issued At claim (`iat` key): This claim identifies the time at which a JWT is issued.

e. JWT ID claim (`jti` key): This claim represents the unique identifier for a JWT.

f. Audience claim (`aud` key): This claims identifies the recipients, which JWT is intended for.

g. Not Before claim (`nbf` key): Represents the time before which a JWT must be rejected.

- **Public claims**: These are defined by JWT issuers and must not collide with registered claims. Therefore, these should either be registered with the IANA JWT Claims Registry or defined as a **Uniform Resource Identifier** (**URI**) with a collision-resistant namespace.

- **Private claims**: These are custom claims defined and used by the issuer and audience. They are neither registered nor public.

Here is a sample JWT string containing the following payload:

```
{
   "sub": "scott",
   "roles": [
     "USER"
   ],
   "iss": "Modern API Development with Spring and Spring
           Boot",
   "exp": 1610520629,
   "iat": 1610198237,
   "jti": "297df8a7-518f-4ed7-ba66-92da494ddd76"
}
```

The preceding payload contains `sub` (subject), `iss` (issuer), `roles` (custom claim roles), `exp` (expires), `iat` (issued at), and `jti` (JWT ID) fields.

Signature

A signature is also a Base64-encoded string—a third part. A signature is there to protect the content of the JWT. The content is visible, but cannot be modified if the token is signed. A Base64-encoded header and payload are passed to the signature's algorithm, along with either a secret or a public key to make the token a signed token. If you wish to include any sensitive or secret information in the payload, then it's better to encrypt it before assigning it to the payload.

A signature makes sure that the content is not modified once it is received back. The use of a public/private key enhances the security step by verifying the sender.

You can use a combination of both a JWT and JWE. The recommended way, however, is to first encrypt the payload using JWE and then sign it.

We'll use the public/private keys to sign the token in this chapter. Let's jump into the code.

Securing REST APIs with JWT

In this section, you'll secure the REST endpoints exposed in *Chapter 4, Writing Business Logic for APIs*. Therefore, we'll use the code from *Chapter 4, Writing Business Logic for APIs* and enhance it to secure the APIs.

The REST APIs should be protected with the following features:

- No secure API should be accessed without JWT.
- A JWT can be generated using sign-in/sign-up or a refresh token.
- A JWT and a refresh token should only be provided for a valid user's username/ password combination or a valid user sign-up.
- The password should be stored in encoded format using a `bcrypt` strong hashing function.
- The JWT should be signed with **RSA** (for **Rivest, Shamir, Adleman**) keys with a strong algorithm.
- Claims in the payload should not store sensitive or secured information. If they do, then these should be encrypted.
- You should be able to authorize API access for certain roles.

We need to include new APIs for the authorization flow. Let's add them first.

Learning new API definitions

You will enhance the existing APIs by adding four new APIs—sign-up, sign-in, sign-out, and a refresh token. Sign-up, sign-in, and sign-out operations are self-explanatory.

The refresh token provides a new access token (JWT) once the existing token expires. This is the reason why the sign-up/sign-in API provides two types of tokens—an access token and a refresh token—as a part of its response. The JWT access token self-expires, therefore a sign-out operation would only remove the refresh token.

Let's add these APIs into the openapi.yaml document.

Modifying the API specification

Apart from adding the new APIs, you also need to add a new user tag for these APIs that will expose all these APIs through the UserApi interface. Let's first add a sign-up endpoint.

Sign-up endpoint

Add the following specification for the sign-up endpoint in openapi.yaml:

```
/api/v1/users:
  post:
    tags:
      - user
    summary: Signup the a new customer (user)
    description: Creates a new customer (user), who can
                 login and do the shopping.
    operationId: signUp
    requestBody:
      content:
        application/xml:
          schema:
            $ref: '#/components/schemas/User'
        application/json:
          schema:
            $ref: '#/components/schemas/User'
    responses:
      201:
        description: For successful user creation.
```

```
      content:
        application/xml:
          schema:
            $ref: '#/components/schemas/SignedInUser'
        application/json:
          schema:
            $ref: '#/components/schemas/SignedInUser'
```

https://github.com/PacktPublishing/Modern-API-Development-with-Spring-and-Spring-Boot/blob/main/Chapter06/src/main/resources/api/openapi.yaml

Add the following new model, SignedInUser, to the schemas. This contains accessToken, refreshToken, username, and user ID fields. The code to add the model is shown in the following snippet:

```
SignedInUser:
  description: Signed-in user information
  type: object
  properties:
    refreshToken:
      description: Refresh Token, a unique secure string
      type: string
    accessToken:
      description: JWT Token aka access token
      type: string
    username:
      description: User Name
      type: string
    userId:
      description: User Identifier
      type: string
```

Now, let's add the sign-in endpoint.

Sign-in endpoint definition

Add the following specification for the sign-in endpoint in `openapi.yaml`:

```yaml
/api/v1/auth/token:
  post:
    tags:
      - user
    summary: Signin the customer (user)
    description: Signin the customer (user) that generates
                 the JWT (access token) and refresh token,
                 which can be used for accessing APIs.
    operationId: signIn
    requestBody:
      content:
        application/xml:
          schema:
            $ref: '#/components/schemas/SignInReq'
        application/json:
          schema:
            $ref: '#/components/schemas/SignInReq'
    responses:
      200:
        description: For user sign-in. Once successful,
                     user
                     receives the access and refresh token.
        content:
          application/xml:
            schema:
              $ref: '#/components/schemas/SignedInUser'
          application/json:
            schema:
              $ref: '#/components/schemas/SignedInUser'
```

We need a new model, `SignInReq`, for the sign-in request payload. It just contains the username and password fields. Let's add it, as follows:

```
SignInReq:
  description: Request body for Sign-in
  type: object
  properties:
    username:
      description: username of the User
      type: string
    password:
      description: password of the User
      type: string
```

Sign-out endpoint

Add the following specification for the sign-out endpoint in `openapi.yaml`:

```
# Under the /api/v1/auth/token
delete:
  tags:
    - user
  summary: Signouts the customer (user)
  description: Signouts the customer (user). It removes the
               refresh
               token from DB. Last issued JWT should be
               removed from
               client end that if not removed last for
               given
               expiration time.
  operationId: signOut
  requestBody:
    content:
      application/xml:
        schema:
          $ref: '#/components/schemas/RefreshToken'
      application/json:
        schema:
```

```
          $ref: '#/components/schemas/RefreshToken'
  responses:
    202:
      description: Accepts the request for logout.
```

In an ideal scenario, you should remove the refresh token of a user received from the request. You can fetch the user ID from the token and then use that ID to remove it from the USER_TOKEN table. However, in that case, you should send a valid access token.

We have opted for an easy way to remove the token, which is for it to be sent by the user as a payload. Therefore, this endpoint needs the following new model, RefreshToken. Here is the code to add the model:

```
RefreshToken:
  description: Contains the refresh token
  type: object
  properties:
    refreshToken:
      description: Refresh Token
      type: string
```

Finally, let's add an endpoint for refreshing the access token.

Refresh token endpoint

Add the following specification for the refresh token endpoint in openapi.yaml:

```
/api/v1/auth/token/refresh:
  post:
    tags:
      - user
    summary: Provides new JWT based on valid refresh token.
    description: Provides new JWT based on valid refresh
                 token.
    operationId: getAccessToken
    requestBody:
      content:
        application/xml:
          schema:
            $ref: '#/components/schemas/RefreshToken'
```

```
        application/json:
          schema:
            $ref: '#/components/schemas/RefreshToken'
    responses:
      200:
        description: For successful operation.
        content:
          application/xml:
            schema:
              $ref: '#/components/schemas/SignedInUser'
          application/json:
            schema:
              $ref: '#/components/schemas/SignedInUser'
```

Here, we have raised an exception by defining the refresh endpoint in terms of forming a URI that represents the refresh token resources. Instead, it generates a new access token.

In the existing code, we don't have a table for storing the refresh token. Therefore, let's add one.

Storing the refresh token using a database table

You can modify the Flyway database script to add a new table, as shown in the following code snippet:

```
create TABLE IF NOT EXISTS ecomm.user_token (
  id uuid NOT NULL DEFAULT random_uuid(),
  refresh_token varchar(128),
  user_id uuid NOT NULL,
  PRIMARY KEY(id),
  FOREIGN KEY (user_id)
    REFERENCES ecomm.user(id)
);
```

https://github.com/PacktPublishing/Modern-API-Development-with-Spring-and-Spring-Boot/blob/main/Chapter06/src/main/resources/db/migration/V1.0.0__Init.sql

Now, you can start writing the implementation code for JWT.

Implementing the JWT manager

Let's add a constant class that contains all the constants related to the security functionality before we implement the JWT manager class, as shown in the following code snippet:

```
public class Constants {
    public static final String ENCODER_ID = "bcrypt";
    public static final String API_URL_PREFIX = "/api/v1/**";
    public static final String H2_URL_PREFIX = "/h2-
      console/**";
    public static final String SIGNUP_URL = "/api/v1/users";
    public static final String TOKEN_URL =
      "/api/v1/auth/token";
    public static final String REFRESH_URL =
                            "/api/v1/auth/token/refresh";
    public static final String AUTHORIZATION =
      "Authorization";
    public static final String TOKEN_PREFIX = "Bearer ";
    public static final String SECRET_KEY = "SECRET_KEY";
    public static final long EXPIRATION_TIME = 900_000;
    public static final String ROLE_CLAIM = "roles";
    public static final String AUTHORITY_PREFIX = "ROLE_";
}
```

https://github.com/PacktPublishing/Modern-API-Development-with-Spring-and-Spring-Boot/blob/main/Chapter06/src/main/java/com/packt/modern/api/security/Constants.java

These constants are self-explanatory, except the EXPIRATION_TIME long value, which represents 15 minutes as a time unit.

Now, we can define the JWT manager class—JwtManager. JwtManager is a custom class that is responsible for generating a new JWT. It uses the java-jwt library from auth0.com. You are going to use public/private keys for signing the token. Let's define this class, as follows:

```
@Component
public class JwtManager {
    private final RSAPrivateKey privateKey;
```

```java
    private final RSAPublicKey pubKey;

    public JwtManager(RSAPrivateKey privateKey, RSAPublicKey
            pubKey) {
        this.privateKey = privateKey;
        this.pubKey = pubKey;
    }

    public String create(UserDetails principal) {
        final long now = System.currentTimeMillis();
        return JWT.create()
            .withIssuer("Modern API Development with Spring…")
            .withSubject(principal.getUsername())
            .withClaim(ROLE_CLAIM,
                    principal.getAuthorities().stream()
                .map(GrantedAuthority::getAuthority).collect(
                    toList())))
            .withIssuedAt(new Date(now))
            .withExpiresAt(new Date(now + EXPIRATION_TIME))
            .sign(Algorithm.RSA256(pubKey, privateKey));
    }
}
```

https://github.com/PacktPublishing/Modern-API-Development-with-Spring-and-Spring-Boot/blob/main/Chapter06/src/main/java/com/packt/modern/api/security/JwtManager.java

Here, JWT is a class from the `java-jwt` library that provides a fluent API for generating the token. It adds issuer (`"iss"`), subject (`"sub"`), issued at (`"iat"`), and expired at (`"exp"`) claims.

It also adds a custom claim, ROLE_CLAIMS ("roles"), which is populated using authorities from UserDetails. UserDetails is an interface provided by Spring Security. You can use the org.springframework.security. core.userdetails.User.builder() method to create a UserBuilder class. UserBuilder is a final builder class that allows you to build an instance of UserDetails.

At the end, it signs the JWT using a SHA256withRSA algorithm, using the provided public and private RSA keys. The JWT header specifies a HS256 value for the algorithm ("alg") claim.

> **RSA**
>
> RSA is an algorithm approved by the **Federal Information Processing Standards (FIPS)** (*FIPS 186*) for digital signatures and in **Special Publication (SP)** (*SP800-56B*) for key establishment.

Signing is done using the public and private RSA keys. Let's add the code for RSA key management in our sample e-commerce application.

Generating the public/private keys

You can use JDK's keytool to create a key store and generate public/private keys, as shown in the following code snippet:

```
$ keytool -genkey -alias "jwt-sign-key" -keyalg RSA -keystore
jwt-keystore.jks -keysize 4096
Enter keystore password:
Re-enter new password:
What is your first and last name?
  [Unknown]:  Modern API Development
What is the name of your organizational unit?
  [Unknown]:  Org Unit
What is the name of your organization?
  [Unknown]:  Packt
What is the name of your City or Locality?
  [Unknown]:  City
What is the name of your State or Province?
  [Unknown]:  State
What is the two-letter country code for this unit?
  [Unknown]:  IN
```

```
Is CN=Modern API Development, OU=Org Unit, O=Packt, L=City,
                    ST=State, C=IN correct?
  [no]:  yes
Generating 4,096 bit RSA key pair and self-signed certificate
(SHA384withRSA) with a validity of 90 days
        for: CN=Modern API Development, OU=Org Unit, O=Packt,
            L=City, ST=State, C=IN
```

The generated key store should be placed under the src/main/resources directory. These keys are valid only for 90 days from the time they got generated. Therefore, make sure that you create a new set of public/private keys when you use this chapter's code before running it.

Required values used in the keytool command should also be configured in the application.properties file, as shown here:

```
app.security.jwt.keystore-location=jwt-keystore.jks
app.security.jwt.keystore-password=password
app.security.jwt.key-alias=jwt-sign-key
app.security.jwt.private-key-passphrase=password
```

https://github.com/PacktPublishing/Modern-API-Development-with-Spring-and-Spring-Boot/blob/main/Chapter06/src/main/resources/application.properties

Now, we can configure the key store and public/private keys in the security configuration class.

Configuring the key store and keys

Let's add a SecurityConfig configuration class to configure the security relation configurations. This class extends the WebSecurityConfigurerAdaptor class. Here's the code to do this:

```
@EnableWebSecurity
public class SecurityConfig extends
WebSecurityConfigurerAdapter {
  @Value("${app.security.jwt.keystore-location}")
  private String keyStorePath;
  @Value("${app.security.jwt.keystore-password}")
  private String keyStorePassword;
  @Value("${app.security.jwt.key-alias}")
```

```
private String keyAlias;
@Value("${app.security.jwt.private-key-passphrase}")
private String privateKeyPassphrase;
…

…
}
```

https://github.com/PacktPublishing/Modern-API-Development-with-Spring-and-Spring-Boot/blob/main/Chapter06/src/main/java/com/packt/modern/api/security/SecurityConfig.java

We have added all the properties defined in application.properties here.

Now, we can make use of the properties defined in application.properties for configuring the KeyStore, RSAPrivateKey, and RSAPublicKey beans in the security configuration class, as shown in the next few sections.

KeyStore bean

You can create a new bean for KeyStore by adding the following method and annotating it with @Bean:

```
@Bean
public KeyStore keyStore() {
  try {
    KeyStore keyStore =
            KeyStore.getInstance(KeyStore.getDefaultType());
    InputStream resourceAsStream =
            Thread.currentThread().getContextClassLoader()
            .getResourceAsStream(keyStorePath);
    keyStore.load(resourceAsStream,
      keyStorePassword.toCharArray());
    return keyStore;
  } catch (IOException | CertificateException |
            NoSuchAlgorithmException |
            KeyStoreException e) {
    LOG.error("Unable to load keystore: {}", keyStorePath, e);
  }

  throw new IllegalArgumentException("Unable to load
```

```
        keystore");
    }
```

This creates a `KeyStore` instance using the `KeyStore` class from the `java.security` package. It loads the key store from the `src/main/resources` package and uses the password configuration in the `application.properties` file.

RSAPrivateKey bean

You can create a new bean for `RSAPrivateKey` by adding the following method and annotating it with `@Bean`:

```
@Bean
public RSAPrivateKey jwtSigningKey(KeyStore keyStore) {
    try {
        Key key = keyStore.getKey(keyAlias,
                    privateKeyPassphrase.toCharArray());
        if (key instanceof RSAPrivateKey) {
            return (RSAPrivateKey) key;
        }
    } catch (UnrecoverableKeyException |
                NoSuchAlgorithmException |
                KeyStoreException e) {
        LOG.error("Unable to load private key from keystore:
                {}", keyStorePath, e);
    }
    throw new IllegalArgumentException("Unable to load
        private key");
}
```

This method uses a key alias and a private key password to retrieve the private key, which is being used to return the `RSAPrivateKey` bean.

RSAPublicKey bean

You can create a new bean for `RSAPublicKey` by adding the following method and annotating it with `@Bean`:

```
@Bean
public RSAPublicKey jwtValidationKey(KeyStore keyStore) {
    try {
        Certificate certificate =
```

```
      keyStore.getCertificate(keyAlias);
   PublicKey publicKey = certificate.getPublicKey();
   if (publicKey instanceof RSAPublicKey) {
     return (RSAPublicKey) publicKey;
   }
 } catch (KeyStoreException e) {
   LOG.error("Unable to load private key from keystore:
          {}", keyStorePath, e);
 }
 throw new IllegalArgumentException("Unable to load public
     key");
}
```

Again, a key alias is used to retrieve the certificate from the key store. Then, the public key is retrieved from the certificate and returned. ·

As you know, JwtManager uses these public and private RSA keys to sign the JWT; therefore, a JWT decoder should use the same public key to decode the token. OAuth 2.0 Resource Server uses the org.springframework.security.oauth2.jwt. JwtDecoder interface to decode the token. Therefore, we need to create an instance of the JwtDecoder implementation and set the same public key in it to decode the token.

Spring OAuth 2.0 Resource Server provides a NimbusJwtDecoder implementation class of JwtDecoder. Let's now create a bean of it with the public key.

JwtDecoder Bean

You can create a new bean for JwtDecoder by adding the following method and annotating it with @Bean:

```
@Bean
public JwtDecoder jwtDecoder(RSAPublicKey rsaPublicKey) {
  return NimbusJwtDecoder.withPublicKey(
    rsaPublicKey).build();
}
```

Now, we can implement the newly added REST APIs.

Implementing new APIs

Let's implement the APIs exposed using `UserApi`. `UserApi` is part of code that has been autogenerated using OpenAPI Codegen. First, we need to add a new entity to the `user_token` table.

Coding user token functionality

You can create a `UserTokenEntity` based on the `user_token` table, as shown in the following code snippet:

```java
@Entity
@Table(name = "user_token")
public class UserTokenEntity {

    @Id
    @GeneratedValue
    @Column(name = "ID", updatable = false, nullable = false)
    private UUID id;

    @NotNull(message = "Refresh token is required.")
    @Basic(optional = false)
    @Column(name = "refresh_token")
    private String refreshToken;

    @ManyToOne(fetch = FetchType.LAZY)
    private UserEntity user;

    ...

    ...
```

https://github.com/PacktPublishing/Modern-API-Development-with-Spring-and-Spring-Boot/blob/main/Chapter06/src/main/java/com/packt/modern/api/entity/UserTokenEntity.java

Similarly, we can expose the following **create, read, update, and delete (CRUD)** repository for `UserTokenEntity` with the following two methods: `deleteByUserId()`, which will remove the `UserToken` table record based on a given user ID, and `findByRefreshToken()`, which will find the `UserToken` table record based on a given refresh token. The code is illustrated in the following code snippet:

```
public interface UserTokenRepository extends
                    CrudRepository<UserTokenEntity, UUID> {
  Optional<UserTokenEntity> findByRefreshToken(
    String refreshToken);
  Optional<UserTokenEntity> deleteByUserId(UUID userId);
}
```

https://github.com/PacktPublishing/Modern-API-Development-with-Spring-and-Spring-Boot/blob/main/Chapter06/src/main/java/com/packt/modern/api/repository/UserTokenRepository.java

Now, let add the new operations into `UserService`.

Enhancing the UserService class

We also need to add new methods into `UserService` for the `UserApi` interface. Let's add new methods into the service, as follows:

```
UserEntity findUserByUsername(String username);
Optional<SignedInUser> createUser(User user);
SignedInUser getSignedInUser(UserEntity userEntity);
Optional<SignedInUser> getAccessToken(RefreshToken
        refreshToken);
void removeRefreshToken(RefreshToken refreshToken);
```

https://github.com/PacktPublishing/Modern-API-Development-with-Spring-and-Spring-Boot/blob/main/Chapter06/src/main/java/com/packt/modern/api/service/UserService.java

Here, each method performs a specific operation, as outlined here:

- `findUserByUsername()`: Finds and returns a user based on a given username.
- `createUser()`: Adds a new signed-up user to the database.

- `getSignedInUser()`: Creates a new model instance of `SignedInUser` that holds the refresh token, access token (JWT), user ID, and username.

- `getAcceessToken()`: Generates and returns a new access token (JWT) for a given valid refresh token.

- `removeRefreshToken()`: Removes the refresh token from the database. It is called when the user wants to sign out.

Let's implement each of these methods in the `UserServiceImpl` class.

findUserByUsername() implementation

First, you can add the implementation for `findUserByUsername()`, as follows:

```
@Override
public UserEntity findUserByUsername(String username) {
  if (Strings.isBlank(username)) {
    throw new UsernameNotFoundException("Invalid user.");
  }
  final String uname = username.trim();
  Optional<UserEntity> oUserEntity =
              repository.findByUsername(uname);
  UserEntity userEntity = oUserEntity.orElseThrow(() ->
              new UsernameNotFoundException(
                String.format("Given user(%s) not found.",
                          uname)));
  return userEntity;
}
```

This is a straightforward operation. You query the database based on a given username. If found, then it returns the user, else it throws a `UsernameNotFoundException` exception.

createUser() implementation

Next, you can add the implementation for the `createUser()` method, as shown in the following code snippet:

```
@Override
@Transactional
public Optional<SignedInUser> createUser(User user) {
  Integer count = repository.findByUsernameOrEmail(
```

```
                            user.getUsername(), user.getEmail());
  if (count > 0) {
    throw new GenericAlreadyExistsException("
      Use different username and email.");
  }
  UserEntity userEntity = repository.save(toEntity(user));
  return Optional.of(createSignedUserWithRefreshToken(
    userEntity));
}
```

https://github.com/PacktPublishing/Modern-API-Development-with-Spring-and-Spring-Boot/blob/main/Chapter06/src/main/java/com/packt/modern/api/service/UserServiceImpl.java

Here, we first check whether there is any existing user that was assigned the same username or email in the sign-up request. If there is, then it simply raises an exception, else it creates a new user in the database and returns a `SignedInUser` instance with refresh and access tokens using the `createSignedUserWithRefreshToken()` method.

First, we can add a private `createSignedUserWithRefreshToken()` method in `UserServiceImpl`, as follows:

```
private SignedInUser createSignedUserWithRefreshToken(
                                  UserEntity userEntity) {
  return createSignedInUser(userEntity)
      .refreshToken(createRefreshToken(userEntity));
}
```

This also uses another private method, `createSignedInUser()`, which returns `SignedInUser`; then, it adds the refresh token by calling the `createRefreshToken()` method.

Let's define these two `createSignedIn()` and `createSignedInUser()` private methods, as follows:

```
private SignedInUser createSignedInUser(
    UserEntity userEntity) {
  String token = tokenManager.create(
      org.springframework.security.core.userdetails.User.
```

```
            builder()
        .username(userEntity.getUsername())
        .password(userEntity.getPassword())
        .authorities(Objects.nonNull(userEntity.getRole()) ?
            userEntity.getRole().name() : "")
        .build());
    return new SignedInUser().username(
            userEntity.getUsername())
        .accessToken(token)
        .userId(userEntity.getId().toString());
}


private String createRefreshToken(UserEntity user) {
    String token = RandomHolder.randomKey(128);
    userTokenRepository.save(new UserTokenEntity()
                    .setRefreshToken(token).setUser(user));
    return token;
}
```

Here, `tokenManager` is used in the `createSignedIn()` method for creating the JWT. `tokenManager` is an instance of `JwtManager`. The `User.builder()` method is used to create a `UserBuilder` class. `UserBuilder`, which is a final builder class, is used to create an instance of `UserDetails`. The `JwtManager.create()` method uses this `UserDetails` instance to create a token.

The `createRefreshToken()` method uses `RandomHolder` private static class to generate a refresh token. This token is not a JWT; however, we can use a longer duration valid token, such as one valid for a day for a refresh token. Saving a JWT as a refresh token in the database removes the sole purpose of using the JWT. Therefore, we should think carefully about using the JWT as a refresh token and then saving it in the database.

Let's add the `RandomHolder` private static class, as follows:

```
// https://stackoverflow.com/a/31214709/109354
private static class RandomHolder {
    static final Random random = new SecureRandom();

    public static String randomKey(int length) {
        return String.format("%"+length+"s",
```

```
              new BigInteger(
                  length*5/*base 32,2^5*/, random)
        .toString(32)).replace('\u0020', '0');
  }
}
```

This class uses a `SecureRandom` instance to generate a random `BigInteger` instance. Then, this random `BigInteger` value is converted to a string with radix size 32. At the end, the space is replaced with 0 if found in a converted string.

You can also use the `org.apache.commons.lang3.RandomStringUtils.randomAlphanumeric()` method to generate a refresh token, or use any other secured random key generator.

We also need to modify the `UserRepository` class to add a new method that returns the count of users having a given username or email.

getSignedInUser() implementation

Implementation of the `getSignedInUser()` method is straightforward, as shown in following code snippet:

```
@Override
@Transactional
public SignedInUser getSignedInUser(UserEntity userEntity) {
   userTokenRepository.deleteByUserId(userEntity.getId());
   return createSignedUserWithRefreshToken(userEntity);
}
```

It first removes the existing token from the database associated with the given user, and then returns the new instance of `SignedInUser` created using `createSignedUserWithRefreshToken()`, defined previously in the *createUser() implementation* subsection.

getAccessToken() implementation

Implementation of the `getAccessToken()` method is again straightforward, as shown in the following code snippet:

```
@Override
public Optional<SignedInUser> getAccessToken(
                            RefreshToken refreshToken) {
   return userTokenRepository
```

```
        .findByRefreshToken(refreshToken.getRefreshToken())
    .map(ut -> Optional.of(createSignedInUser(
            ut.getUser())
        .refreshToken(refreshToken.getRefreshToken())))
    .orElseThrow(() -> new InvalidRefreshTokenException(
                                    "Invalid token."));
}
```

First, it finds the user's token entity using the `UserTokenRepository` instance. Then, it populate the `SignedInUser` POJO using the retrieved `UserToken` entity. The `createSignedInUser()` method does not populate the refresh token, therefore we assign the same refresh token back. If it does find the user token entry in the database based on the refresh token, it throws an exception.

You can also add a time validation logic for the refresh token—for example, store the refresh token creation time in the database and use the configured valid time for refresh token validation: a kind of expiration logic of JWT.

removeRefreshToken() implementation

Implementation of the `removeRefreshToken()` method is shown in the following code snippet:

```
@Override
public void removeRefreshToken(RefreshToken refreshToken) {
    userTokenRepository
        .findByRefreshToken(refreshToken.getRefreshToken())
        .ifPresentOrElse(userTokenRepository::delete, () -> {
            throw new InvalidRefreshTokenException(
                                    "Invalid token.");
        });
}
```

First, it finds the given refresh token in the database. If this is not found, then it throws an exception. If the given refresh token is found in the database, then it deletes it.

Enhancing the UserRepository class

Let's add `findByUsername()` and `findByUsernameOrEmail()` methods to `UserRepository`, as follows:

```
public interface UserRepository
                extends CrudRepository<UserEntity, UUID> {
  Optional<UserEntity> findByUsername(String username);
  @Query(value = "select count(u.*) from ecomm.user u
                where u.username = :username or u.email =
                :email",
                nativeQuery = true
  Integer findByUsernameOrEmail(String username, String
                                email);
}
```

https://github.com/PacktPublishing/Modern-API-Development-with-Spring-and-Spring-Boot/blob/main/Chapter06/src/main/java/com/packt/modern/api/repository/UserRepository.java

This returns a count of records matching the given username or email.

Now, we can implement the `UserApi` interface to write REST controllers.

Implementing the REST controllers

You have already developed and enhanced services and repositories required to implement the APIs defined in the `UserApi` interface generated by OpenAPI Codegen in the previous section. The only pending dependency is `PasswordEncoder`. `PasswordEncoder` is required for encoding the password before storing and matching the password given as part of the sign-in request.

Adding a bean for PasswordEncoder

You should expose the `PasswordEncoder` bean because Spring Security needs to know which encoding you want to use for password encoding, as well as for decoding the passwords. Let's add a `PasswordEncoder` bean in `AppConfig`, as follows:

```
@Bean
public PasswordEncoder passwordEncoder() {
  Map<String, PasswordEncoder> encoders = Map.of(
```

```
    "bcrypt", new BCryptPasswordEncoder(),
    "pbkdf2", new Pbkdf2PasswordEncoder(),
    "scrypt", new SCryptPasswordEncoder());
  return new DelegatingPasswordEncoder("bcrypt", encoders);
}
```

https://github.com/PacktPublishing/Modern-API-Development-with-Spring-and-Spring-Boot/blob/main/Chapter06/src/main/java/com/packt/modern/api/AppConfig.java

You can directly create a new instance of `BCryptPasswordEncoder` and return it for bcrypt encoding. However, use of `DelegatingPasswordEncoder` not only allows you to support existing passwords but also facilitates migration to a new, better encoder if one is available in the future. This code uses *Bcrypt* as a default password encoder, which is the best among the current available encoders.

For `DelegatingPasswordEncoder` to work, you need to add a hashing algorithm prefix such as `{bcrypt}` to encoded passwords—for example, add `{bcrypt}$2a$10$neR0EcYY5./tLVp4litNyuBy/kfrTsqEv8hiyqEKX0TXIQQwC/5Rm` in the persistent store if you already have a hashed password in the database or if you're adding any seed/test users in the database script. The new password would store the password with a prefix anyway, as configured in the `DelegatingPasswordEncoder` constructor. You have passed bcrypt into the constructor, therefore all new passwords will be stored with a `{bcrypt}` prefix.

`PasswordEncoder` reads the password from the persistence store and removes the prefix before matching. It uses the same prefix to find out which encoder it needs to use for matching. Now, you can start implementing the new APIs based on `UserApi`.

Implementing the Controller class

First, create a new `AuthController` class, as shown in the following code snippet:

```
@RestController
public class AuthController implements UserApi {

  private final UserService service;
  private final PasswordEncoder passwordEncoder;

  public AuthController(UserService service,
                        PasswordEncoder passwordEncoder, ) {
```

```
    this.service = service;
    this.passwordEncoder = passwordEncoder;
  }
  ...
  ...
}
```

https://github.com/PacktPublishing/Modern-API-Development-with-Spring-and-Spring-Boot/blob/main/Chapter06/src/main/java/com/packt/modern/api/controller/AuthController.java

The `AuthController` class is annotated with `@RestController` to mark it as a REST controller. Then, it uses two beans, `UserService` and `PasswordEncoder`, which will be injected at the time of the `AuthController` construction.

Let's first add the sign-in operation, as follows:

```
@Override
public ResponseEntity<SignedInUser> signIn(
                              @Valid SignInReq signInReq) {
  UserEntity userEntity = service
              .findUserByUsername(signInReq.getUsername());
  if (passwordEncoder.matches(
          signInReq.getPassword(),
              userEntity.getPassword())) {
    return ok(service.getSignedInUser(userEntity));
  }
  throw new InsufficientAuthenticationException(
        "Unauthorized.");
}
```

It first finds the user and matches the password using the `PasswordEncoder` instance. If everything goes through successfully, it returns the `SignedInUser` instance with refresh and access tokens; else, it throws an exception.

Let's add other operations to `AuthController`, as follows:

```
@Override
public ResponseEntity<Void> signOut(
                          @Valid RefreshToken refreshToken) {
```

```
    service.removeRefreshToken(refreshToken);
    return accepted().build();
}


@Override
public ResponseEntity<SignedInUser> signUp(
                                @Valid User user) {
    return status(HttpStatus.CREATED)
            .body(service.createUser(user).get());
}


@Override
public ResponseEntity<SignedInUser> getAccessToken(
                                @Valid RefreshToken refreshToken)
{
    return ok(service.getAccessToken(refreshToken)
            .orElseThrow(InvalidRefreshTokenException::new));
}
```

All operations such as `signOut()`, `signUp()`, and `getAccessToken()` are straightforward, as outlined here:

- `signOut()` uses the user service to remove the given refresh token.
- `signUp()` creates a valid new user and returns the `SignedInUser` instance as a response.
- `getAccessToken()` returns the `SignedInUser` with a new access token if the given refresh token is valid.

We are done with coding the controllers. Let's configure security in the next subsection.

Configuring web-based security

You have already learned about how to configure Spring Security in one of the previous subsections, *Configuring Spring Security*. In that subsection, we defined a new class, `SecurityConfig`, which extends `WebSecurityConfigurerAdaptor` and is also annotated with `@EnableWebSecurity`.

Let's modify its overridden `configure()` method, which will allow us to configure `HttpSecurity` using DSL (fluent methods). Let's make the following configurations:

- Disable HTTP **Basic authentication (BA)** using `httpBasic().disable()`.

- Disable the form login using `formLogin().disable()`.

- Restrict access based on URL patterns using `authorizeRequests()`.

- Configure URL patterns and respective HTTP methods using `antMatchers()`, which allows you to use `ant` (build tool) pattern matching styles. You can also use `mvcMatchers()`, which uses the same pattern matching style as Spring **Model-View-Controller (MVC)**.

- All URLs except those configured explicitly by `authorizeRequests()` should be allowed by any authenticated user (by using `anyRequest().authenticated()`).

- Enable JWT bearer token support for OAuth 2.0 Resource Server (`oauth2ResourceServer(OAuth2ResourceServerConfigurer::jwt`).

- Enable the `STATELESS` session creation policy (that is, it won't create any `HTTPSession`).

Let's add these to `HttpSecurity`, as follows:

```
@Override

protected void configure(HttpSecurity http) throws
    Exception {
  http.httpBasic().disable().formLogin().disable()
      .and()
      .headers().frameOptions().sameOrigin() // for H2
                                             // Console
      .and()
      .authorizeRequests()
      .antMatchers(HttpMethod.POST, TOKEN_URL).permitAll()
      .antMatchers(HttpMethod.DELETE,
          TOKEN_URL).permitAll()
      .antMatchers(HttpMethod.POST, SIGNUP_URL).permitAll()
      .antMatchers(HttpMethod.POST,
          REFRESH_URL).permitAll()
      .antMatchers(H2_URL_PREFIX).permitAll()
```

```
    .anyRequest().authenticated()
    .and()
    .oauth2ResourceServer(OAuth2ResourceServerConfigurer:
        :jwt)
    .sessionManagement().sessionCreationPolicy(
            SessionCreationPolicy.STATELESS);
}
```

https://github.com/PacktPublishing/Modern-API-Development-with-Spring-and-Spring-Boot/blob/main/Chapter06/src/main/java/com/packt/modern/api/security/SecurityConfig.java

Here, we have added one more important configure for the H2 console app. The H2 console **user interface** (**UI**) is based on HTML frames. The H2 console UI won't display in browsers because by default, the security header (X-Frame-Options) is not sent with the permission to allow frames with the same origin. Therefore, you need to configure headers().frameOptions().sameOrigin().

Now, you can configure CORS and CSRF.

Configuring CORS and CSRF

Browsers restrict cross-origin requests from scripts for security reasons. For example, a call from http://mydomain.com to http://mydomain-2.com can't be made using a script. Also, an origin not only indicates a domain—in fact, it includes a scheme and a port too.

Before hitting to any endpoint, the browser sends a preflight request using the HTTP method option to check whether the server would permit the actual request. This request contains the following headers:

- Actual request's headers (Access-Control-Request-Headers)
- A header containing the actual request's HTTP method (Access-Control-Request-Method)

- A `Origin` header that contains the requesting origin (scheme, domain, and port)

- If the response from the server is successful, then only the browser allows the actual request to fire. The server responds with other headers, such as `Access-Control-Allow-Origin`, which contains the allowed origins (an asterisk `*` value means any origin), `Access-Control-Allow-Methods` (allowed methods), `Access-Control-Allow-Headers` (allowed headers), and `Access-Control-Max-Age` (allowed time in seconds).

You can configure CORS to take care of cross-origin requests. For that, you need to make the following two changes:

- Add a `CorsConfigurationSource` bean that takes care of the CORS configuration using a `CorsConfiguration` instance.

- Add the `cors()` method into `HTTPSecurity` in the `configure()` method. It uses `CorsFilter` if a `corsFilter` bean is added, else it uses `CorsConfigurationSource`. If neither is configured, then it uses the Spring MVC pattern inspector handler.

Let's now add the `CorsConfigurationSource` bean to the `SecurityConfig` class.

The default permitted values (`new CorsConfiguraton().applyPermitDefaultValues()`) configure CORS for any origin (`*`), all headers, and simple methods (`GET`, `HEAD`, `POST`) with allowed max age is 30 minutes.

You need to allow mostly all of the HTTP methods, including the `DELETE` method, and need more custom configuration; therefore, we will use the following bean definition:

```
@Bean
CorsConfigurationSource corsConfigurationSource() {
  CorsConfiguration configuration = new
      CorsConfiguration();
  configuration.setAllowedOrigins(Arrays.asList("*"));
  configuration.setAllowedMethods(Arrays.asList("HEAD",
      "GET", "PUT", "POST", "DELETE", "PATCH"));
   // For CORS response headers
  configuration.addAllowedOrigin("*");
  configuration.addAllowedHeader("*");
  configuration.addAllowedMethod("*");
  UrlBasedCorsConfigurationSource source = new
      UrlBasedCorsConfigurationSource();
```

```
source.registerCorsConfiguration("/**", configuration);
    return source;
}
```

https://github.com/PacktPublishing/Modern-API-Development-with-Spring-and-Spring-Boot/blob/main/Chapter06/src/main/java/com/packt/modern/api/security/SecurityConfig.java

Here, you are creating a `CorsConfiguration` instance using the default constructor, and then setting the allowed origins, allowed methods, and response headers. Finally, you are passing it as an argument while registering it to the `UrlBasedCorsConfigurationSource` instance and returning it.

Let's add `cors()` to `HttpSecurity`, as follows:

```
@Override
protected void configure(HttpSecurity http) throws Exception {
    http.httpBasic().disable().formLogin().disable()
        .csrf().ignoringAntMatchers(API_URL_PREFIX,
            H2_URL_PREFIX)
        .and()
        .headers().frameOptions().sameOrigin() // for H2 Console
        .and()
        .cors()
        .and()
        .authorizeRequests()
        .antMatchers(HttpMethod.POST, TOKEN_URL).permitAll()
        .antMatchers(HttpMethod.DELETE,
            TOKEN_URL).permitAll()
        .antMatchers(HttpMethod.POST, SIGNUP_URL).permitAll()
        .antMatchers(HttpMethod.POST,
            REFRESH_URL).permitAll()
        ...
        ...
}
```

Here, we have also configured CSRF using `csrf()` DSL. We have applied CSRF protection to all URLs except URLs starting with `/api/v1` and the `/h2-console` H2 database console URL. You can change the configuration based on your requirement.

Let's first understand what CSRF/XSRF is. **CSRF** or **XSRF** stands for **Cross-Origin Request Forgery**, which is a web security vulnerability. Let's assume you are a bank customer and are currently signed in to the bank. You may get an email and you click on a link in the email, or click on any malicious website's link that may contain a malicious script. This script then sends a request to your bank for a fund transfer. The bank then transfers the funds to a perpetrator's account because the bank thinks that the request has been sent by you as you are signed in. This is just an example.

To prevent such attacks, the application sends new unique CSRF tokens associated with the signed-in user for each new request. These tokens are stored in hidden form fields. When a user submits a form, the same token should be sent back with the request. The application then verifies the CSRF token and only processes the request if the verification is successful. This works because malicious scripts can't read the token due to the same origin policy,

However, if a perpetuator also tricks you into revealing the CSRF token, then it is very difficult to prevent such attacks. You can disable CSRF protection for this web service by using `csrf().disable()` because it only provides REST endpoints.

Now, let's move on to the final section, where you will configure the authorization based on the user's role.

Understanding authorization

Your valid username/password or access token for authentication allows you access to secure resources such as URLs, web resources, or secure web pages. Authorization is one step ahead; it allows you to configure access security further with scopes such as read, write, or roles such as Admin, User, Manager, and so on. Spring Security allows you to configure any custom authority.

We will configure three types of roles for our sample e-commerce app—namely, Customer (user), Admin, and **Customer Support Representative (CSR)**. Obviously, each user would have their own specific authority. For example, a user can place an order and buy stuff online, but should not be able to access the CSR or Admin resources. Similarly, a CSR should not be able to have access to Admin-only resources. A security configuration that allows authority or role-based access to resources is known authorization. A failed authentication should return HTTP (status 401 unauthorized), and a failed authorization should return HTTP status 403 (forbidden), which means the user is authenticated but does not have the required authority/role to access the resource.

Let's introduce these three roles in a sample e-commerce app, as shown in the following code snippet:

```
public enum RoleEnum implements GrantedAuthority {
  USER(Const.USER), ADMIN(Const.ADMIN), CSR(Const.CSR);
  private String authority;
  RoleEnum(String authority) { this.authority = authority; }
  @Override
  @JsonValue
  public String getAuthority() { return authority; }
  @JsonCreator
  public static RoleEnum fromAuthority(String authority) {
    for (RoleEnum b : RoleEnum.values()) {
      if (b.authority.equals(authority)) { return b; }
    }
    throw new IllegalArgumentException("Unexpected value");
  }

  @Override
  public String toString() { return String.valueOf(authority);
}

  public class Const {
    public static final String ADMIN = "ROLE_ADMIN";
    public static final String USER = "ROLE_USER";
    public static final String CSR = "ROLE_CSR";
  }
}
```

Here, we have declared an `enum` that implements Spring Security's `GrantedAuthority` interface to override the `getAuthority()` method. `GrantedAuthority` is an authority granted to an `Authentication` (interface) object. As you know, `BearerTokenAuthenticationToken` is a type of `AbstractAuthenticationToken` class that implements the `Authentication` interface, which represents the token/principal for an authenticated request. We have used the `String` constants for user's roles in this `enum` as we need these when we configure the role-based restriction at a method level.

Let's discuss role and authority in detail.

Role and authority

You can take authorities for finer-grained control, whereas roles should be applied to large sets of permissions. A role is an authority that has the `ROLE_` prefix. This prefix is configurable in Spring Security.

Spring Security provides `hasRole()` and `hasAuthority()` methods for applying role- and authority-based restrictions. `hasRole()` and `hasAuthority()` are almost identical, but the `hasRole()` method maps with `Authority` without the `ROLE_` prefix. If you use `hasRole('ADMIN')`, your Admin enum must be `ROLE_ADMIN` instead of `ADMIN` because a role is an authority and should have a `ROLE_` prefix, whereas if you use `hasAuthority('ADMIN')`, your `ADMIN` enum must be only `ADMIN`.

OAuth 2.0 Resource Server by default populates the authorities based on the scope (`scp`) claim. If you provide access to user's resources such as order history and so on for integration to other application, then you can limit an application's access to a user's account before granting access to other applications for third-party integration. Third-party applications can request one or more scopes; this information is then presented to the user in the consent screen, and the access token issued to the application will be limited to the scopes granted. However, in this chapter, we are not providing OAuth 2.0 authorization flows and are limiting security access to REST endpoints.

If the JWT contains a claim with the name "scope" (`scp`), then Spring Security will use the value in that claim to construct the authorities by prefixing each value with `"SCOPE_"`. For example, if a payload contains a `scp=["READ", "WRITE"]` claim, this means that a list of `Authority` will consist of `SCOPE_READ` and `SCOPE_WRITE`.

We need to change the default authority mapping behavior because a scope (scp) claim is the default authority for OAuth2.0 Resource Server in Spring. We can do that by adding a custom authentication converter to JwtConfigurer in OAuth2ResourceServer in your security configuration. Let's add a method that returns the converter, as follows:

```java
private Converter<Jwt, AbstractAuthenticationToken>
                        getJwtAuthenticationConverter() {
  JwtGrantedAuthoritiesConverter authorityConverter =
                new JwtGrantedAuthoritiesConverter();
  authorityConverter.setAuthorityPrefix(AUTHORITY_PREFIX);
  authorityConverter.setAuthoritiesClaimName(ROLE_CLAIM);
  JwtAuthenticationConverter converter =
                new JwtAuthenticationConverter();
  converter.setJwtGrantedAuthoritiesConverter(
      authorityConverter);
  return converter;
}
```

https://github.com/PacktPublishing/Modern-API-Development-with-Spring-and-Spring-Boot/blob/main/Chapter06/src/main/java/com/packt/modern/api/security/SecurityConfig.java

Here, first create a new instance of JwtGrantedAuthorityConverter and then assign an authority prefix (ROLE_) and authority claim name (key of the claim in JWT) as roles.

Now, we can use this private method for configuring the OAuth 2.0 resource server. You can now modify the existing configuration with the following code. We can also add configuration for adding role-based restriction on the POST /api/v1/addresses API call in the following code snippet:

```java
@Override
protected void configure(HttpSecurity http) throws
    Exception {
  http.httpBasic().disable().formLogin().disable()
      .csrf().ignoringAntMatchers(API_URL_PREFIX,
        H2_URL_PREFIX)
      .and()
      .headers().frameOptions().sameOrigin() // for H2
                                        // Console
```

```
        .and()
        .cors()
        .and()
        .authorizeRequests()
        .antMatchers(HttpMethod.POST, TOKEN_URL).permitAll()
        .antMatchers(HttpMethod.DELETE,
          TOKEN_URL).permitAll()
        .antMatchers(HttpMethod.POST, SIGNUP_URL).permitAll()
        .antMatchers(HttpMethod.POST,
          REFRESH_URL).permitAll()
        .antMatchers(H2_URL_PREFIX).permitAll()
        .mvcMatchers(HttpMethod.POST, "/api/v1/addresses/**")
        .hasAuthority(RoleEnum.ADMIN.getAuthority())
        .anyRequest().authenticated()
        .and()
        .oauth2ResourceServer(oauth2ResourceServer ->
            oauth2ResourceServer.jwt(jwt ->
                jwt.jwtAuthenticationConverter(
                    getJwtAuthenticationConverter())))
        .sessionManagement().sessionCreationPolicy(
            SessionCreationPolicy.STATELESS);
}
```

After setting this configuration to add an address (POST /api/v1/addresses), it now requires both authentication and authorization. This means the logged-in user must have the ADMIN role to call this endpoint successfully. Also, we have changed the default claim from scope to role.

Now, we can proceed further with method-level role-based restrictions. Spring Security provides a feature that allows you to place authority-/role-based restrictions on public methods of Spring beans using a set of annotations such as @PreAuthorize, @Secured, and @RolesAllowed. By default, these are disabled, therefore you need to enable them explicitly.

Let's enable these by adding @EnableGlobalMethodSecurity(prePostEnabled = true) annotation to the Spring Security configuration class, as follows:

```
@EnableWebSecurity
@EnableGlobalMethodSecurity(prePostEnabled = true)
```

```
public class SecurityConfig extends
WebSecurityConfigurerAdapter {
```

Now, you can use @PreAuthorize (the given access-control expression would be evaluated before the method invocation) and @PostAuthorize (the given access-control expression would be evaluated after the method invocation) annotations to place restrictions on public methods of Spring beans because you have set the prePostEnabled property to true when enabling the global method-level security.

@EnableGlobalMethodSecurity also supports the following properties:

- securedEnabled: This allows you to use @Secured annotation on public methods.

- jsr250Enabled: This allows you to use JSR-250 annotations such as @RolesAllowed. @RolesAllowed can be applied to both public classes and methods. As the name suggests, you can use a list of roles for access restrictions.

@PreAuthorize/@PostAuthorize are more powerful than the other security annotations because these can not only be configured for authorities/roles, but you can also use any valid **Spring Expression Language (SpEL)** expression.

For demonstration purposes, let's add @PreAuthorize annotation to the deleteAddressesById() method, which is associated with DELETE /v1/auth/addresses/{id} in AddressController, as shown in the following code snippet:

```
@PreAuthorize("hasRole('" + Const.ADMIN + "')")
@Override
public ResponseEntity<Void> deleteAddressesById(String id) {
  service.deleteAddressesById(id);
  return accepted().build();
}
```

- Here, hasRole() is a built-in SpEL expression. We need to pass a valid *SpEL* expression, and it should be a String. Any variable used to form this *SpEL* expression should be final. Therefore, we have declared final string constants in the RoleEnum enum (for example, Const.ADMIN).

- Now, the DELETE /api/v1/addresses/{id} REST API can only be invoked if the user has the ADMIN role.

- Spring Security provides various built-in SpEL expressions, such as `hasRole()`. Here are some others:

a. `hasAnyRole(String... roles)`: Returns `true` if principal's role matches any of the given roles.

b. `hasAuthority(String authority)`: Returns `true` if principal has given authority. Similarly, you can also use `hasAnyAuthority(String... authorities)`.

c. `permitAll`: Returns `true`.

d. `denyAll`: Returns `false`.

- `isAnonymous()`: Returns `true` if current user is anonymous.

- `isAuthenticated()`: Returns `true` if current user is not anonymous.

A full list of these expressions is available at `https://docs.spring.io/spring-security/site/docs/current/reference/html5/#el-access`.

Similarly, you can apply access restrictions for other APIs. Let's test security in the next section.

Testing security

You can clone the code and build it using the following command:

```
Run it from project home
$ gradlew clean build
```

This code is tested with Java 15.

> **Important**
> Make sure to generate the keys again, as keys generated by the JDK keytool are only valid for 90 days.

Then, you can run the application from your project home, as shown in the following code snippet:

```
$ java -jar build/libs/Chapter06-0.0.1-SNAPSHOT.jar
```

Now, you must be looking forward to testing. Let's test our first use case.

Let's hit the `GET /api/vi/addresses` API without the `Authorization` header, as shown in the following code snippet:

```
$ curl -v 'http://localhost:8080/api/v1/addresses' -H 'Content-
Type: application/json' -H 'Accept: application/json'

< HTTP/1.1 401
< Vary: Origin
< Vary: Access-Control-Request-Method
< Vary: Access-Control-Request-Headers
< WWW-Authenticate: Bearer
< X-Content-Type-Options: nosniff
< X-XSS-Protection: 1; mode=block
< Cache-Control: no-cache, no-store, max-age=0, must-revalidate
< Other information is removed for brevity
```

This returns HTTP Status 401 (unauthorized) and a `WWW-Authenticate: Bearer` response header, which suggests the request should be sent with an `Authorization` header.

Let's send the request again with an invalid token, as follows:

```
$  curl -v 'http://localhost:8080/api/v1/addresses' -H
'Content-Type: application/json' -H 'Accept: application/json'
-H 'Authorization: Bearer eyJ0eXAiOiJKV1QiLCJhbGciOiJSUzI1NiJ9…
rest of the JWT string removed for brevity'

< HTTP/1.1 401
< Vary: Origin
< Vary: Access-Control-Request-Method
< Vary: Access-Control-Request-Headers
< WWW-Authenticate: Bearer error="invalid_token", error_
description="An error occurred while attempting to decode the
Jwt: Jwt expired at 2021-01-09T14:19:49Z", error_uri="https://
tools.ietf.org/html/rfc6750#section-3.1"
< Other information is removed for brevity
```

The server again responds with 401 (unauthorized), but this time with an error message and description that tells us that a given token has expired. It can also send an `invalid_token` error with HTTP status 401 based on a given bearer token, as outlined here:

- If the token is not a well-formatted JWT, then it shows `An error occurred while attempting to decode the Jwt: Invalid JWT serialization: Missing dot delimiter(s)`.

- If the token doesn't have a valid signature, then it shows `An error occurred while attempting to decode the Jwt: Signed JWT rejected: Invalid signature`.

We have created two users using a Flyway database migration script: `scott/tiger` and `scott2/tiger`. Now, let's perform a sign-in with the username `scott` to get the valid JWT, as follows:

```
$ curl -X POST 'http://localhost:8080/api/v1/auth/token' -H
'Content-Type: application/json' -H 'Accept: application/json'
-d '{
    "username": "scott",
    "password": "tiger"
}'


{
    "refreshToken": "3i2tlrmdqnp60drl6i9c2kdm36s48qg5vm2ucgt
flsk0cjo4dthhjan9aj1ck83det8m8hkl461cqkfl57puk81ct6j09ilpo
ranf1jj414ht4ob7dkcakq6lk92cnct",
    "accessToken": "eyJ0eXAiOiJKV1QiLCJhbGciOiJSUzI1NiJ9.
eyJzdWIiOiJzY290dCIsInJvbGVzIjpbIlVTRVVI
iXSwiaXNzIjoiTW9kZXJuIEFQSSBEZXZlbG9wbW
VudCB3aXRoIFNwcmluZyBhbmQgU3ByaW5nIEJvb
3QiLCJleHAiOjE2MTA5NTQ3NzMsImlhdCI6MTYxMDk1Mzg3M30.GOE2WwgN-
1s82KqU2U-hd7rcrhdblrfV59HXTL9B7BL2eAgshjxtJGVhj9CtR_
LQwA54fZo0yVwYFyMUrQBBFkn_2fDRU_8j1LD91N
HDO1wqpiVx9kRzB9nUIR0OcpT2OdMaPD_HpmRiQOchwZSxsi7_
c5dO59-URJn17ahXeBDJoAFrYQGmetjvuZtGwd9nLAvdSq9KK
OL_gLle1wqsjJOYqJ9l_djLzeaO3Xgg-Kva5rmyZP0tWws7A95H2S
i2tIqrRGESZUCAQ3GbezpZB2OO_YgyCkQSuJkFTuQWc1MFbqtgeRcR
iklX53BBngcHCfAeOAsBtKL17yXnd-IQSPn1GBLmCJh1-nMgrwAKS-
1bx1k55FI93qGVoXDFFnVRUgjf_mA5aKNx9VECDtaXLDR7TA7LgjXiDXJ3
ZPRNsF3-8fagHNKq42BjPdGH62XtWBve_Ide00DXNtSffHAlo2ukjGpN_
zdmuZu7-UNeObg3g_dD6vvSnfupylJbVJooVDOQctR0u-
ausMNKvh32NBG4-IQS2pW5Xo3i3l0GEtTP8AIy0vtafnFWBJI_OwTKVM
```

```
8s966cgliswmeahLxIpLPeSo4Q0NxdE7MDnGg8wUbnzxoiq-ExcUjm_
e7M2N7LMgdlsk0asQQYOJDe50EwMr2oE9ZDQepTtqwfSjcpKdKQ",
    "username": "scott",
    "userId": "a1b9b31d-e73c-4112-af7c-b68530f38222"
}
```

It returns with both a refresh and an access token. Let's use this access token to call the
`GET /api/v1/addresses` API again (please note that the `Bearer` token value in the
`Authorization` header is taken from the response of the previous `GET /api/v1/`
`auth/token` API call). The code is shown in the following snippet:

```
$ curl -v 'http://localhost:8080/api/v1/addresses' -H
'Content-Type: application/json' -H 'Accept: application/json'
-H 'Authorization: Bearer eyJ0eXAiOiJKV1QiLCJhbGciOiJSUzI1NiJ9.
ey… rest of the JWT string removed for brevity'
< Response
[
    {
        "links": [
            {
                "rel": "self",
                "href": "http://localhost:8080/api/v1/
                addresses/a731fda1-aaad-42ea-bdbc-a27eeebe2cc0"
            }
        ],
        "id": "a731fda1-aaad-42ea-bdbc-a27eeebe2cc0",
        "number": "9I-999",
        "residency": "Fraser Suites Le Claridge",
        "street": "Champs-Elysees",
        "city": "Paris",
        "state": "Île-de-France",
        "country": "France",
        "pincode": "75008"
    }
]
```

This time, the call is successful. Now, let's use the refresh token to get a new access token, as follows:

```
$ curl -X POST 'http://localhost:8080/api/v1/auth/token/
refresh'
 -H 'Content-Type: application/json' -H 'Accept: application/
json'
 -d '{
    "refreshToken":
"3i2tlrmdqnp60drl6i9c2kdm36s48qg5vm2ucgtflsk0cjo4dth
hjan9aj1ck83det8m8hkl461cqkfl57puk81ct6j09ilporanf1j
j414ht4ob7dkcakq6lk92cnct"
 }'
```

```
< Response
{
    "refreshToken":
"3i2tlrmdqnp60drl6i9c2kdm36s48qg5vm2ucgtflsk0cjo4dthhjan9
aj1ck83det8m8hkl461cqkfl57puk81ct6j09ilporanf1jj4
14ht4ob7dkcakq6lk92cnct",
    "accessToken": "eyJ0eXAiOiJKV1QiLCJhbGciOiJSUzI1NiJ9.
               Rest of token truncated for brevity",
    "username": "scott",
    "userId": "a1b9b31d-e73c-4112-af7c-b68530f38222"
}
```

This time, it returns a new access token with the same refresh token given in the payload.

If you pass an invalid refresh token while calling the refresh token API, it would provide the following response:

```
{"errorCode":"PACKT-0010",
"message":"Requested resource not found. Invalid token.",
"status":404,"url":"http://localhost:8080/api/v1/auth/
token/refresh","reqMethod":"POST","timestamp":"2021-01-
18T07:20:35.846649200Z"}
```

After testing the authentication using JWT, we can now test the authorization. Let's create an address using a token created by user SCOTT. SCOTT has a USER role. Here is the code to do this:

```
$ curl -v -X POST 'http://localhost:8080/api/v1/addresses' -H
'Content-Type: application/json' -H 'Accept: application/json'
-H 'Authorization: Bearer eyJ0eXAiOiJKV1QiLCJhbGciOiJSUzI1NiJ9.
ey
Rest of the token is truncated for brevity'
-d '{
    "number": "9I-999",
    "residency": "Fraser Suites Le Claridge",
    "street": "Champs-Elysees",
    "city": "Paris",
    "state": "Ile-de-France",
    "country": "France",
    "pincode": "75008"
}'
```

```
< HTTP/1.1 403
< Vary: Origin
< Vary: Access-Control-Request-Method
< Vary: Access-Control-Request-Headers
< WWW-Authenticate: Bearer error="insufficient_scope", error_
description="The request requires higher privileges than
provided by the access token.", error_uri="https://tools.ietf.
org/html/r
fc6750#section-3.1"
< output truncated for brevity
```

The API responded with 403 (forbidden) because SCOTT has a USER role and we have configured this API to only be allowed to be accessed by a user with an ADMIN role.

Let's create a token again, using the SCOTT2 user who has an ADMIN role, with the following code:

```
$ curl -X POST 'http://localhost:8080/api/v1/auth/token' -H
'Content-Type: application/json' -H 'Accept: application/json'
-d '{
    "username": "scott2",
```

```
    "password": "tiger"
}'
```

Now, let's call the `create address` API again using the access token received from the
SCOTT2 sign-in, as shown in the following code snippet:

```
$ curl -X POST 'http://localhost:8080/api/v1/addresses'
-H 'Content-Type: application/json' -H 'Accept: application/
json'
-H 'Authorization: Bearer eyJ0eXAiOiJKV1QiLCJhbGciOiJSUzI1NiJ9.
                            eyJzd
Rest of the token is truncated for brevity' -d '{
    "number": "9I-999",
    "residency": "Fraser Suites Le Claridge",
    "street": "Champs-Elysees",
    "city": "Paris",
    "state": "Ile-de-France",
    "country": "France",
    "pincode": "75008"
}'
```

```
< Response
```

```
{"_links":{"self":[{"href":"http://localhost:8080/
b78d485e-16a0-4b11-98d2-6e4dadbc60e7"},{"href":"http://
localhost:8080/api/v1/addresses/b78d485e-16a0-4b11-98d2-
6e4dadbc60e7"}]},"id":"b78d48
5e-16a0-4b11-98d2-6e4dadbc60e7","number":"9I-
999","residency":"Fraser Suites Le Claridge","street":"Champs-
Elysees","city":"Paris","state":"Ile-de-France","country":"Fran
ce","pincode":"75008
"}
```

https://github.com/PacktPublishing/Modern-API-Development-with-Spring-and-Spring-Boot/blob/main/Chapter06/Chapter06.postman_collection.json

Similarly, you can try to delete address operations using the REST API. This would only
allow an ADMIN role to perform the operation.

You can find the Postman (API client) collection of this chapter APIs at the following location, which is based on Postman Collection version 2.1. You can import it and then test the APIs.

Summary

In this chapter, you have learned about JWTs, Spring Security, authentication using filters, and JWT token validation using filter and authentication with Spring OAuth 2.0 Resource Server. You have also learned how you can add CORS and CSRF protection and why these are necessary.

You have also learned about access protection based on roles and authorities. You have now got the skills to implement JWTs, Spring Security, and Spring Security OAuth 2.0 Resource Server to protect your web resources.

In the next chapter, you will develop a sample e-commerce app's UI using the Spring Security framework and APIs used in this chapter. This integration will allow you to understand the UI flows and how to consume REST APIs using JavaScript.

Questions

1. What is a security context and principal?

2. Which is the preferred way to secure a JWT—signing or encrypting a token?

3. What are the best practices for using a JWT?

Further reading

- Hands-On Spring Security 5.x (video course):

  `https://www.packtpub.com/product/hands-on-spring-security-5-x-video/9781789802931`

- Spring Security documentation:

  `https://docs.spring.io/spring-security/site/docs/current/reference/html5/`

- List of filters available in Spring Security:

  `https://docs.spring.io/spring-security/site/docs/current/reference/html5/#servlet-security-filters`

- JWT:

  `https://tools.ietf.org/html/rfc7519`
- JWS:

  `https://www.rfc-editor.org/info/rfc7515`
- JWE:

  `https://www.rfc-editor.org/info/rfc7516`
- Spring Security in-built SpEL expressions:

  `https://docs.spring.io/spring-security/site/docs/current/reference/html5/#el-access`

7

Designing a User Interface

In the previous chapter, you implemented authentication and authorization using Spring Security, which also includes all the sample e-commerce app **application programming interfaces** (**APIs**). In this chapter, you will develop the frontend of a sample e-commerce app using the React library. This **user interface** (**UI**) app will then consume the APIs developed in the previous chapter, *Chapter 6, Security (Authorization and Authentication)*. This UI app will be a **single-page application** (**SPA**) that consists of interactive components such as **Login**, **Product Listing**, **Product Detail**, **Cart**, and **Order Listing**. This chapter will conclude the end-to-end development and communication between different layers of an online shopping app. By the end of the chapter, you will have learned about SPAs, UI component development using React, and consuming the **REpresentational State Transfer** (**REST**) APIs using the browser built-in `Fetch` API.

This chapter will cover the following topics:

- Learning React fundamentals
- Exploring React components and other features
- Designing e-commerce app components
- Consuming APIs using Fetch
- Implementing authentication

Technical requirements

You need the following prerequisites for developing and executing the code:

- You should be familiar with JavaScript—data types, variables, functions, loops, and array methods such as `map()`, `Promises`, and `async`, and so on.

- Node.js 14.x with **Node Package Manager (npm)** 6.x (and optionally, `yarn`, which you can install using `npm install yarn -g`).

- **Visual Studio Code** (**VS Code**)—a free source code editor.

- React 17 libraries that will be included when you use `create-react-app`.

Let's get the ball rolling!

Please visit the following link to check the code: `https://github.com/PacktPublishing/Modern-API-Development-with-Spring-and-Spring-Boot/tree/main/Chapter07/ecomm-ui`

Learning React fundamentals

React is a declarative library to build interactive and dynamic UIs, including isolated small components. Sometimes, it is also referred to as a framework because it is capable and comparable with other JavaScript frameworks, such as AngularJS. However, React is a library that works with other supported React libraries such as React Router, React Redux, and so on. You normally use it to develop SPAs, but it can also be used to develop full-stack applications.

React is used for building the view layer of the application, as per a **Model-View-Controller** (**MVC**) architecture. You can build reusable UI components with their own state. You can either use plain JavaScript with **HyperText Markup Language** (**HTML**) or **JavaScript Syntax Extension** (**JSX**) for templating. We'll be using JSX in this chapter. It uses a **virtual Document Object Model** (**VDOM**) for dynamic changes and interactions.

Let's create a new React app using the `create-react-app` utility. It scaffolds and provides the basic app structure that you'll use to develop the sample e-commerce app frontend.

Creating a React app

You can configure and build a React UI app from scratch. However, React provides a `create-react-app` utility that bootstraps and builds a basic running sample app. You can further use it to build a full-fleshed UI application.

Its syntax is shown here:

```
npx create-react-app <app name>
```

npm package executor (NPX) is a tool that allows you to use **command-line interface (CLI)** tools and other executables available in the npm registry. It is by default available with npm 5.2.0, else you can install it using npm i npx. Therefore, it executes the create-react-app React package directly.

> **Using npm in place of yarn**
>
> By default, create-react-app uses the yarn package as a package manager. However, if you want, you can also use npm with the following command:
>
> npx create-react-app ecomm-ui --use-npm

Now, let's create an ecomm-ui application using this command:

```
$ npx create-react-app ecomm-ui
Creating a new React app in C:\modern-api-with-spring-and-
sprint-boot\Chapter07\ecomm-ui.
Installing packages. This might take a couple of minutes.
Installing react, react-dom, and react-scripts with cra-
template...
yarn add v1.22.5
[1/4] Resolving packages...
[2/4] Fetching packages...
[3/4] Linking dependencies...
[4/4] Building fresh packages...
success Saved lockfile.
success Saved 297 new dependencies.
info Direct dependencies
├─ cra-template@1.1.1
├─ react-dom@17.0.1
├─ react-scripts@4.0.1
└─ react@17.0.1
info All dependencies
├─ @babel/compat-data@7.12.7
├─ @babel/core@7.12.10
```

```
├─ <Output truncated for brevity>
├─ yargs-parser@18.1.3
└─ yocto-queue@0.1.0
Done in 644.56s.
```

After it has installed all the required dependent packages, it continues by installing the template dependencies, as follows:

```
Installing template dependencies using yarnpkg...
yarn add v1.22.5
[1/4] Resolving packages...
[2/4] Fetching packages...
[3/4] Linking dependencies...
[4/4] Building fresh packages...
success Saved lockfile.
success Saved 15 new dependencies.
info Direct dependencies
├─ @testing-library/jest-dom@5.11.9
├─ @testing-library/react@11.2.3
├─ @testing-library/user-event@12.6.0
└─ web-vitals@0.2.4
info All dependencies
├─ @testing-library/dom@7.29.4
├─ @testing-library/jest-dom@5.11.9
├─ <Output truncated for brevity>
├─ strip-indent@3.0.0
└─ web-vitals@0.2.4
Done in 109.90s.
```

It may also ask you to add some testing dependencies— you can install these using the following command:

```
yarnpkg add @testing-library/jest-dom@^5.11.4 @testing-library/
react@^11.1.0 @testing-library/user-event@^12.1.10 web-
vitals@^0.2.4
```

Once it is installed successfully, you can go to the app directory and start the application installed using `create-react-app` by running the following code:

```
$ cd ecomm-ui
$ code .
```

The `code .` command opens the `ecomm-ui` app project in VS Code. You can then use the following command to start the development server:

```
$ yarn start
```

Once the server has started successfully, it will open a new tab on your default browser with `localhost:3000`, as shown in the following screenshot:

Figure 7.1 – Default UI app created by the create-react-app utility

Our bootstrapped React UI is up and running, but you now need to understand the basic concepts and files generated by `create-react-app` before you build an e-commerce UI app on top of it.

Exploring basic structures and files

A scaffolded React app contains the following directories and files inside the root project directory:

```
ecomm-ui
├── README.md
├── node_modules
```

```
├── package.json
├── .gitignore
├── public
│   ├── favicon.ico
│   ├── index.html
│   ├── logo192.png
│   ├── logo512.png
│   ├── manifest.json
│   └── robots.txt
└── src
    ├── App.css
    ├── App.js
    ├── App.test.js
    ├── index.css
    ├── index.js
    ├── logo.svg
    ├── reportWebVitals.js
    └── setupTests.js
```

Let's understand the main parts, as follows:

- `node_modules`: You don't make any changes here. Node-based applications keep a local copy of all the dependent packages here.

- `public`: This directory contains all the static assets of an app here, including `index.html`, images, favicon icon, and `robots.txt`.

- `src`: This directory contains all the dynamic code, including React code and **Cascading Style Sheets (CSS)** (including **Synthetically Awesome Style Sheets (Sass)**, **Leaner Style Sheets (Less)**, and so on). It also contains the test code.

- `package.json`: This **JavaScript Object Notation (JSON)** file contains all the metadata, commands (inside `scripts`), and dependent packages (inside `dependencies` and `dev-dependencies`).

You can remove the `serviceWorker.js` file (if generated), the `logo.svg` file, and test files from the `src` directory for now as we are not going to use them in this chapter.

Let's understand the `package.json` file in the next subsection.

Understanding the package.json file

You can also view the `package.json` file that contains all the dependencies under the `dependencies` and `dev-dependencies` fields. It is similar in nature to the `build.gradle` file.

The main React libraries are `react` and `react-dom`, mentioned in the `dependencies` field; these are for React and the virtual DOM respectively.

`package.json` also contains a `script` field that contains all the commands you can execute on this application. We have used the `yarn start` command to start the application in development mode. Similarly, you can execute other commands, as shown in the following code block, with `yarn` and npm:

```
"scripts": {
    "start": "react-scripts start",
    "build": "react-scripts build",
    "test": "react-scripts test",
    "eject": "react-scripts eject"
},
```

`react-scripts` is a CLI package installed by the `create-react-app` utility. It contains many dependencies, and few of the primary dependencies are listed here:

- **Webpack** (`https://webpack.js.org/`): This is a module bundler that bundles JavaScript, CSS, images, HTML, and so on. CSS and images may require extra loaders as dependencies. For example, it will pick all JavaScript files and bundle them into a single JavaScript file, though you can customize the way it bundles them by using a `webpack.config.js` configuration.

- **Jest** (`https://jestjs.io/`): Jest is a JavaScript testing framework maintained by Facebook.

- **ESLint** (`https://eslint.org/`): ESLint is a linter that allows you to maintain code quality. It is very similar to Checkstyle in the Java world.

- **Babel** (`https://babeljs.io/`): Babel is a JavaScript transcompiler tool that converts JavaScript code to backward-compatible JavaScript code. The latest JavaScript draft version is **ECMAScript 2020**, which is also referred as **ES10**. The latest JavaScript stable version is **ECMAScript 2018 (ES9)**. Babel allows you to generate optimized backward-compatible code from JavaScript code written using the latest versions.

You can find `react-scripts` under the `dependencies` field in `package.json`. Let's understand each of these commands, as follows:

- `start`: This command allows you to start the development server in a node environment. It also provides the hot reload feature, which means any changes to the React code would be reflected in the application, without a restart being required. Therefore, if there are any linting or code issues, this would show up accordingly in the console (terminal window) and web browser.

- `build`: This command packages the React application code for production deployment. It does the bundling of the JavaScript files in one CSS file into another and also minifies and optimizes the code files. You can then use this bundle to deploy on any web server.

- `test`: This command executes a test using the test runner (Jest tool). It executes all test files having extensions such as `.test.js` or `.spec.js`.

- `eject`: React comes with default build configurations such as `webpack`, `Babel`, and so on. The build configuration has the best practices implemented for optimizing the built app. This command helps you to eject the hidden configuration, after which you can override and customize the build configuration. However, you should do this with the utmost care because this is a one-way activity and you can't reverse it.

Let's understand how React works, in the next subsection.

Understanding how React works

A web page is nothing but an HTML document. HTML documents contain the DOM, a tree-like structure of HTML elements. Any changes to the DOM are reflected in the rendering of the HTML document in the browser. Making changes in the actual DOM—and, specifically to the nth level—is a heavy operation in terms of traversal and rendering the DOM, because each change is done on the whole DOM, and this is a time-and memory-consuming operation.

React uses a **VDOM** to make these operations lightweight. A VDOM is an in-memory copy of the actual DOM. React maintains the VDOM using the `react-dom` package. Therefore, when you initialize the React app, you first pass the root HTML element ID to the `ReactDOM` object's `render` function. React writes the VDOM under this root element after its first render.

After the first render, only the necessary changes are written to the actual DOM based on changes to React components and their state. The React components' `render` function returns the markup in JSX syntax. Then, React transforms it to HTML markup and compares the generated VDOM with the actual HTML DOM, and only makes the necessary changes to the actual DOM. This process then continues till the components get changed. Let's explore how the first render takes place.

Bootstrapping of the React app

The `index.html` file under the `public` directory contains the main HTML file. It's an application skeleton that contains the site `title`, `meta` elements, a `body` element, and a `div` element under the `body` with an ID of `root`. You pass this root element to the `render` function of `ReactDOM` in `index.js`, in the `src` directory. **This is the entry point of the React app**. Let's have a look at its code, as follows:

```
import React from 'react';
import ReactDOM from 'react-dom';
import './index.css';
import App from './App';
import reportWebVitals from './reportWebVitals';

ReactDOM.render(
  <React.StrictMode>
    <App />
  </React.StrictMode>,
  document.getElementById('root')
);
```

https://github.com/PacktPublishing/Modern-API-Development-with-Spring-and-Spring-Boot/blob/main/Chapter07/ecomm-ui/src/index.js

Here, React uses the `ReactDOM` object from the `react-dom` package to render the page. The `render()` function contains two arguments: `element` and `container`. It can also have a third (optional) argument for the callback.

You are passing an `<App />` tag component wrapped with React's strict mode component as an `element` argument and a `<div id= "root">` inside the `<body>` element of `index.html` as a `container` argument to the `render` function.

App components can be a single component or a parent component with single or multilayer child components. A single component won't contain any other React component; it would simply contain the JSX, and that's it. However, parent components may contain one or more child component, and those child components may contain one or more child components, and so on. For example, an App component may have header, footer, and content components. A content component may have a cart component, and then the cart component may have items inside it.

A <React.StrictMode> component is a special React component that gets rendered twice in development mode to check the best practices, deprecated methods, and potential risk in your React components, and prints warnings and suggestions in the console log. It has no impact on production build because it works only in development mode.

The render function transforms the JSX of the app component to HTML and adds it inside the <div id="root"> tag, then it compares the VDOM with the real DOM and makes the necessary changes in the real DOM. This is how React components get rendered on the browser.

You now understand that React components are key here. Let's deep dive into them in the next section.

Exploring React components and other features

Each page is built up using React components—for example, the **Product Listing** page of Amazon can broadly be divided into Header, Footer, Content, Product List, Filter and Sorting options, and Product Card components. You can create components in React in two ways: by using JavaScript classes or by using functions.

Let's create a sample header component in React with both a function and a class.

You can either write a plain old JavaScript function or **ECMAScript 6 (ES6)** arrow functions. We'll mostly use arrow functions. In the following code snippet, check out the Header component using a JavaScript arrow function:

```
export const Header = (props) => {
  return (
    <div>
      <h1>{props.title}</h1>
    <div>
  )
}
```

Let's create the same `Header` component using a JavaScript class, as follows:

```
export default class Header extends React.Component {
  render() {
    return (
      <div>
        <h1>{this.props.title}</h1>
      <div>
      )
    }
}
```

Let's understand both of these functions point by point, as follows:

- Both are returning the JSX that looks similar to HTML, which actually gets rendered after transformation.

- Both are exporting the function and class respectively so that they can be imported by other components.

- Both are having props—one as an argument and one bound with a `this` scope, which is part of `React.Component`. Props represent the attributes and their values—for example, here, a `title` attribute is used. When it gets rendered, it is replaced by the `title` attribute's value.

- The class needs a `render()` function, whereas the function simply needs a `return` statement.

Let's see how the `Header` component could be used. You can use this `Header` component as you would use any other HTML tag in your JSX code, as shown next:

```
<Header title="Sample Ecommerce App" />
```

When this `Header` component gets rendered, it will show the title wrapped in an `<H1>` element.

Let's explore the JSX next. This is how you use the props: you add an attribute (such as title) to its value while using the component. Inside the component, you can access these attributes (properties) by using `props` directly or using the `{ title }` destructuring form in functional components and by using `this.props` in class components.

Exploring JSX

React components would return the JSX. You can write HTML code to design the components because JSX is very similar to HTML, except for the HTML attributes. Therefore, you need to make sure to update attributes such as class to className, for to htmlFor, fill-rule to fillRule, and so on. The advantage of using the React.StrictMode component is that you get a warning and a suggestion to use the correct JSX attribute names if you use HTML attributes or have a typo.

You can also put any JavaScript expressions inside JSX or an element's attributes to make the component dynamic by using the expression wrapped in curly braces ({ }).

Let's have a look at some sample code to understand both JSX and expressions. The following JSX code snippet has been taken from the CartItem component. Check out the highlighted code for expressions; the rest of the code is JSX, which is very similar to HTML:

```
<div className="w-32">
    <img className="h-24" src={item?.imageUrl} alt="" />
</div>
<div className="flex flex-col justify-between ml-4 flex-
        grow">
    <Link to={"/products/" + item.id} className="font-bold
        text-sm text-indigo-500 hover:text-indigo-700">
        {item?.name}
    </Link>
    <span className="text-xs">Author: {author}</span>
    <button className="font-semibold hover:text-red-500
        text-indigo-500 text-xs text-left" onClick={() =>
        removeItem(item.id) }>
        Remove
    </button>
</div>
```

The preceding code fragment represents a cart item that shows the product image, product name, author, and **Remove** button. The product name is also a link that links to the product detail page. You can design using JSX (read HTML) as shown. Please also note that the class attribute name is changed to className because it is a JSX. Link is a part of the react-router-dom library.

You are done with the cart item's design part. Now, you need a mechanism to populate the values and add the event handling in it. This is where a JSX expression helps you.

You use `item`—an object that represents the cart item, and `author`—a variable that contains the author name. Both are part of the React component's state. You will learn more about the state in the next subsection, but for the time being you can think of them as variables defined in the `CartItem` component. Once you write the JSX (read HTML), dynamic values (from variables) and interaction (for events) can be defined using the expressions wrapped inside curly braces (`{}`).

Let's understand each of the expressions next, as follows:

- `src={item?.imageUrl}`: You get the item (product) image **Uniform Resource Locator (URL)** as part of the API response. You simply assign it to the `src` attribute of the `img` tag. Note that the dot operator (`.`) allows you to access the property of an object. The code may throw an error if you try to read the property of any null or undefined object. You can avoid that by using the `?.` operator. Then, the property (in this case, `imageUrl`) would only be read if an object (in this case, `item`) is not null or undefined.

- `to={"/products/" + item.id}`: Here, links to an attribute are formed by using the object item's `id` property.

- `{item?.name}`: Here, the name of the product is displayed using the `name` property of the `item` object.

- `Author: {author}`: The author value is displayed using the `author` variable.

- `onClick={() => removeItem(item.id)}`: This is the way you associate a user-defined function with an event. Here, `removeItem()` would be called by passing the item object's `id` property on the click of a button. If you are not passing any argument or using multiple statements, then you can directly pass the function name instead of using the arrow function—for example, `onClick={removeItem}`.

Next, we will deep dive into the state of React components. Let's see how this works.

Exploring a component's state

Components are dynamic and contain a state. The state represents the data and metadata held by the component at a given point in time. There are two levels of state: a global (app-level) state and a local (component-level) state.

Earlier (prior to React version 16.8), the state was only supported in components defined using classes. Now, React supports the state in both functional and class components. React supports the state in functional components using hooks such as `useState()`, `useContext()`, and so on.

React introduced hooks (a set of functions) in the 16.8 version, which introduced many features to the functional component that were earlier not supported, such as state and an event similar to `componentDidMount` (a lifecycle method in the class that indicates a component was mounted), and you can now perform certain operations such as loading data using APIs, and so on.

Let's understand the React hooks next.

Hooks

Hooks are special React functions that are provided in React version 16.8 onward. Each hook represents a special feature that you can use in functional components. Let's understand the most popular and common hooks one by one, as follows:

- `useState`: `useState` allows you to define and maintain the state. Let's see how you can use this hook. First, you import the `useState` hook at the top of the component code file, as follows:

  ```
  import {useState} from "react";
  ```

  Next, inside your component's arrow function code, define the state before the `return` statement, as shown next:

  ```
  const [total, setTotal] = useState(0);
  ```

  You need to define both state and state setter functions in an array while declaring the state. Here, the `total` state is defined with its `setter` function. You can use any type of state, such as an object, array, string, or number. The `total` state is of type `number`, therefore it is initialized with `0`. `setTotal` is a setter function. The `setter` function allows you to update the state (`total` here)—for example, you can update the total state by calling `setTotal(100)`, then the `total` state would be changed from `0` to `100`.

  React tracks the state's `setter` function and whenever it is called, React updates the state of the component and re-renders the component. The naming convention of the `setter` function is to prefix the state name with `set` and make the state's first letter a capital letter. Therefore, we have used the `setTotal` name for the `total` state. You'll use `useState` for local state management in most components.

- `useEffect`: You use a `useEffect()` hook when you want to do something after rendering a component. This gets called after each render. You can also use it when you want to load the initial data from an API or add an event listener. However, if an API call should be made once, then you can pass the empty array (`[]`) dependency while calling it. You'll find multiple instances of `useEffect` in `ecomm-ui` code when an empty array is passed for a single call.

- React recommends using multiple `useEffect` functions inside components for separating the concern. Also, make sure it returns an arrow function for cleanup. For example, when you add the event listener for any component, it should return an arrow function that removes the event listener.

- `useContext`: You can pass props from one component to another. Sometimes, you have to use props drilling to the *n*th level. React also provides an alternative way to define these props so that they can be used in any component in a tree without using prop drilling. You would use it for props that are common across components, such as `theme` or `isUserLoggedIn`.

- React provides a `createContext()` function to create a context. It returns a provider and consumer to provide access to its values and changes respectively (see the next code block). However, `useContext` can easily make use of the context by removing usage of the consumer. The following code snippet depicts `useContext` usage:

```
import {createContext} from "react";
import ReactDOM from "react-dom";


const LoggedInContext = createContext();
const App = () => {
    return (
        <LoggedInContext.Provider isUserLoggedIn=true>
            <ProductList/>
        <LoggedInContext.Provider/>
    );
}
const ProductList = () => {
    return (
        <LoggedInContext.Consumer> { (isUserLoggedIn) =>
            <div>Is user logged-in: {isUserLoggedIn}</div>
        } <LoggedInContext.Consumer>
    );
```

```
}
ReactDOM.render(<App/>,
    document.getElementById("root"));
```

You can simplify the `ProductList` component's `return` block in the previous code snippet (check the highlighted code) with `useContext`, as follows:

```
import {createContext, useContext} from "react";
import ReactDOM from "react-dom";
const LoggedInContext = createContext();
const App = () => {
    return (
        <LoggedInContext.Provider isUserLoggedIn=true>
            <ProductList/>
        <LoggedInContext.Provider/>
    );
}
const ProductList = () => {
    const isUserLoggedIn = useContext(LoggedInContext);
    return (
        <div>Is user logged-in: {isUserLoggedIn}</div>
    );
}
ReactDOM.render(<App/>, document.getElementById("root"));
```

This is how you can use `createContext` and `useContext` hooks.

- `useReducer`: This is an advanced version of the `useState` hook that not only allows you to use a component's state but also provides better controls to manage its state by taking the `reducer` function as a first argument. It takes the initial state as a second argument. Check out its syntax, as seen in the following code block:

```
const [state, dispatch] = useReducer(reducer,
initialState);
```

The `reducer` function is a special function that takes state and action as arguments and returns a new state. We'll explore this more when we build the `CartContext` component later in this chapter.

Now that you have learned the basic concepts of React, let's add some styling to the `ecomm-ui` application using `TailwindCSS`.

Styling components using Tailwind

Tailwind CSS is a utility CSS framework that helps you to design a responsive UI. It supports theming, animation, pre-defined padding and margins, flex, grids, and so on. You can install Tailwind and its peer packages using `yarn`, as shown in the following code snippet (executing it from the project root directory):

```
$ yarn add -D tailwindcss@npm:@tailwindcss/postcss7-compat @
tailwindcss/postcss7-compat postcss@^7 autoprefixer@^9
```

`create-react-app` doesn't support PostCSS 8 at the time of writing this chapter, so you need to install the Tailwind CSS v2.0 with PostCSS 7 compatibility build for now, as shown in the previous code snippet. However, it can be changed to the appropriate version of PostCSS once `create-react-app` starts supporting it (version 8+).

Configuring the Tailwind build needs the **Create React App Configuration Override (CRACO)** package.

Installing and configuring CRACO

You also need to install CRACO to be able to configure the Tailwind build because `create-react-app` doesn't let you override the PostCSS configuration natively. CRACO allows you to override the configuration created by `create-react-app`. Let's install it by executing the following command from the project root directory:

```
$ yarn add -D @craco/craco
```

Once installation is done successfully, you can update `scripts` in the `package.json` file to replace `react-scripts` with `craco` for all scripts except `eject`, as follows:

```json
{
    ...
    ...
    "scripts": {
      "start": "craco start",
      "build": "craco build",
      "test": "craco test",
      "eject": "react-scripts eject"
    },
}
```

Next, create a file named `craco.config.js` at the root of the project and add `tailwindcss` and `autoprefixer` as `postcss` plugins, as follows:

```
module.exports = {
  style: {
    postcss: {
      plugins: [
        require('tailwindcss'),
        require('autoprefixer'),
      ],
    },
  },
}
```

Now, you can create a Tailwind configuration file.

Creating a Tailwind configuration file

You can create and initialize a `tailwind.config.js` file using the following command:

```
npx tailwindcss init
```

This will create a default `tailwind.config.js` file at the root of the project with minimal configuration, as illustrated in the following code snippet:

```
module.exports = {
  purge: [],
  darkMode: false, // or 'media' or 'class'
  theme: {
    extend: {},
  },
  variants: {
    extend: {},
  },
```

```
    plugins: [],
}
```

Now, we can add configuration to purge unused styles in production.

Configuration to remove unused styles in production

You would like to keep the style sheet size down in a production environment because this improves the performance of the application. You can purge unnecessary styles by adding the following purge block in the `tailwind.config.js` file. Then, Tailwind can tree-shake unused styles while building the production build. You can set the `PURGE_CSS` environment variable to `production` for production builds. The code is illustrated in the following snippet:

```
module.exports = {
  purge: {
    enabled: process.env.PURGE_CSS === "production" ? true
    : false,
    content: ["./src/**/*.{js,jsx,ts,tsx}",
    "./public/index.html"],
  },
  darkMode: false, // or 'media' or 'class'
  theme: {
    extend: {},
  },
  variants: {
    extend: {},
  },
  plugins: [],
}
```

Next, we will add Tailwind to React.

Including Tailwind in React

Open the `src/index.css` file that `create-react-app` generates for you by default and import Tailwind's base, components, and utilities styles, replacing the original file contents, as follows:

```
@import "tailwindcss/base";
@import "tailwindcss/components";
@import "tailwindcss/utilities";
```

https://github.com/PacktPublishing/Modern-API-Development-with-Spring-and-Spring-Boot/blob/main/Chapter07/ecomm-ui/src/index.css

These statements import the styles generated by the build based on the Tailwind configuration when you execute the build.

Finally, make sure that the CSS file is being imported in the `src/index.js` file by running the following code:

```
import React from 'react';
import ReactDOM from 'react-dom';
import './index.css';
import App from './App';
import reportWebVitals from './reportWebVitals';

ReactDOM.render(
  <React.StrictMode>
    <App />
  </React.StrictMode>,
  document.getElementById('root')
);
...
```

https://github.com/PacktPublishing/Modern-API-Development-with-Spring-and-Spring-Boot/blob/main/Chapter07/ecomm-ui/src/index.js

Done! Next, you can execute `yarn run start` to use Tailwind CSS in the `ecomm-ui` app.

Adding basic components

First, remove the following files created by `create-react-app`:

- `App.css`
- `logo.svg`

Don't forget to remove these file references from `/src/App.js` too.

Then, create a new `components` directory under `/src`. You will create all new components under this directory, as shown in *Figure 7.2*. Let's create three new components, as follows:

- `Header`: This will display at the top and contains header items such as app name and **Login/Logout** button.
- `Container`: This will contain the main content, such as a product list.
- `Footer`: This will display at the bottom and contains footer items such as copyright information.
- The basic structure can be seen in the following screenshot:

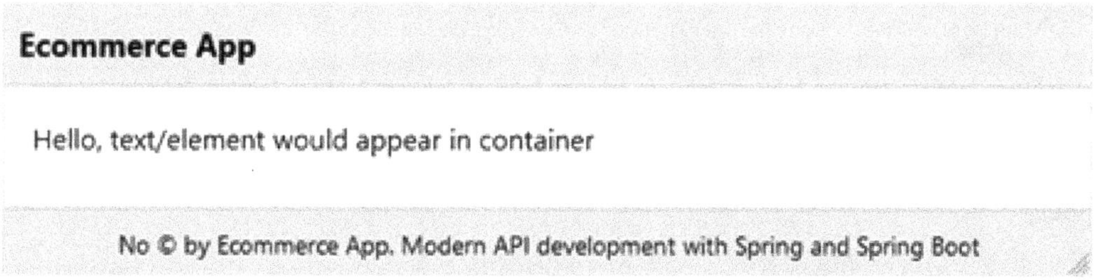

Ecommerce App

Hello, text/element would appear in container

No © by Ecommerce App. Modern API development with Spring and Spring Boot

Figure 7.2 – Basic structure of app containing Header, Footer, and Container components

- Let's add these containers. First, we'll create a `Header` component, as shown in the following code snippet:

```
const Header = () => {
  return (
    <div>
      <header className="p-2 border-b-2 border-gray-300 bg-
          gray-200">
        <h1 className="text-lg font-bold">Ecommerce
            App</h1>
      </header>
    </div>
```

```
    );
};
export default Header;
```

Similarly, you can create a `Footer` component, as shown in the following code snippet:

```
const Footer = () => {
  return (
    <div>
      <footer className="text-center p-2 border-t-2 bg-
        gray-200 border-gray-300 text-sm">
        No &copy; by Ecommerce App.{" "}
        <a href="https://github.com/PacktPublishing/Modern-
           API-Development-with-Spring-and-Spring-Boot">
          Modern API development with Spring and Spring Boot
        </a>
      </footer>
    </div>
  );
};
export default Footer;
```

Similarly, you can create a `Container` component, as shown in the following code snippet:

```
const Container = () => {
  return (
    <div className="flex-grow flex-shrink-0 p-4">
      <p>Hello, text/element would appear in container</p>
    </div>
  );
};
export default Container;
```

And finally, you can modify the `/src/App.js` file, as shown in the following code snippet:

```
import Header from "./components/Header";
import Footer from "./components/Footer";
import Container from "./components/Container";
function App() {
  return (
    <div className="flex flex-col min-h-screen h-full">
      <Header />
      <Container />
      <Footer />
    </div>
  );
}
export default App;
```

This is how you can create and use new components. These components are in their simplest form and are kept as such to understand these more easily. However, you can find refined and improved versions of these components on GitHub, as follows:

- **Header component source**: `https://github.com/PacktPublishing/ Modern-API-Development-with-Spring-and-Spring-Boot/blob/ main/Chapter07/ecomm-ui/src/components/Header.js`

- **Footer component source**: `https://github.com/PacktPublishing/ Modern-API-Development-with-Spring-and-Spring-Boot/blob/ main/Chapter07/ecomm-ui/src/components/Footer.js`

- The `Container` component (which contains the actual content in the center) would be replaced with the component `switch` from `react-router-dom` that would display the components based on a given route, such as cart, orders, and login.

Now, you can start the actual `ecomm-ui` development.

Designing e-commerce app components

Design is not only a key part of the **user experience** (**UX**)/a UI, but is also important for frontend developers. Based on the design, you can create reusable and maintainable components. However, a sample e-commerce app is a simple application that does not need much attention. You will create the following components in this application:

- **Product listing component**: A component that displays all the products and also acts as a home page. Each product in the listing will be displayed as a card with the product name, price, and two buttons—**Buy now** and **Add to bag**. The following screenshot displays the **Product listing** page, which shows product information along with an image of the product:

Figure 7.3 – Product listing page (home page)

- **Product detail component**: This is a component that displays details of the clicked product. It displays the product image, product name, product description, tags, and **Buy now** and **Add to bag** buttons, as shown next:

Figure 7.4 – Product detail page

- **Login component**: Login components allow a user to log in to an app by using their username and password, as illustrated in the following screenshot. It displays an error message when a login attempt fails. Click on **Cancel** to go back to the to the **Product listing** page. The **Product listing** page shows a list of products a customer can buy:

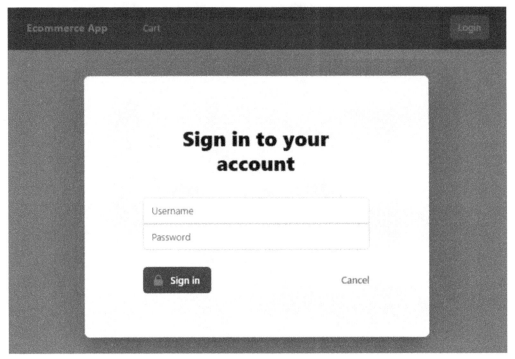

Figure 7.5 – Login page

- **Cart component**: A Cart component lists all the items that have been added to the cart. Each item displays product image, name, description, price, quantity, and total. It also provides a button to decrease and increase quantity, and a button to remove an item from the cart.

Product name is a link that takes the user back to the **Product detail** page. **The Continue shopping** button takes the user to the **Product listing** page. The **CHECKOUT** button performs the checkout. On successful checkout, an order is generated and the user is redirected to the **Orders** page, as shown next:

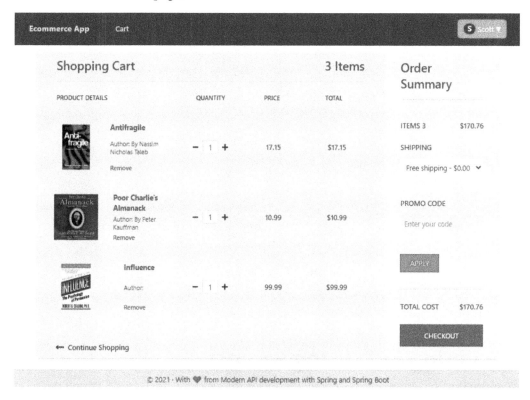

Figure 7.6 – Cart page

Orders component: The **Orders** page shows all orders placed by the user in a tabular form. The **Orders** table displays the order date, ordered items, order status, and order amount for each order.

The order date willbe displayed in the user's local time, but on the server it will be in **Universal Coordinated Time** (**UTC**) format. Order items would be displayed in an order list, with their quantity and unit price in brackets, as illustrated in the following screenshot:

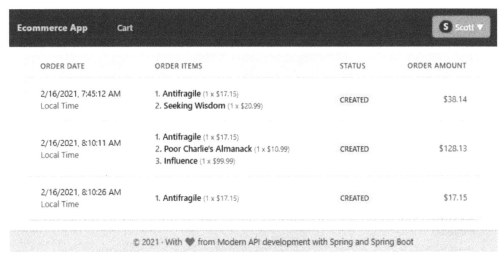

Figure 7.7 – Order page

Let's start coding these components. First, you will code the **Product listing** page, which fetches the products from the backend server using the REST API.

Consuming APIs using Fetch

Let's create the first component—that is, the `Product Listing Page`. Create a new file in the `src/components` directory with the name `ProductList.js`. This is the parent component of the **Product Listing** page.

This component fetches the products from the backend server and passes them to the child component, `Products` (It create a new `Products.js` file under the `components` directory).

Products contain the logic of fetched product list iterations. Each iteration renders the card UI for each product. `ProductCard` is another component, therefore you'll create another file, `ProductCard.js`, under `src/components`. You can write the product card logic inside products, but to single out the responsibility it's better to create a new component.

The `ProductCard` component has a **Buy now** button and an **Add to bag** link. These links should only work if the user is logged in, else it should redirect the user to the login page.

You now have an idea about the `Product Listing Page` component tree structure. Now, our first task is to have an API client that fetches products we can render in these components.

Writing the product API client

You are going to use the `Fetch` browser library as a REST API client. You can also use a third-party library such as `axios`. However, this means you need to include another dependency. When you can do the same job using a built-in browser API, why include extra dependencies?

You'll create a configuration file for all API clients to use. Let's name it `Config.js` so that you can create under the `src/api` directory.

`Config` is a JavaScript class that contains constants such as URLs and common methods such as `DefaultHeaders()` and `tokenExpired()`. Check out its code in the following snippet:

```
class Config {
  SCHEME = process.env.SCHEME ? process.env.SCHEME :
      "http";
  HOST = process.env.HOST ? process.env.HOST : "localhost";
  PORT = process.env.PORT ? process.env.PORT : "8080";
  LOGIN_URL = `${this.SCHEME}://${this.HOST}:${this.PORT}
    /api/v1/auth/token`;
  PRODUCT_URL = `${this.SCHEME}://${this.HOST}:${
    this.PORT}/api/v1/products`;
  // other constants removed for brevity
  defaultHeaders() {
    return { "Content-Type": "application/json",
            Accept: "application/json",
    };
  }
  headersWithAuthorization() {
    return {...this.defaultHeaders(),
      Authorization: localStorage.getItem(
          this.ACCESS_TOKEN),
```

```
    };
  }
// continue...
```

https://github.com/PacktPublishing/Modern-API-Development-with-Spring-and-Spring-Boot/blob/main/Chapter07/ecomm-ui/src/api/Config.js

Here, you can see that we have created constants that are formed using environment variables. The defaultHeaders() function returns the common headers being used in all API calls, and headersWithAuthorization() returns common headers with the Authorization header. headersWithAuthorization() uses object destruction for retrieving the default headers. The Authorization header is fetched from the local storage, which is set when a user is logged in successfully and is removed once the user logs out.

It also has a tokenExpired() function that simply checks the expiration time for a token stored in local storage. This expiration time is extracted from the access token (**JSON Web Token**, or **JWT**). It returns true if the expiration time is past the current time. Check out this function's code in the following code snippet:

```
// Config.js continue
  tokenExpired() {
    const expDate = Number(localStorage.getItem(
        this.EXPIRATION));
    if (expDate > Date.now()) {
      return false;
    }
    return true;
  }
  storeAccessToken(token) {
    localStorage.setItem(this.ACCESS_TOKEN, `Bearer
        ${token}`);
    localStorage.setItem(this.EXPIRATION,
        this.getExpiration(token));
  }
  getExpiration(token) {
    let encodedPayload = token ? token.split(".")[1] : null;
    if (encodedPayload) {
      encodedPayload =
```

```
                encodedPayload.replace(/-/g,
                    "+").replace(/_/g, "/");
        const payload = JSON.parse(window.atob(
                            encodedPayload));
        return payload?.exp ? payload?.exp * 1000 : 0;
    }
    return 0;
  }
}
```

The `Config` class also contains a `storeAccessToken()` function that simply stores the access token and expiration time in local storage. It uses a `getExpiration()` function to extract the expiration time from the access token. This function simply first extracts the payload from the token string and then decodes the payload and converts it to JSON. At the end, it returns the expiration time if a payload is a valid object, else it returns 0.

Now, let's use this `Config` class in the `ProductClient.js` file as shown in the following code block:

```
import Config from "./Config";
class ProductClient {
  constructor() { this.config = new Config(); }
  async fetchList() {
    return fetch(this.config.PRODUCT_URL, {
      method: "GET",
      mode: "cors",
      headers: {
        ...this.config.defaultHeaders(),
      },
    })
    .then((response) => Promise.all([response,
                        response.json()]))
    .then(([response, json]) => {
      if (!response.ok) {
        return { success: false, error: json };
      }
      return { success: true, data: json };
```

```
        })
        .catch((e) => { return this.handleError(e); });
    }
// continue...
```

https://github.com/PacktPublishing/Modern-API-Development-with-Spring-and-Spring-Boot/blob/main/Chapter07/ecomm-ui/src/api/ProductClient.js

ProductClient is a class, and a config instance is instantiated in its constructor. This class contains two asynchronous functions for fetching the products: fetchList() and fetch(). The former fetches all products, and the latter is for fetching a single product based on its ID. fetchList() makes use of the fetch browser function to fetch the product list. You pass the URL as the first argument input, and request an initialization object that contains the HTTP method, mode, and headers as the second argument. The fetch browser call returns a promise that you use to handle the request. First, you resolve the promise to response and response JSON and then check if response. ok is true or not. response.ok returns true for a status in the 200 to 299 range. Upon a successful response, the fetchList() method returns an object with data and success fields as true. Upon an unsuccessful response, it returns success as false and shows an error response in the data field.

Similarly, you can write a function to retrieve the product by ID. Everything will be the same except the URL, as you can see in the following code block:

```
// ProductClient.js continue...
async fetch(prodId) {
  return fetch(this.config.PRODUCT_URL + "/" + prodId, {
    method: "GET",
    mode: "cors",
    headers: {
      ...this.config.defaultHeaders(),
    },
  })
    .then((response) => Promise.all([response,
                                    response.json()]))
    .then(([response, json]) => {
      if (!response.ok) {
        return { success: false, error: json };
      }
```

```
            return { success: true, data: json };
        })
        .catch((e) => { this.handleError(e); });
    }
    handleError(error) {
        const err = new Map([
            [TypeError, "There was a problem fetching the
                response."],
            [SyntaxError, "There was a problem parsing the
                response."],
            [Error, error.message],
        ]).get(error.constructor);
        console.log(err);
        return err;
    }
}
export default ProductClient;
```

The handleError() function checks the type of the error (using error. constructor) and, based on that, returns the appropriate error message.

Please note that other API clients such as CartClient, CustomerClient, and OrderClient are developed in a similar fashion. The code is available at the following locations:

- CartClient: https://github.com/PacktPublishing/Modern-API-Development-with-Spring-and-Spring-Boot/blob/main/Chapter07/ecomm-ui/src/api/CartClient.js

- CustomerClient: https://github.com/PacktPublishing/Modern-API-Development-with-Spring-and-Spring-Boot/blob/main/Chapter07/ecomm-ui/src/api/CustomerClient.js

- OrderClient: https://github.com/PacktPublishing/Modern-API-Development-with-Spring-and-Spring-Boot/blob/main/Chapter07/ecomm-ui/src/api/OrderClient.js

Now, we can use ProductClient to fetch the products. Let's code the ProductList component and its child components.

Coding the Product Listing page

`ProductList` is a straightforward component that loads the products after their first render using `ProductClient`. You know that for this purpose, `useEffect` hooks should be used. Let's code it, as follows:

```
import { useEffect, useState } from "react";
import CartClient from "../api/CartClient";
import ProductClient from "../api/ProductClient";
import { updateCart, useCartContext } from
    "../hooks/CartContext";
import Products from "./Products";
const ProductList = ({ auth }) => {
  const [productList, setProductList] = useState();
  const [noRecMsg, setNoRecMsg] = useState("Loading...");
  const { dispatch } = useCartContext();
  useEffect(() => {
    async function fetchProducts() {
      const res = await new ProductClient().fetchList();
      if (res && res.success) {
        setProductList(res.data);
      } else {
        setNoRecMsg(res);
      }
    }
    async function fetchCart(auth) {
      const res = await new CartClient(auth).fetch();
      if (res && res.success) {
        console.log(res.data);
        dispatch(updateCart(res.data.items));
        if (res.data?.items && res.data.items?.length < 1) {
          setNoRecMsg("Cart is empty.");
        }
      } else {
        setNoRecMsg(res && typeof res === "string" ? res :
                                    res?.error?.message);
      }
    }
```

```
    if (auth?.token) fetchCart(auth);
    fetchProducts();
}, []);
// Continue…
```

https://github.com/PacktPublishing/Modern-API-Development-with-Spring-and-Spring-Boot/blob/main/Chapter07/ecomm-ui/src/components/ProductList.js

The `ProductList` component uses `auth` as a prop. It contains authentication information such as a token. The `ProductList` component is used as the main `App` component, and `auth` is passed to the `ProductList` component by it.

Please note that you have passed an empty array (`[]`) as a dependency to make sure that the API is called only once. You are using a `useState` hook to store the product list (`productList`) and message states (`noRecMsg`—no record) by using setter methods.

Reason for fetching the cart in ProductList

The `ProductList` component and its child components are available for non-authenticated users. Once the user clicks on the **Buy now** button or the **Add to bag** link, it will ask the user to log in. Once logged in, the user can add items to the cart. It is quite possible that the user might already have some items in the cart. Therefore, when you add an item to the cart, the quantity of existing products should be increased, and if a clicked item does not exist in the cart, then it should be added to the cart.

`Cart` is a separate component altogether; it means you can't access the `cart` unless you do cart prop drilling from the `App` component to both the `Cart` and `ProductCard` components, or have a `useContext` hook for the cart. We have built a custom store to maintain the cart state, very similar to **Redux** (a library to maintain the state in React app). We'll learn more about this library later in this chapter. `Dispatch` is an action that updates the cart items received from the backend server to the cart context.

Next, you create a JSX template and pass the fetched `productList` component to the child component, `Products`, for further rendering, as illustrated in the following code snippet:

```
// ProductList.js continue…
return (
    <div className="max-w-7xl mx-auto px-4 sm:px-6 lg:px-
        8">
        {productList ? (
            <div className="flex flex-wrap -mx-1 lg:-mx-4">
```

```
        <Products auth={auth} productList={productList ?
                productList : []} />
      </div>
    ) : (
      <div className="text-lg font-
          semibold">{noRecMsg}</div>
    )}
    </div>
  );
};
export default ProductList;
```

Here, it also passes the `auth` object as a prop to `Products`.

Let's have a look at the `Products` code, as follows:

```
import ProductCard from "./ProductCard";
const Products = ({ auth, productList }) => {
  return (
    <>
      {productList.map((item) => (
        <ProductCard key={item.id} product={item}
            auth={auth} />
      ))}
    </>
  );
};
export default Products;
```

https://github.com/PacktPublishing/Modern-API-Development-with-Spring-and-Spring-Boot/blob/main/Chapter07/ecomm-ui/src/components/Products.js

It simply does the job of iterating the product list passed by the `ProductList` component and passes each item with `product` props to the `ProductCard` component along with the `auth` object.

You can observe the usage of two React concepts here, as follows:

- It uses a `<></ >` fragment, which is an empty tag. Ideally, this is used when a component returns more than one top-level tag because React needs only one top-level tag in each component. Then, you can wrap those tags with a fragment. You can also use `<React.Fragment>` in place of an empty tag after importing `React` from the `react` package.

- Another usage is for the `key` props in the `ProductCard` component. When you generate components based on a collection, then React requires the `key` props to uniquely identify them. This will allow React to identify which item is changed, removed, or added. We have used the item ID here. If you don't have an ID in your collection, you can also use the index, as shown in the following code example:

```
{productList.map((item, index) => (
    <ProductCard key={index} product={item} auth={auth} />
))}
```

Now, let's have a look at the last child component of the `ProductList` component: `ProductCard`. The `ProductCard` component simply passes `Product` values to JSX template expressions for rendering.

We have added some extra code to add the functionality associated with **Add to bag** and **Buy now** click events.

Configuring routing

You are creating a SPA. Here, routing is not available by default. Routing is the mechanism to providing the routing to a single page, which means that with each new page, the browser URL would reflect the change and allows you to bookmark the page. It also maintains the URL history. You are going to use the `react-router-dom` package for routing management. You need to add the `react-router-dom` package to use routing, as shown in the following code snippet. Make sure to execute it from the project root directory:

```
$ yarn add react-router-dom
```

You are going to configure the routing in the App component because it is the root component of the ecomm-ui application. In the ProductList component, you are going to use the Link component and the useHistory() hook from the react-router-dom package. Let's understand them, as follows:

- Link: This is similar to a <a> HTML anchor tag. Instead of a href attribute, it uses a to attribute to link the URL. The route library maintains the links, therefore it knows which component to render when a link is passed in with a to attribute when Link is clicked.

- useHistory(): This allows navigation inside the component and accesses the state of the router. You would use the push("/path ") function of it to navigate, as shown in the checkLogin() function of the ProductList component.

Let's continue with the development of the next product-based component: ProductCard.

Developing the ProductCard component

First, you import the required packages. Then, declare the state (using useCartContext and useState) and variables. Please note in the following code snippet that it has auth and product as props:

```
import { useState } from "react";
import { Link, useHistory } from "react-router-dom";
import CartClient from "../api/CartClient";
import { updateCart, useCartContext } from "../hooks/
CartContext";
const ProductCard = ({ auth, product }) => {
  const history = new useHistory();
  const cartClient = new CartClient(auth);
  const { cartItems, dispatch } = useCartContext();
  const [msg, setMsg] = new useState("");
  // continue...
```

https://github.com/PacktPublishing/Modern-API-Development-with-Spring-and-Spring-Boot/blob/main/Chapter07/ecomm-ui/src/components/ProductCard.js

First, you write the `add()` asynchronous function that adds the product to the cart. It first checks whether the user is logged in or not. If not, it redirects the user to the login page. `checkLogin()` uses the `useHistory` hook's push method to redirect. The `token` property of `auth` is used to identify whether the user is logged in or not.

Once it has identified that the user is logged in, it calls the `callAddItemApi` function to add a product to the cart. The `callAddItemApi` function first finds out whether the product exists in the cart or not. If it exists, it finds out the quantity and adds one more to it. The `callAddItemApi` function then calls the REST API using the `CartClient` to add a new item or update the quantity in the existing cart item.

At the end, the `add` function calls the `dispatch` to update the state of `cartItems` in the cart context.

The following code snippet contains the same logic:

```
// ProductCard.js continue…
const add = async () => {
  const isLoggedIn = checkLogin();
  if (isLoggedIn && product?.id) {
    const res = await callAddItemApi();
    if (res && res.success) {
      if (res.data?.length > 0) {
        setMsg("Product added to bag.");
        dispatch(updateCart(res.data));
      }
    } else {
      setMsg(res && typeof res === "string" ? res :
                                    res.error.message);
    }
  }
};
const checkLogin = () => {
  if (!auth.token) {
    history.push("/login");
    return false;
  }
  return true;
};
```

```
const callAddItemApi = async () => {
  const qty = findQty(product.id);
  return cartClient.addOrUpdate({
    id: product.id, quantity: qty + 1, unitPrice:
        product.price
  });
};
const findQty = (id) => {
  const idx = cartItems.findIndex((i) => i.id === id);
  if (~idx) { return cartItems[idx].quantity; }
  return 0;
};
// continue...
```

The add function is called on a click of the **Add to bag** link. Similarly, the buy function shown in the following code snippet will be called when the user clicks on the **Buy now** button:

```
// ProductCard.js continue...
const buy = async () => {
  const isLoggedIn = checkLogin();
  if (isLoggedIn && product?.id) {
    const res = await callAddItemApi();
    if (res && res.success) {
      history.push("/cart");
    } else {
      setMsg(res && typeof res === "string" ? res :
                                    res.error.message);
    }
  }
};
// continue...
```

This is very similar to the add function. However, on a successful response from callAddItemApi, it redirects the user to the cart page, using a useHistory hook instance.

Let's have a look at a JSX template. In the following code snippet, the `className` attribute values have been stripped for better readability:

```
// ProductCard.js continue…
return (
  <div id={product.id} className="…">
    <figure className="…">
      <img src={product.imageUrl} alt={product.name}/>
      <div className="…">
        <form className="…">
          <div className="…">
            <h1 className="…">
              <Link to={`/products/${product.id}`}>
              {product.name}</Link>
            </h1>
            <div className="…">{"$"}{
                product.price.toFixed(2)}
            </div>
            <div className="…">In stock</div>
          </div>
          <div className="…">
            <div className="…">
              <button className="…" type="button"
                onClick={buy}
              >Buy now</button>
              <button className="…" type="button"
                onClick={add}
              >Add to bag</button>
            </div>
          </div>
          <p className="…">Free shipping on all local
              orders.</p>
        </form>
      </div>
    </figure>
  </div>
);
```

```
};
export default ProductCard;
```

https://github.com/PacktPublishing/Modern-API-Development-with-Spring-and-Spring-Boot/blob/main/Chapter07/ecomm-ui/src/components/ProductCard.js

The onClick event has been bound to buy and add for the **Buy now** button and the **Add to bag** link respectively. Also, the product name is a link created using Link. The to attribute of Link contains the path that points to the ProductDetail component. This path also contains the path parameter ID. You can use this parameter to perform certain operations on it. Similarly, you can also pass the query parameters the way you do in the browser URL.

When the user clicks on the product name, the user is redirected to the ProductDetail component (ProductDetail.js). Let's develop this next.

Developing the ProductDetail component

The ProductDetail component is similar to the ProductCard component, except that it loads the product details from the backend by using the ID from the path.

Let's see how this is done. Only code related to the Fetch product has been shown in the following snippet. The rest of the code is the same as for the ProductCard component. However, you can refer to the full code in the GitHub repository:

```
import { Link, useParams, useHistory } from "react-router-dom";
import ProductClient from "../api/ProductClient";
// Other imports removed for brevity
const ProductDetail = ({ auth }) => {
  const { id } = useParams();
  // Other declaration removed for brevity
  // Other functions removed for brevity
  useEffect(() => {
    async function getProduct(id) {
      const client = new ProductClient();
      const res = await client.fetch(id);
      if (res && res.success) {
        setProduct(res.data);
      }
    }
```

```
    // rest of code removed from brevity
    getProduct(id);
  }, [id]);
  return ( /* JSX Template */  );
};
export default ProductDetail;
```

https://github.com/PacktPublishing/Modern-API-Development-with-Spring-and-Spring-Boot/blob/main/Chapter07/ecomm-ui/src/components/ProductDetail.js

You have used `useParams()` from the `react-router-dom` package to retrieve the product ID passed from the `ProductCard` component. This `id` property is then used to fetch the product from the backend server using the `ProductClient` component. Upon a successful response, the retrieved product detail is set in the state product using the `setProduct` state function.

We are done with the development of product-based components such as `ProductList`, `Products`, `ProductCard`, and `ProductDetail`. We will now focus on authentication functionality so that we can later work on the `cart` and `orders` components, which require an authenticated user.

Implementing authentication

Before you jump into the `Login` component development, you will want to figure out how to manage a token received from a successful login response and how to make sure that if the access token has expired, then a refresh token request should be fired before making any call that requires authentication.

The browser allows you to store tokens or any other information in cookies, session storage, and local storage. From the server side, we haven't opted for cookie or stateful communication, therefore we are left with the remaining two options. Session storage is preferable for more secure applications because it is specific to the same tab and it gets cleared as soon as you click on the **Refresh** button or close the tab. We want to manage login persistence between different tabs and page refresh, therefore we'll opt for local storage of the browser.

On top of that, you can also store them in the state in the same way you are going to manage the cart state. However, this will be very similar to session storage. Let's leave that option for now.

Creating a custom useToken hook

You have now used different React hooks. Let's move a step forward and **create a custom hook**. First, create a new hooks directory under the src directory, and create a useToken.js file in it.

Then, add the following code to it:

```javascript
import { useState } from "react";
export default function useToken() {
  const getToken = () => {
    const tokenResponse = localStorage.getItem(
        "tokenResponse");
    const userInfo = tokenResponse ?
                        JSON.parse(tokenResponse) : "";
    return userInfo;
  };
  const [token, setToken] = useState(getToken());
  const saveToken = (tokenResponse) => {
    localStorage.setItem("tokenResponse",
        JSON.stringify(tokenResponse));
    setToken(tokenResponse);
  };
  return { setToken: saveToken, token };
}
```

https://github.com/PacktPublishing/Modern-API-Development-with-Spring-and-Spring-Boot/blob/main/Chapter07/ecomm-ui/src/hooks/useToken.js

Here, you are using a useState hook to maintain the token state. The token state is initialized while declaring the token state by calling the getToken function in the constructor of useState. Now, you need to provide a mechanism that should update the initial token state whenever there is a change in action, such as login or logout. You can create a new function, saveToken, for this purpose.

Both the getToken and saveToken functions use localStorage to retrieve and update the token respectively. Finally, both the token state and the saveToken function (in the form of setToken) are returned for their usage.

Next, you create another REST API client for authentication. Let's add another client, `Auth.js` (`https://github.com/PacktPublishing/Modern-API-Development-with-Spring-and-Spring-Boot/blob/main/Chapter07/ecomm-ui/src/api/Auth.js`), under the `src/api` directory.

This `Auth.js` client is very similar to other API clients. It has three functions that perform login, logout, and refresh access token operations by using the backend server REST APIs, outlined as follows:

- The login operation sets the access token, refresh token, user ID, and username in the `responseToken` key of the local storage by using the state arguments passed by the App component. The App component, as usual, uses the `useToken` custom hook. The login operation also sets the access token's expiration time.

- The refresh access token operation updates the access token and its expiration time.

- The logout operation removes the tokens and sets the expiration time to zero.

- You are done with the prerequisite work for implementing the login functionality and can now move on to creating the `Login` component.

Writing the Login component

Let's create a new `Login.js` file under the `src/components` directory and then run the following code:

```
import { useHistory } from "react-router-dom";
import { useState } from "react";
import PropTypes from "prop-types";
Login.propTypes = {
  auth: PropTypes.object.isRequired,
};
const Login = ({ uri, auth }) => {
  const [username, setUserName] = useState();
  const [password, setPassword] = useState();
  const [errMsg, setErrMsg] = useState();
  const history = useHistory();
  const cancel = () => {
    const l = history.length;
    l > 2 ? history.goBack() : history.push("/");
  };
  const handleSubmit = async (e) => {
```

```
    e.preventDefault();
    const res = await auth.loginUser({username, password});
    if (res && res.success) {
      setErrMsg(null);
      history.push(uri ? uri : "/");
    } else {
      setErrMsg(
        res && typeof res === "string" ? res : "Invalid
                               Username/Password");
    }
  };
```

https://github.com/PacktPublishing/Modern-API-Development-with-Spring-and-Spring-Boot/blob/main/Chapter07/ecomm-ui/src/components/Login.js

Before you start understanding the code, it's useful to know that PropTypes provides a way to check the type of passed props. Here, we have made sure that the auth prop is an object and a required prop. You may see messages in the console if it fails. Normally, you add this props check at the end of a file (in source code, it is at the bottom), but here it has been added at the top for better readability.

This component contains two props: auth and uri. The auth prop represents the authentication client, and uri is a string that sends the user back to the appropriate page after a successful login.

It has two functions: handleSubmit and cancel. The cancel function just sends back the user back to the previous page or the home page. The handleSubmit function makes use of the authentication client and calls the login API with the username and password.

Let's have a look at its JSX template, as follows:

```
    return (
      <div className="..."><div className="...">
        <div className="..." role="dialog" aria-modal="true">
        <div className="..."><div className="...">
        <div className="..."><div className="...">
        <h2 className="...">Sign in to your account</h2>
        <form className="..." onSubmit={handleSubmit}>
        <div className="..."><div>
```

```
          <span className="…" style={{ dispay: errMsg ? "block" :
                              "none" }} >{errMsg}</span>
          <label htmlFor="username" className="…">Username
          </label>
          <input id="username" name="username" type="username"
           autoComplete="username" placeholder="Username"
               required
           className="…" onChange={(e) =>
               setUserName(e.target.value)}/>
          </div><div>
          <label htmlFor="password" className="…">Password
          </label>
          <input id="password" name="password" type="password"
           autoComplete="password" placeholder="Password" required
           className="…" onChange={(e) => setPassword(e.target.
             value)}/>
          </div></div><div className="…"><div>
          <button type="submit" className="…">
           <span className="…">
           <svg className="…" xmlns=http://www.w3.org/2000/svg
            viewBox="0 0 20 20" fill="currentColor" aria-
            hidden="true"><path fillRule="evenodd" d="M5 9V7a5 5 0
            0110
            0v2a2 2 0 012 2v5a2 2 0 01-2 2H5a2 2 0 01-2-2v-5a2 2 0
            012-
            2zm8-2v2H7V7a3 3 0 016 0z" clipRule="evenodd"/>
           </svg>
          </span><span className="…">Sign in</span>
          </button>
          </div><div className="…">
          <button type="button" onClick={cancel}
          >Cancel</button>
          </div></div>
          </form>
          </div></div></div></div></div></div>
      );};
export default Login;
```

https://github.com/PacktPublishing/Modern-API-Development-with-Spring-and-Spring-Boot/blob/main/Chapter07/ecomm-ui/src/components/Login.js

The `handleSubmit` function is called when a form is submitted (when the user clicks on the **Sign in** button). The `cancel` function is called when the user clicks on the **Cancel** button. Another noticeable point relates to setting the username and password states. These are set on `onChange` events respectively. The `e.target.value` argument represents the typed value in the respective input field. The `e` instance represents the event and `target` represents the target input field for the respective event.

So, now you know the complete flow: the user logs in and the app sets the required token and information in local storage. The API client uses this information to call the authenticated APIs. The logout operation, which is a part of the `Header` component (`https://github.com/PacktPublishing/Modern-API-Development-with-Spring-and-Spring-Boot/blob/main/Chapter07/ecomm-ui/src/components/Header.js`), calls the `Auth` client's logout function, which calls the remove refresh token backend server's REST API and removes the authentication information from the local storage.

After authentication implementation, you need to write one more piece of code before you jump to writing the `Cart` component: cart context. Let's do that now.

Writing the custom cart context

You can use the Redux library for centralizing and maintaining an application's global state. However, you would write a Redux-like custom hook to maintain the state for the cart. This uses `createContext`, `useReducer`, and `useContext` hooks from the React library.

You already know that `createContext` returns the `Provider` and `Consumer`. Therefore, when you create a `CartContext` using `createContext`, it would provide the `CartContext.Provider`. You won't use the `Consumer`, as you are going to use a `useContext` hook.

Next, you need a cart state (`cartItems`) that you pass to the value in `CartContext.Provider` so that it will be available in the component that use the `CartContext`. Now, we just need a `reducer` function. A `reducer` function accepts two arguments: `state` and `action`. Based on the provided action, it updates (mutates) the state and returns the updated state.

Now, let's jump into the code and see how it turns out. Have a look at the following snippet:

```
import React,{ createContext, useReducer, useContext } from
"react";
export const CartContext = createContext();
function useCartContext() {
  return useContext(CartContext);
}
export const UPDATE_CART = "UPDATE_CART";
export const ADD_ITEM = "ADD_ITEM";
export const REMOVE_ITEM = "REMOVE_ITEM";

export function updateCart(items) {
  return { type: UPDATE_CART, items };
}
export function addItem(item) {
  return { type: ADD_ITEM, item };
}
export function removeItem(index) {
  return { type: REMOVE_ITEM, index };
}
export function cartReducer(state, action) {
  switch (action.type) {
    case UPDATE_CART:
      return [...action?.items];
    case ADD_ITEM:
      return [...state, action.item];
    case REMOVE_ITEM:
      const list = [...state];
      list.splice(action.index, 1);
      return list;
    default:
      return state;
  }
}
const CartContextProvider = (props) => {
```

```
    const [cartItems, dispatch] = useReducer(cartReducer,
        []);
    const cartData = { cartItems, dispatch };
    return <CartContext.Provider value={cartData} {...props}
        />;
};
export { CartContextProvider, useCartContext };
```

https://github.com/PacktPublishing/Modern-API-Development-with-Spring-and-Spring-Boot/blob/main/Chapter07/ecomm-ui/src/hooks/CartContext.js

First, we have created a `CartContext` with a `createContext` hook. Then, we have declared a function that uses a `useContext` hook and returns the `value` field's value declared in the `CartContext.Provider` tag.

Next, you need a `reducer` function that uses the action and state. Therefore, we first define action types such as `UPDATE_CART` and then write functions that return an action object that contains both action type and argument value, such as `updateCart`. Finally, you can write a `reducer` function that takes `state` and `action` as arguments and, based on the passed action type, it mutates the state and returns the updated state.

Next, you define a `CartContextProvider` function that returns the `CartContext.Provider` component. Here, you use the `reducer` function in `useReducer` hook, and in its second argument, you pass the empty array as an initial state. The `useReducer` hook returns to the `state` and `dispatch` functions. The `dispatch` function takes the `action` object as an argument. You can use the function that returns the `action` object, such as `updateCart` and `addItem`. You wrap the `state` (`cartItems`) and `dispatcher` functions (`dispatch`) in the `cartData` object and pass it to the `value` attribute in the `CartContext.Provider` component. At the end, it exports both the `CartContextProvider` and `useCartContext` functions.

You are going to use `CartContextProvider` as a component wrapper in the `App` component. This makes `cartData` (`cartItems` and `dispatch`) available to all components inside `CartContextProvider`, which can be accessed and used using `useCartContext`.

Now, finally, you can write the `Cart` component in the next subsection.

Writing the Cart component

The Cart component is a parent component because it can have multiple items (CartItem component) in it. Let's create a new cart.js file in the src/components directory and add the following code to it:

```
import { useEffect, useState } from "react";
import { Link, useHistory } from "react-router-dom";
import CartClient from "../api/CartClient";
import CustomerClient from "../api/CustomerClient";
import OrderClient from "../api/OrderClient";
import { removeItem, updateCart, useCartContext }
                            from "../hooks/CartContext";
import CartItem from "./CartItem";
const Cart = ({ auth }) => {
  const [grandTotal, setGrandTotal] = useState(0)
  const [noRecMsg, setNoRecMsg] = useState("Loading...");
  const history = useHistory();
  const cartClient = new CartClient(auth);
  const orderClient = new OrderClient(auth);
  const customerClient = new CustomerClient(auth);
  const { cartItems, dispatch } = useCartContext();
  // Continue...
```

https://github.com/PacktPublishing/Modern-API-Development-with-Spring-and-Spring-Boot/blob/main/Chapter07/ecomm-ui/src/components/Cart.js

You use useCartContext here and import functions such as updateCart that return the action object (consumed by the dispatch function). Apart from CartClient, you also use OrderClient and CustomerClient here for checkout operations.

Let's add functions for calculating the total (calTotal) and increasing the quantity (increaseQty) of a given product ID, as shown next:

```
// Cart.js Continue...
  const calTotal = (items) => {
    let total = 0;
    items?.forEach((i) => (total = total + i?.unitPrice *
                                        i?.quantity));
    return total.toFixed(2);
```

```
};
const increaseQty = async (id) => {
  const idx = cartItems.findIndex((i) => i.id === id);
  if (~idx) {
    cartItems[idx].quantity = cartItems[idx].quantity + 1;
    const res = await cartClient.addOrUpdate(
      cartItems[idx]);
    if (res && res.success) {
      refreshCart(res.data);
      if (res.data?.length < 1) { setNoRecMsg("Cart is
        empty."); }
    } else {
      setNoRecMsg(res && typeof res === "string" ? res :
                                    res.error.message);
    }
  }
}; // Continue...
```

The increaseQty function first finds whether the given ID exists in cart items or not. If it exists, then it increases the quantity of a product by 1. Finally, it calls the REST API to update the cart items and uses the response to update the cart by calling the refreshCart function.

Let's add a decreaseQty function, which is similar to increaseQty but decreases the quantity by one. Also, the deleteItem function would remove the cart item from the cart. The code is shown in the following snippet:

```
// Cart.js Continue...
const decreaseQty = async (id) => {
  const idx = cartItems.findIndex((i) => i.id === id);
  if (~idx && cartItems[idx].quantity <= 1) {
    return deleteItem(id);
  } else if (cartItems[idx]?.quantity > 1) {
    cartItems[idx].quantity = cartItems[idx].quantity -
      1;
    const res = await cartClient.addOrUpdate(
      cartItems[idx]);
    if (res && res.success) {
      refreshCart(res.data);
```

```
            if (res.data?.length < 1) { setNoRecMsg("Cart is
                                        empty.");}
        return;
    } else { setNoRecMsg(res && typeof res === "string" ?
                            res : res?.error?.message); }
    }
};
const deleteItem = async (id) => {
  const idx = cartItems.findIndex((i) => i.id === id);
  if (~idx) {
    const res = await cartClient.remove(
        cartItems[idx].id);
    if (res && res.success) {
      dispatch(removeItem(idx));
      if (res.data?.length < 1) { setNoRecMsg("Item is
        removed.");}
    } else { setNoRecMsg(res && typeof res === "string" ?
          res
        : "There is an error performing the remove."); }
  }
}; // Continue…
```

The decreaseQty function does one extra step in comparison to increaseQty— it removes the item if the existing quantity is 1 by calling the deleteItem. function.

The deleteItem function first finds the product based on a given ID. If it exists, then it calls the REST API to remove the product from the cart and updates the cart item state by calling the dispatch function with the action object returned by the removeItem function.

Let's define refreshCart and useEffect functions, as shown in the following code snippet:

```
// Cart.js Continue…
const refreshCart = (items) => {
  setGrandTotal(calTotal(items));
  dispatch(updateCart(items));
};
useEffect(() => {
```

```
  async function fetch() {
    const res = await cartClient.fetch();
    if (res && res.success) {
      refreshCart(res.data.items);
      if (res.data?.items && res.data.items?.length < 1) {
        setNoRecMsg("Cart is empty.");
      }
    } else {
      setNoRecMsg(res && typeof res === "string" ? res
                              : res.error.message);
    }
  }
  fetch();
}, []);// Continue...
```

The refreshCart function updates the total and dispatches the updateCart action. The useEffect loads the cart items from the backend server and calls refreshCart to update the cartItems global state.

Let's add the last function of the Cart component to perform the checkout operation, as shown in the following code snippet:

```
// Cart.js Continue...
const checkout = async () => {
  const res = await customerClient.fetch();
  if (res && res.success) {
    const payload = {
      address: { id: res.data.addressId },
      card: { id: res.data.cardId },
    };
    const orderRes = await orderClient.add(payload);
    if (orderRes && orderRes.success) {
      history.push("/orders");
    } else {
      setNoRecMsg(orderRes && typeof orderRes ===
        "string"
          ? orderRes: "Couldn't process checkout."
      );
```

```
        }
    } else {
        setNoRecMsg( res && typeof res === "string" ? res
                            : "error retreiving customer");
    }
};
```

The checkout function first fetches the customer information and forms a payload for placing the order. On a successful POST order API response, the user is redirected to the Orders component.

Finally, let's add a JSX template, which is used from codepen user abdelrhman for the Cart component as shown in the next code block (Code and className values have been stripped for brevity):

```
// Cart.js Continue...
return (
  <div className="...">
    <!-- code stripped for brevity  -->
        <div className="...">
            <h1 className="...">Shopping Cart</h1>
            <h2 className="...">{cartItems?.length}
            Items</h2>
        </div>
        <div className="...">
            <h3 className="...">Product Details</h3>
            <h3 className="...">Quantity</h3>
            <h3 className="...">Price</h3>
            <h3 className="...">Total</h3>
        </div>
        {cartItems && cartItems.length > 0 ? (
          cartItems?.map((i) => (
            <CartItem item={i} key={i.id}
                removeItem={deleteItem}
              increaseQty={
                  increaseQty}decreaseQty={decreaseQty}/>

    <!-- code stripped for brevity  -->
```

```
            <div className="...">
              <div className="...">
                <span>Total cost</span>
                <span>${grandTotal}</span>
              </div>
              <button className="..." onClick={checkout}
                disabled={grandTotal == 0 ? true : false} >
                Checkout
              </button>
          <!-- code stripped for brevity  -->
      </div>
    );
  };
export default Cart;
```

https://github.com/PacktPublishing/Modern-API-Development-with-Spring-and-Spring-Boot/blob/main/Chapter07/ecomm-ui/src/components/Cart.js

Here, you can see that on a click of the **Checkout** button, it calls the checkout function to place the user order. Cart items are rendered using the CartItem component that you create next. You pass the removeItem, increaseQty, and decreaseQty functions as props to it.

Let's write the CartItem component by creating a new file (src/components/CartItem.js) and adding the following code:

```
import { useEffect, useState } from "react";
import { Link } from "react-router-dom";
const CartItem = ({ item, increaseQty, decreaseQty, removeItem
}) => { const des = item ? item.description?.split(".") : [];
  const author = des && des.length > 0
      ? des[des.length - 1] : "";
  const [total, setTotal] = useState();
  const calTotal = (item) => {
    setTotal((item?.unitPrice *
            item?.quantity)?.toFixed(2));
  };
  const updateQty = (qty) => {
    if (qty === -1) { decreaseQty(item?.id); }
```

```
    else if (qty === 1) { increaseQty(item?.id); }
    else { return false; }
    calTotal(item);
  };

  useEffect(() => {
    calTotal(item);
  }, []);
```

https://github.com/PacktPublishing/Modern-API-Development-with-Spring-and-Spring-Boot/blob/main/Chapter07/ecomm-ui/src/components/CartItem.js

Here, you maintain the state of the total that is a product of the quantity and the unit price (`calTotal` function) and the `updateQty` helper function to perform the increase/decrease quantity operations. The `useEffect` hook also calls `calTotal` to update the total on the **Cart** page.

Let's add the last piece of the JSX template for the `CartItem` component, as shown in the next code block (`className` values have been stripped for brevity):

```
return (
  <div className="…">
    <div className="…">
      <div className="…">
        <img className="…" src={item?.imageUrl} alt="" />
      </div>
      <div className="…">
        <Link to={"/products/" + item.id} className="…">
          {item?.name}
        </Link>
        <span className="…">Author: {author}</span>
        <button className="…" onClick={() =>
            removeItem(item.id)}>
          Remove
        </button>
      </div>
    </div>
    <div className="…">
```

```
          <span className="…" onClick={() => updateQty(-1)}>
            <svg className="…" viewBox="0 0 448 512">
              <path d="M416 208H32c-17.67 0-32 14.33-32 32v32c0
              17.67
              14.33 32 32 32h384c17.67 0 32-14.33
              32-32v-32c0-17.67-
              14.33-32-32-32z" /></svg>
          </span>
          <input type="text" readOnly value={item?.quantity} />
          <span className="…" onClick={() => updateQty(1)}>
            <svg className="…" viewBox="0 0 448 512">
              <path d="M416 208H272V64c0-17.67-14.33-32-32-32h-
              32c-
              17.67 0-32 14.33-32 32v144H32c-17.67 0-32 14.33-32
              32v32c0 17.67 14.33 32 32 32h144v144c0 17.67 14.33
              32 32
              32h32c17.67 0 32-14.33 32-32V304h144c17.67 0 32-
              14.33
              32-32v-32c0-17.67-14.33-32-32-32z" /></svg>
          </span>
        </div>
        <span className="…">{item?.unitPrice?.toFixed(2)}</span>
        <span className="…">${total}</span>
      </div>
    );
  };
  export default CartItem;
```

https://github.com/PacktPublishing/Modern-API-Development-with-Spring-and-Spring-Boot/blob/main/Chapter07/ecomm-ui/src/components/CartItem.js

The `state` item is used in an expression to generate the values. The `removeItem` and `updateQty` functions are bound to the `onClick` event for the respective JSX elements.

Now, you can write the last component (page) of this application in the next subsection: the `Order` Component.

Writing the Order component

The Order component contains the order details fetched from the backend server. It shows date, status, amount, and items in a tabular format. It loads the order details on the first render with the useEffect hook and then the orders state is used in the JSX expression to display it.

Let's create a new file, Orders.js, in the src/components directory and add the following code to it:

```
import { useEffect, useState } from "react";
import OrderClient from "../api/OrderClient";
const Orders = ({ auth }) => {
  const [orders, setOrders] = useState([]);
  const formatDate = (dt) => {
    return dt && new Date(dt).toLocaleString();
  };
  useEffect(() => {
    async function fetchOrders() {
      const client = new OrderClient(auth);
      const res = await client.fetch();
      if (res && res.success) {
        setOrders(res.data);
      }
    }
    fetchOrders();
  }, []);
```

https://github.com/PacktPublishing/Modern-API-Development-with-Spring-and-Spring-Boot/blob/main/Chapter07/ecomm-ui/src/components/Orders.js

Next, add the following JSX template. Here, the code and className have been stripped for brevity:

```
  return (
    <div className="…">
      <!-- code stripped for brevity  -->
            <table className="…">
              <thead className="…">
                <tr>
```

```
        <th scope="col" className="…">Order
    Date</th>
        <th scope="col" className="…">Order
    Items</th>
        <th scope="col"
    className="…">Status</th>
        <th scope="col" className="…">Order
    Amount</th>
    </tr>
</thead>
<tbody className="…">{orders &&
orders.length < 1 ?

    ( <tr className="px-6 py-4 whitespace-
    nowrap">
        Found zero order</tr>
    ) : (
    orders?.map((order) => (
    <tr>
        <td className="…">
          <div className="…">{formatDate(
              order?.date)}
          </div>
          <div className="…">Local Time</div>
        </td>
        <td className="…">
          <!-- code stripped for brevity   -->
              {order?.items.map((o, idx) => (
                <div>
                    <span className="…">
                    {idx + 1}.
                    </span>{" "}{o.name}{" "}
                    <span className="…">
                        ({o?.quantity +" x $" +
                        o?.unitPrice?.toFixed(2)}
                    )
```

```
                      </span><br/>
                    </div>
                  ))}
                </div>
              </div>
            </div>
          </td>
          <td className="…"><span className="…">{
              order?.status}</span>
          </td>
          <td className="…">${
              order?.total?.toFixed(2)}</td>
            </tr>
          ))
        )}
        </tbody>
      </table>
    <!-- code stripped for brevity  -->
    </div>
  </div>
  );
};
export default Orders;
```

https://github.com/PacktPublishing/Modern-API-Development-with-Spring-and-Spring-Boot/blob/main/Chapter07/ecomm-ui/src/components/Orders.js

It simply displays the information fetched from the `orders` state.

Now, we can update the root component to complete the flow and test the application after starting again with the `yarn start` command.

Writing the root (App) component

The App component is a root component of the React application. It contains routing information and the application layout with all the parent components, such as the product list and orders components.

Update the App.js file available in the project src directory with the following code:

```
import { BrowserRouter as Router, Route, Switch } from "react-
router-dom";
import Header from "./components/Header";
import Footer from "./components/Footer";
import ProductList from "./components/ProductList";
import Login from "./components/Login";
import useToken from "./hooks/useToken";
import Cart from "./components/Cart";
import ProductDetail from "./components/ProductDetail";
import NotFound from "./components/NotFound";
import Auth from "./api/Auth";
import { CartContextProvider } from "./hooks/CartContext";
import Orders from "./components/Orders";
function App() {
  const { token, setToken } = useToken();
  const auth = new Auth(token, setToken);
  const loginComponent = (props) => (
    <Login {...props} uri="/login" auth={auth} />
  );
  const productListComponent= (props)=> <ProductList
auth={auth}/>;
  // continue…
```

This contains all the imports required for the `App` component. Then, you use the `useToken()` hook and the `Auth` authentication REST API client for authentication purposes. You create functions that return `loginComponent` and `productListComponent`.

Its JSX template is different from what we have used till now. It uses the `BrowserRouter` (`Router`), `Route`, and `Switch` components from the `react-router-dom` package. You define all the `Route` components inside the `BrowserRouter` component. Here, we are also using the `Switch` component because we want to render components exclusively. It also allows you to render the `NotFound` component (the typical `404 - not found` page) if no path matches. The `Route` component allows you to define the path and component to be rendered. You have used the arrow function as a `render` property value because we can then use the expressions too. The following code snippet contains the logic explained here:

```
// App.js continue...
return (
  <div className="flex flex-col min-h-screen h-full ">
    <Router>
      <Header userInfo={token} auth={auth} />
      <div className="flex-grow flex-shrink-0 p-4">
        <CartContextProvider>
          <Switch>
            <Route path="/" exact render={() =>
                                 productListComponent()} />
            <Route path="/login" render={(props) => token ?
               productListComponent() : loginComponent(props)} />
            <Route path="/cart" render={(props) => token ?
              <Cart auth={auth} /> : loginComponent(props)} />
            <Route path="/orders" render={(props) => token ?
              <Orders auth={auth} /> : loginComponent(props)} />
            <Route path="/products/:id" render={() =>
              <ProductDetail auth={auth} />} />
            <Route path="*" exact component={NotFound} />
          </Switch>
        </CartContextProvider>
      </div>
      <Footer />
```

```
    </Router>
    </div>
  );
}
export default App;
```

https://github.com/PacktPublishing/Modern-API-Development-with-Spring-and-Spring-Boot/blob/main/Chapter07/ecomm-ui/src/App.js

All components are wrapped inside `CartContextProvider` to allow `cartItems` and `dispatch` to be accessible in all components provided they use the `useCartContext` custom hook.

Running the application

You can start the backend server by using code from *Chapter 6, Security (Authorization and Authentication)*. Then, you can start the `ecomm-ui` app by executing a `yarn start` command from the project root directory. You can log in with `scott/tiger` and perform all the operations.

Summary

In this chapter, you have learned React basic concepts and created different types of components using them. You have also learned how to use the browser's built-in `Fetch` API to consume the REST APIs. You acquired the following skills in React: developing a component-based UI, implementing routing, consuming REST APIs, implementing functional components with hooks, writing custom hooks, and building a global state store with a React context API and a `useReducer` hook. The concepts and skills you acquired in this chapter lay a solid foundation for modern frontend development and give you an edge to gain the perspective of 360-degree application development.

In the next chapter, you will learn about writing tests for REST-based web services.

Questions

1. What is the difference between `props` and `state`?
2. What is an event and how you can bind events in a React component?
3. What is a higher-order component?

Further reading

- *Mastering React Test Driven Development:*

  `https://www.packtpub.com/product/mastering-react-test-driven-development/9781789133417`

- React documentation:

  `https://reactjs.org/docs/`

- React Router guide:

  `https://reactrouter.com/web/guides/quick-start`

8
Testing APIs

In this chapter, we will learn about how to manually and automatically test APIs. First, you will learn about automating unit and integration tests. After learning about these forms of automation, you will be able to make both types of testing an integral part of any build. You will also learn how to set up the Java Code Coverage tool to calculate different code coverage metrics.

In this chapter, we will cover the following topics:

- Manual testing
- Integration testing automation

This chapter will help you learn about test automation by showing you how to implement unit and integration test automation. You will also learn how to set up code coverage using the **Java Code Coverage (JaCoCo)** tool.

Technical requirements

You will need the following for developing and executing the code in this chapter:

- Any Java IDE, such as NetBeans, IntelliJ, or Eclipse
- **Java Development Kit (JDK)** 15+
- An internet connection to clone the code and download the dependencies and Gradle
- Postman/cURL (for API testing)

The code present in this chapter can be found on GitHub at `https://github.com/PacktPublishing/Modern-API-Development-with-Spring-and-Spring-Boot/tree/main/Chapter08`

Testing APIs and code manually

Testing is a continuous process in software development and maintenance cycles. You need to do full testing that covers all possible use cases and the respective code for each change. Different types of testing can be performed for APIs, including the following:

- **Unit testing**: Unit testing is performed by the developers to test the smallest unit (such as a class method) of code.
- **Integration testing**: Integration testing is performed by the developers to test the integration of different layers of components.
- **Contract testing**: Contract testing is performed by the developers to make sure any changes that are made to the API won't break the consumer code. The consumer code should always comply with the producer's contract (API). It is primarily required in microservices-based development.
- **End-to-End (E2E) testing**: E2E testing is performed by the **Quality Assurance (QA)** team to test end-to-end scenarios, such as from the UI (consumer) to the backend.
- **User Acceptance Testing (UAT)**: This is performed by the business users from a business perspective, and may overlap with E2E testing.

You performed manual API testing by using the cURL and Postman tools earlier in this book. Every change requires the APIs to be completely tested – not only the impacted APIs. There is a reason for this. You may assume that it only impacts certain APIs, but what if your underlying assumptions are wrong? It may impact the other APIs that you skipped, which would lead to production issues. This can create panic and may require a release to be rolled over or a patch to be released with fix.

You don't want to be in such situations, so products have a separate **QA** team that ensures releases are delivered with the best possible quality. QA teams do the separate end-to-end and acceptance testing (along with business/domain users), apart from the testing that's done by the development team.

This extra assurance for high-quality deliverables need more time and effort. Therefore, software development cycles used to be huge in comparison to today's. **Time to market (TTM)** is a huge factor in today's competitive software industry. Today, you need faster release cycles. Moreover, quality checks, also known as testing, is an important and major part of release cycles.

You can reduce testing time by automating the testing process and making it an integral part of the CI/CD pipeline. **CI** stands for **continuous integration**, which means *build > test > merge* in a code repository. **CD** stands for **continuous delivery** and/or **continuous deployment**, both of which may be used interchangeably. Continuous delivery is a process where code is automatically tested and released (read and uploaded) to an artifact repository or container registry. Then, it can be picked and deployed to a production environment. Continuous deployment is one step ahead of continuous delivery in the pipeline, and code is deployed to the production environment once the previous steps have been successful. Products that don't release their code for public access use this approach, such as Facebook and Twitter. On the other hand, products/services that is available publicly, such as the Spring Framework and Java, use continuous delivery pipelines.

We'll automate the manual testing we have done so far in the next section.

Testing automation

Whatever testing you are doing manually can be automated and made part of the build. This means that any change or code commit will run the test suite part as a part of the build. A build will only be successful if all the tests pass.

You can add automated integration tests for all the APIs. So, instead of firing each API manually using cURL or Postman, the build will fire them, and the test result will be available at the end of the build.

In this section, you are going to write an integration test that will replicate the REST client call and test all the application layers, starting from the controller, all the way down to the persistence layer, including the database (H2).

But before that, you will add the necessary unit tests. Ideally, these unit tests should have been added alongside the development process, or before the development process in the case of **test-driven development (TDD)**.

Unit tests are tests that validate the expected results of small units of code, such as a class's methods. You can avoid most bugs if you have proper tests in place with good code (90% or above) and branch coverage (80% and above). Code coverage refers to metrics such as the number of lines and branches (such as if-else), which is validated when the tests are executed.

Some classes or methods have dependencies on other classes or infrastructure services. For example, controller classes have dependencies on service and assembler classes, while repository classes have dependencies on Hibernate APIs. You can create mocks to replicate dependencies behaviors and assume these are working as expected or behave as per the defined tests. This allows you test the actual code unit (such as a method) and validate its behavior.

In the next section, we'll explore how to add unit tests before writing the integration tests.

Unit testing

I advise you to go back to *Chapter 6, Security (Authorization and Authentication)*, as a base for this chapter's code. You don't have to add any additional dependencies for unit tests. You already have the following dependency in `build.gradle` (`https://github.com/PacktPublishing/Modern-API-Development-with-Spring-and-Spring-Boot/blob/main/Chapter08/build.gradle`):

```
testImplementation('org.springframework.boot:spring-boot-
starter-test')
```

`spring-boot-starter-test` adds all the required test dependencies, not only for the unit tests, but also for the integration tests. You are going to primarily use the following libraries for testing:

- **JUnit 5**: JUnit 5 is a bundle of modules – the JUnit Platform, JUnit Jupiter, and JUnit Vintage:

    a. The **JUnit Platform** allows you to launch tests on JVM and its engine provides APIs for writing testing frameworks that run on the platform. The JUnit Platform consists of `junit-platform-engine` and `junit-platform-commons`.

    b. **JUnit Jupiter** provides the programming and extension models for writing tests and extensions. It has a separate library called `junit-jupiter-engine` that allows you to run Jupiter-based tests on JUnit Platform. It also provides the `junit-jupiter`, `junit-jupitor-api`, and `junit-jupiter-params` libraries.

c. **JUnit Vintage** supports older versions of JUnit, such as versions 3 and 4. You are going to use latest version, which is 5, so you don't need it.

You can find out more about JUnit at `https://junit.org/`.

- **AssertJ**: AssertJ is a test assertion library that simplifies assertion writing by providing fluent APIs. It is also extendable. You can write custom assertions for your domain objects. You can find more about it at `https://assertj.github.io/doc/`.

- **Hamcrest**: Hamcrest is another assertion library that provides assertions based on matchers. It also allows you to write custom matchers. You'll find an example of both of these in this chapter. Though AssertJ is preferable, you can choose one of them or both based on your use cases and liking. You can find out more about it at `http://hamcrest.org/`.

- **Mockito**: Mockito is a mocking framework that allows you to mock objects (read dependencies) and stubs method calls. You can find out more about it at `https://site.mockito.org/`.

You already know that unit tests test the smallest testable code unit. But how can we write a unit test for controller methods? Controller runs on web servers and have the Spring web application context. If you write a test that uses `WebApplicationContext` and is running on top of a web server, then you can call it an integration test rather than a unit test.

Unit tests should be lightweight and must be executed quickly. Therefore, you must use `MockMvc`, a special class provided by the Spring test library, to test the controllers. You can use the standalone setup for `MockMvc` for unit testing. You can also use `MockitoExtension` to run the unit test on the JUnit Platform (JUnit 5 provides an extension for runners) that supports object mocking and method stubbing. You will also use the Mockito library to mock the required dependencies. These tests are really fast and help developers build faster.

Let's write our test using AssertJ assertions.

Testing using AssertJ assertions

Let's write our first unit test for `ShipmentController`. The following code can be found in `src/test/java/com/packt/modern/api/controller/ShipmentControllerTest.java`:

```
@ExtendWith(MockitoExtension.class)
public class ShipmentControllerTest {
  private static final String id = "a1b9b31d-e73c-4112-
```

```
                                    af7c-b68530f38222";
private MockMvc mockMvc;
@Mock
private ShipmentService service;
@Mock
private ShipmentRepresentationModelAssembler assembler;
@Mock
private MessageSource msgSource;
@InjectMocks
private ShipmentController controller;
private ShipmentEntity entity;
private Shipment model = new Shipment();
private JacksonTester<List<Shipment>> shipmentTester;
// continue…
```

https://github.com/PacktPublishing/Modern-API-Development-with-Spring-and-Spring-Boot/blob/main/Chapter08/src/test/java/com/packt/modern/api/controller/ShipmentControllerTest.java

Here, our test is using a Jupiter-based annotation (`ExtendWith`) that registers the extension (`MockitoExtension`) for running tests and supporting Mockito-based mocks and stubbing.

Here, the Spring test library provides the `MockMvc` class, which allows you to mock the Spring MVC. As a result, you can execute the controller methods by calling the associated API endpoints' URI. The dependencies of the `ShipmentController` controller class such as the service and assembler are marked with `@Mock` annotations to create the mock instances of its dependencies. You can also use `Mockito.mock(classOrInterface)` to create the mock objects.

Another noticeable annotation is `@InjectMocks` on the controller declaration. It finds out all the declared mocks that are required for a testing class and injects them automatically. `ShipmentController` uses the `ShipmentService` and `ShipmentRepresentationModelAssembler` instances, which are injected using its constructor. The Mockito-based `InjectMocks` annotation finds the dependencies in the `ShipmentController` class (service and assembler). Then, it looks for mocks of the service and assembler in the test class. Once it finds them, it injects these mock objects into the `ShipmentController` class. If required, you can also create an instance of the testing class using a constructor instead of using `@InjectsMocks`, as shown here:

```
controller = new ShipmentController(service, assembler);
```

A mock of `MessageSource` is created for `RestApiHandler`, which is being used in the setup method. You'll explore it further in the following code block.

The last part of the declaration is `JacksonTester`, which is part of the Spring testing library. `JacksonTester` is a custom JSON assertion class that's created using the AssertJ and Jackson libraries.

The JUnit Jupiter API provides the `@BeforeAll` and `@BeforeEach` method annotations, which can be used to set up the prerequisites. As their names suggest, `@BeforeAll` is run once per test class, while `@BeforeEach` gets executed before each test execution. `@BeforeEach` can be placed on public non-static methods, whereas `@BeforeAll` should be used to annotate public static methods.

Similarly, JUnit provides the `@AfterAll` and `@AfterEach` annotations, which execute the associated methods after each test is executed and after each test is executed, respectively.

Let's use the `@BeforeEach` annotation to set up the prerequisites for the `ShipmentControllerTest` class, as shown here:

```
@BeforeEach
public void setup() {
    ObjectMapper mapper = new AppConfig().objectMapper();
    JacksonTester.initFields(this, mapper);
    MappingJackson2HttpMessageConverter mappingConverter =
        new MappingJackson2HttpMessageConverter();
    mappingConverter.setObjectMapper(mapper);
    mockMvc = MockMvcBuilders.standaloneSetup(controller)
        .setControllerAdvice(new
            RestApiErrorHandler(msgSource))
```

```
        .setMessageConverters(mappingConverter)
        .build();
    final Instant now = Instant.now();
    entity = // entity initialization code
    BeanUtils.copyProperties(entity, model);
    // extra model property initialization
}
```

https://github.com/PacktPublishing/Modern-API-Development-with-Spring-and-Spring-Boot/blob/main/Chapter08/src/test/java/com/packt/modern/api/controller/ShipmentControllerTest.java

First, we initialize the JacksonTester fields with the object mapper instance received from AppConfig. This creates a custom message converter instance (MappingJackson2HttpMessageConverter).

Next, you can create a mockMvc instance using the standalone setup and initialize the controller advice using its setter method. The RestApiErrorHandler instance uses the mock object of the MessageResource class. You can also set the message converter to mockMvc before building it.

Finally, you initialize the instances of ShipmentEntity and Shipment (model).

Next, you are going to write the test for the GET /api/v1/shipping/{id} call, which uses the getShipmentByOrderId() method of the Shipment Controller class. Tests are marked with @Test. You can also use @DisplayName to customize a test's name in the test reports:

```
@Test
@DisplayName("returns shipments by given order ID")
public void testGetShipmentByOrderId() throws Exception {
    // given
    given(service.getShipmentByOrderId(id))
        .willReturn(List.of(entity));
    given(assembler.toListModel(List.of(entity)))
        .willReturn(List.of(model));
    // when
    MockHttpServletResponse response = mockMvc.perform(
        get("/api/v1/shipping/" + id)
            .contentType(MediaType.APPLICATION_JSON)
```

```
            .accept (MediaType.APPLICATION_JSON))
        .andDo (print())
        .andReturn().getResponse();
    // then
    assertThat (response.getStatus())
        .isEqualTo (HttpStatus.OK.value());
    assertThat (response.getContentAsString())
        .isEqualTo (shipmentTester.write(
            List.of (model)).getJson());
}
```

Here, you are using the **behavior-driven development** (**BDD**) test style. You can find out more about BDD at https://cucumber.io/docs/bdd/. BDD tests are written using the Gherkin *Given > When > Then* language (https://cucumber.io/docs/gherkin/), which can be defined as follows:

- Given: Context of the test
- When: Test action
- Then: Test result, followed by validation

Let's read this test from a BDD perspective:

- Given: The service is available and returns the list of shipments based on the given order ID and an assembler, which converts the list of entities into a list of models. It also adds HATEOAS links.
- When: The user calls the API via GET /api/shipping/a1b9b31d-e73c-4112-af7c-b68530f38222.
- Then: The test validates the received shipments associated with the given order ID.

Mockito's MockitoBDD class provides the given() fluent API to stub the mock objects methods. When mockMvc.perform() is called, internally, it calls the respective service and assembler mocks, which, in turn, call the stubbed methods and return the values defined in the stub (using given()).

The andDo(MockMvcResultHandlers.print()) method logs the request and response trace, including the payload and response body. If you want to trace all the mockMvc logs inside a test class, then you can configure them directly while initializing mockMvc instead of defining them individually in mockMvc.perform() calls, as shown here (highlighted code):

```
mockMvc = MockMvcBuilders.standaloneSetup(controller)
        .setControllerAdvice(new
            RestApiErrorHandler(msgSource))
        .setMessageConverters(
            mappingJackson2HttpMessageConverter)
            .alwaysDo(print())
        .build();
```

At the end, you perform assertions (whether the status is 200 OK or not and returned JSON object matching the expected object or not) using AssertJ fluent APIs. First, you use the Asserts.assertThat() function, which takes the actual object and compares it with the expected object using the isEqualTo() method.

So far, you have used AssertJ assertions. Similarly, you can also use Spring and Hamcrest assertions.

Testing using Spring and Hamcrest assertions

At this point, you know how to write JUnit 5 tests using MockitoExtension. You'll use the same approach to write a unit test, except with assertions. This time, you will write an assertion using Hamcrest assertions, as shown here:

```
@Test
@DisplayName("returns address by given existing ID")
public void getAddressByOrderIdWhenExists() throws
      Exception {
  given(service.getAddressesById(id))
        .willReturn(Optional.of(entity));


  // when
  ResultActions result = mockMvc.perform(
        get("/api/v1/addresses/a1b9b31d-e73c-4112-af7c-
            b68530f38222")
          .contentType(MediaType.APPLICATION_JSON)
```

```
            .accept(MediaType.APPLICATION_JSON));

  // then
  result.andExpect(status().isOk());
  verifyJson(result);
}
```

https://github.com/PacktPublishing/Modern-API-Development-with-Spring-and-Spring-Boot/blob/main/Chapter08/src/test/java/com/packt/modern/api/controller/AddressControllerTest.java

You have captured the `MockHttpResponse` instance from the `mockMvc.perform()` call in the previous test example; that is, `testGetShipmentByOrderId()`. This time, you will directly use the returned value of the `mockMvc.perform()` call rather than calling an extra `andReturn().getResponse()` on it.

The `ResultAction` class provides the `andExpect()` assertion method, which takes `ResultMatcher` as an argument. The `StatusResultMatchers.status().isOk()` result matcher evaluates the HTTP status returned by the `perform()` call. The `VerifyJson()` method evaluates the JSON response object, as shown in the following code:

```
private void verifyJson(final ResultActions result)
                                          throws Exception {
  final String BASE_PATH = "http://localhost";
  result
      .andExpect(jsonPath("id",
          is(entity.getId().toString())))
      .andExpect(jsonPath("number",
          is(entity.getNumber())))
      .andExpect(jsonPath("residency",
          is(entity.getResidency())))
      .andExpect(jsonPath("street",
          is(entity.getStreet())))
      .andExpect(jsonPath("city", is(entity.getCity())))
      .andExpect(jsonPath("state", is(entity.getState())))
      .andExpect(jsonPath("country",
          is(entity.getCountry())))
      .andExpect(jsonPath("pincode",
```

```
        is(entity.getPincode()))))
    .andExpect(jsonPath("links[0].rel", is("self")))
    .andExpect(jsonPath("links[0].href",
                    is(BASE_PATH + "/" +
                        entity.getId()))))
    .andExpect(jsonPath("links[1].rel", is("self")))
    .andExpect(jsonPath("links[1].href",
                is(BASE_PATH + URI + "/" +
                    entity.getId()))));
}
```

Here, the `MockMvcResultMatchers.jsonPath()` result matcher takes two arguments – a JSON path expression and a matcher. Therefore, first, you must pass the JSON field name and then the Hamcrest matcher known as `Is.is()`, which is a shortcut for `Is.is(equalsTo(entity.getCity()))`.

Writing the unit test for a service is much easier compared to writing one for the controller because you don't have to deal with `MockMvc`.

You will learn how to test private methods in the next subsection.

Testing private methods

Unit testing a private method is a challenge. The Spring test library provides the `ReflectionTestUtils` class, which provides a method called `invokeMethod`. This method allows to you invoke private methods. The `invokeMethod` method takes three argument – the target class, the method's name, and the method's arguments (using variable arguments). Let's use it to test the `AddressServiceImpl.toEntity()` private method, as shown in the following code block:

```
@Test
@DisplayName("returns an AddressEntity when private method
                    toEntity() is called with Address
                    model")
public void convertModelToEntity() {
    // given
    AddressServiceImpl srvc = new AddressServiceImpl(
        repository);

    // when
```

```
AddressEntity e =
    ReflectionTestUtils.invokeMethod(
    srvc, "toEntity",addAddressReq);

// then
then(e)
  .as("Check address entity is returned and not
      null").isNotNull();
then(e.getNumber())
  .as("Check house/flat no is
      set").isEqualTo(entity.getNumber());
then(e.getResidency())
  .as("Check residency is
      set").isEqualTo(entity.getResidency());
then(e.getStreet())
  .as("Check street is
      set").isEqualTo(entity.getStreet());
then(e.getCity())
  .as("Check city is set").isEqualTo(entity.getCity());
then(e.getState())
  .as("Check state is set").isEqualTo(entity.getState());
then(e.getCountry())
  .as("Check country is
      set").isEqualTo(entity.getCountry());
then(e.getPincode())
  .as("Check pincode is
      set").isEqualTo(entity.getPincode());
}
```

https://github.com/PacktPublishing/Modern-API-Development-with-Spring-and-Spring-Boot/blob/main/Chapter08/src/test/java/com/packt/modern/api/service/AddressServiceTest.java

Here, you can see that when you call `ReflectionTestUtils.invokeMethod()` with the given arguments, it returns the `AddressEntity` instance, which has been converted using the passed argument's `AddAddressReq` model instance.

Here, you are using a third kind of assertion using AssertJ's BDDAssertions class. The BDDAssertions class provides methods that resonate with the BDD style. BDDAssertions.then() takes the actual value that you want to verify. The as() method describes the assertion and should be added before you perform the assertion. Finally, you perform verification using AssertJ's assertion methods, such as isEqualTo().

You will learn how to test void methods in the next subsection.

Testing void methods

A method that returns a value can easily be stubbed, but how can we stub a method that returns nothing? Mockito provides the doNothing() method for this. It has a wrapper willDoNothing() method in the BDDMockito class that internally uses doNothing().

This is very handy, especially when you want such methods to do nothing while you're spying, as shown here:

```
List linkedList = new LinkedList();
List spyLinkedList = spy(linkedList);
doNothing().when(spyLinkedList).clear();
```

Here, linkedList is a real object and not a mock. However, if you want to stub a specific method, then you can use spy(). Here, when the clear() method is called on spyLinkedList, it will do nothing.

Let's use willDoNothing to stub the void method and see how it helps test void methods:

```
@Test
@DisplayName("delete address by given existing id")
public void deleteAddressesByIdWhenExists() {
  given(repository.findById(UUID.fromString(nonExistId)))
      .willReturn(Optional.of(entity));
  willDoNothing().given(repository)
      .deleteById(UUID.fromString(nonExistId));

  // when
  service.deleteAddressesById(nonExistId);
```

```
  // then
  verify(repository, times(1))
      .findById(UUID.fromString(nonExistId));
  verify(repository, times(1))
      .deleteById(UUID.fromString(nonExistId));
}
```

https://github.com/PacktPublishing/Modern-API-Development-with-Spring-and-Spring-Boot/blob/main/Chapter08/src/test/java/com/packt/modern/api/service/AddressServiceTest.java

Here, `AddresRepository.deleteById()` is being stubbed using Mockito's `willDoNothing()` method. Now, you can use the `verify()` method of Mockito, which takes two arguments – the mock object and its verification mode. Here, the `times()` verification mode is being used, which determines how many times a method is invoked.

We'll learn how to unit test exceptional scenarios in the next subsection.

Testing exceptions

Mockito provides `thenThrow()` for stubbing methods with exceptions. BDDMockito's `willThrow()` is a wrapper that uses it internally. You can pass the `Throwable` argument and test it like so:

```
@Test
@DisplayName("delete address by given non-existing id,
                       should throw
                       ResourceNotFoundException")
public void deleteAddressesByNonExistId() throws Exception {
  given(repository.findById(UUID.fromString(nonExistId)))
    .willReturn(Optional.empty())
    .willThrow(new ResourceNotFoundException(
        String.format("No Address found with id %s.",
            nonExistId)));

  // when
  try {
    service.deleteAddressesById(nonExistId);
```

```
} catch (Exception ex) {
// then
   assertThat(ex)
      .isInstanceOf(ResourceNotFoundException.class);
   assertThat(ex.getMessage())
      .contains("No Address found with id " + nonExistId);
}
// then
verify(repository, times(1))
   .findById(UUID.fromString(nonExistId));
verify(repository, times(0))
   .deleteById(UUID.fromString(nonExistId));
}
```

https://github.com/PacktPublishing/Modern-API-Development-with-Spring-and-Spring-Boot/blob/main/Chapter08/src/test/java/com/packt/modern/api/service/AddressServiceTest.java

Here, you basically catch the exception and perform assertions on it.

With that, you have explored the unit tests that you can perform for both controllers and services. You can make use of these examples and write unit tests for the rest of the classes.

Executing unit tests

You can run the `gradlew clean test` command to execute our tests. This will generate the unit test reports at `Chapter08/build/reports/tests/test/index.html`.

A generated test report will look like this:

Test Summary

16	0	0	4.514s	100%
tests	failures	ignored	duration	successful

Packages   Classes

Package	Tests	Failures	Ignored	Duration	Success rate
com.packt.modern.api	1	0	0	0 065s	100%
com.packt.modern.api.controller	8	0	0	4 258s	100%
com.packt.modern.api.service	7	0	0	0 191s	100%

Figure 8.1 – Unit tests report

You can click on the links to drill down further. If the test fails, it also shows the cause of the error.

Let's move on to the next section to learn how to configure code coverage for unit tests.

Code coverage

Code coverage provides important metrics, including line and branch coverage. You are going to use the **JaCoCo** tool to perform and report your code coverage.

First, you need to add the `jacoco` Gradle plugin to the `build.gradle` file, as shown in the following code:

```
plugins {
    id 'org.springframework.boot' version '2.4.3'
    id 'io.spring.dependency-management' version
        '1.0.10.RELEASE'
    id 'java'
    id 'org.hidetake.swagger.generator' version '2.18.2'
    id 'jacoco'
}
```

https://github.com/PacktPublishing/Modern-API-Development-with-Spring-and-Spring-Boot/blob/main/Chapter08/build.gradle

Next, configure the `jacoco` plugin by providing its version and reports directory:

```
jacoco {
    toolVersion = "0.8.6"
    reportsDir = file("$buildDir/jacoco")
}
```

Next, create a new task called `jacocoTestReport` that depends on the `test` task. You don't want to calculate coverage for auto-generated code, so add the `exclude` block. Exclusion can be added by configuring `afterEvaluate`, as shown in the following code block:

```
jacocoTestReport {
    dependsOn test // tests should be run before generating
                   // report
    afterEvaluate {
```

```
classDirectories.setFrom(
    files(classDirectories.files.collect {
    fileTree(
        dir: it,
        exclude: [
            'com/packt/modern/api/model/*',
            'com/packt/modern/api/*Api.*',
            'com/packt/modern/api/security/UNUSED/*',
        ])
    }))
    }
}
```

Next, you need to configure `jacocoTestCoverageVerification`, which defines the violation rules. We have added instructions to cover the ratio rule in the following code block. This will set the expected ratio to a minimum of 90%. If the ratio is below 0.9, then it will fail the build. You can find out more about such rules at `https://docs.gradle.org/current/userguide/jacoco_plugin.html#sec:jacoco_report_violation_rules`:

```
jacocoTestCoverageVerification {
    violationRules {
        rule {
            limit {
                minimum = 0.9
            }
        }
    }
}
```

Next, add `finalizedBy(jacocoTestReport)` to the test task, which ensures that the `jacocoTestReport` task will execute after performing the tests:

```
test {
    jvmArgs '--enable-preview'
    useJUnitPlatform()
    finalizedBy(jacocoTestReport)
}
```

Once you've run `gradlew clean build`, it will not only run the test but also generate the code coverage report, along with the test reports. The code coverage report will be available at `Chapter08/build/jacoco/test/html` and look as follows:

Chapter08

Element	Missed Instructions	Cov.	Missed Branches	Cov.	Missed	Cxty	Missed	Lines	Missed	Methods	Missed	Classes
com.packt.modern.api.service		15%		0%	77	93	148	194	56	72	1	11
com.packt.modern.api.entity		23%		0%	149	200	218	302	130	181	1	15
com.packt.modern.api.exception		29%		33%	37	60	172	239	34	57	4	9
com.packt.modern.api.hateoas		25%		3%	36	48	83	126	21	33	0	8
com.packt.modern.api.controller		37%		0%	27	43	41	89	26	42	0	9
com.packt.modern.api.repository		10%		0%	6	7	28	36	4	5	0	1
com.packt.modern.api.security		78%		50%	4	17	22	88	2	15	1	3
com.packt.modern.api		59%		n/a	3	8	11	20	3	8	1	3
Total	3,313 of 4,674	29%	121 of 126	3%	339	476	723	1,094	276	413	8	59

Figure 8.2 – Code coverage report

Here, you can see that our instruction coverage is only at 29%, while our branch coverage is only at 3%. You can add more tests and increase these percentages.

You will learn about integration testing in the next section.

Integration testing

Once you have the automated integration tests in place, you can ensure that any changes you make won't produce bugs, provided you cover all the testing scenarios. You don't have to add any additional plugins or libraries to support integration testing in this chapter. The Spring test library provides all the libraries required to write and perform integration testing.

Let's add the configuration for integration testing in next subsection.

Configuring the Integration testing

First, you need a separate location for your integration tests. This can be configured in `build.gradle`, as shown in the following code block:

```
sourceSets {
    integrationTest {
        java {
            compileClasspath += main.output + test.output
            runtimeClasspath += main.output + test.output
            srcDir file('src/integration/java')
        }
        resources.srcDir file('src/integration/resources')
```

```
    }
}
```

Next, you can configure the integration test's implementation and runtime so that that it's extended from the test's implementation and runtime, as shown in the following code block:

```
configurations {
    integrationTestImplementation.extendsFrom
        testImplementation
    integrationTestRuntime.extendsFrom testRuntime
}
```

Finally, create a task called `integrationTest` that will not only use the JUnit Platform, but also use our `classpath` and test `classpath` from `sourceSets. integrationTest`.

Finally, configure the `check` task so that it depends on the `integrationTest` task and run `integrationTest` after the `test` task. You can remove the last line in the following code block if you want to run `integrationTest` separately:

```
task integrationTest(type: Test) {
    useJUnitPlatform()
    description = 'Runs the integration tests.'
    group = 'verification'
    testClassesDirs = sourceSets.integrationTest.
        output.classesDirs
    classpath = sourceSets.integrationTest.runtimeClasspath
}

check.dependsOn integrationTest
integrationTest.mustRunAfter test
```

Now, we can start writing the integration tests. Before writing integration tests, first let's write the supporting Java classes in next subsection. First, let's create the `TestUtils` class. This will contain a method that returns an instance of `ObjectMapper`. It will contain a method to check whether JWT has expired.

Writing supporting classes for Integration test

The `ObjectMapper` instance was retrieved from the `AppConfig` class and added an extra configuration so that we can accept a single value as an array. For example, a JSON string field value might be { [{...}, {...}] }. If you take a closer look at it, you will see that it is an array wrapped as a single value. When you convert this value into an object, `ObjectMapper` treats it as an array. The complete code for this class is as follows:

```java
public class TestUtils {

  private static ObjectMapper objectMapper;
  public static ObjectMapper objectMapper() {
    if (Objects.isNull(objectMapper)) {
      objectMapper = new AppConfig().objectMapper();
      objectMapper.configure(
        DeserializationFeature.ACCEPT_SINGLE_VALUE_AS_ARRAY,
          true);
    }
    return objectMapper;
  }
  public static boolean isTokenExpired(String jwt)
                             throws JsonProcessingException {
    var encodedPayload = jwt.split("\\.")[1];
    var payload = new String(Base64.getDecoder()
                              .decode(encodedPayload));
    JsonNode parent = new ObjectMapper().readTree(payload);
    String expiration = parent.path("exp").asText();
    Instant expTime = Instant.ofEpochMilli(
                        Long.valueOf(expiration) * 1000);
    return Instant.now().compareTo(expTime) < 0;
  }
}
```

https://github.com/PacktPublishing/Modern-API-Development-with-Spring-and-Spring-Boot/blob/main/Chapter08/src/integration/java/com/packt/modern/api/TestUtils.java

Next, you need a client that lets you log in so that you can retrieve the JWT. RestTemplate is an HTTP client in Spring that provides support for making HTTP calls. The AuthClient class makes use of TestRestTemplate, which is a replica of RestTemplate from a testing perspective.

Let's write this AuthClient, as follows:

```
public class AuthClient {

  private TestRestTemplate restTemplate;
  private ObjectMapper objectMapper;
  public AuthClient(TestRestTemplate restTemplate,
                             ObjectMapper objectMapper) {
    this.restTemplate = restTemplate;
    this.objectMapper = objectMapper;
  }
  public SignedInUser login(String username, String
        password) {
    SignInReq signInReq = new SignInReq()
                 .username(username).password(password);
    return restTemplate
      .execute("/api/v1/auth/token", HttpMethod.POST,
        request -> {
            objectMapper.writeValue(request.getBody(),
                                    signInReq);
            request.getHeaders()
                .add(HttpHeaders.CONTENT_TYPE,
                    MediaType.APPLICATION_JSON_VALUE);
            request.getHeaders().add(HttpHeaders.ACCEPT,
                    MediaType.APPLICATION_JSON_VALUE);
        },
        response -> objectMapper.readValue(
                        response.getBody(),
                        SignedInUser.class));
  }
}
```

https://github.com/PacktPublishing/Modern-API-Development-with-Spring-and-Spring-Boot/blob/main/Chapter08/src/integration/java/com/packt/modern/api/AuthClient.java

The Sprint test library provides `MockMvc`, `WebTestClient`, and `TestRestTemplate` for performing integration testing. You have already used `MockMvc` in unit testing. The same approach can be used for integration testing as well. However, instead of using mocks, you can use the actual objects by adding the `@SpringBootTest` annotation to the test class. `@SpringBootTest`, along with `SpringExtension`, provides all the necessary Spring context, such as the actual application.

`WebTestClient` is used to test the reactive applications. However, to test REST services, you must use `TestRestTemplate`, which is a replica of `RestTemplate`.

The integration test you are going to write is fully fleshed out test that doesn't contain any mocks. It will use flyway scripts, similar to the actual application, which we added to `src/integration/resources/db/migration`. The integration test will also have its own `application.properties` located in `src/integration/resources`.

Therefore, the integration test will be as good as long as you are hitting the REST endpoints from REST clients such as cURL or Postman. These flyway scripts create the tables and data required in the H2 memory database. This data will then be used by the RESTful web service. You can also use other databases, such as Postgres or MySQL, using their test containers.

Let's create a new integration test called `AddressControllerIT` in `src/integration/java` in an appropriate package and add the following code:

```
@ExtendWith(SpringExtension.class)
@SpringBootTest(webEnvironment = WebEnvironment.RANDOM_PORT)
@TestMethodOrder(OrderAnnotation.class)
public class AddressControllerIT {

    private static ObjectMapper objectMapper;
    private static AuthClient authClient;
    private static SignedInUser signedInUser;
    private static Address address;
    private static String idOfAddressToBeRemoved;
    @Autowired
    private AddressRepository repository;
    @Autowired
```

```java
private TestRestTemplate restTemplate;

@BeforeAll
public static void init() {
    objectMapper = TestUtils.objectMapper();
    address = new Address().id(
            "a731fda1-aaad-42ea-bdbc-
                a27eeebe2cc0").number("9I-999")
        .residency("Fraser Suites Le Claridge")
        .street("Champs-Elysees").city("Paris")
        .state("Île-de-France").country("France")
            .pincode("75008");
}


@BeforeEach
public void setup() throws JsonProcessingException {
    if (Objects.isNull(signedInUser)
        || Strings.isNullOrEmpty(
            signedInUser.getAccessToken())
        || isTokenExpired(signedInUser.getAccessToken())) {
        authClient = new AuthClient(restTemplate,
                    objectMapper);
        signedInUser = authClient.login("scott", "tiger");
    }
}
```

https://github.com/PacktPublishing/Modern-API-Development-with-Spring-and-Spring-Boot/blob/main/Chapter08/src/integration/java/com/packt/modern/api/controller/AddressControllerIT.java

Here, `SpringExtension` is now being used to run the unit test on the JUnit Platform. The `SpringBootTest` annotation provides all the dependencies and context for the test class. A random port is being used to run the test server. You are also using @ `TestMethodOrder`, along with the `@Order` annotation, to run the test in a particular order. You are going to execute the test in a particular order so that the `POST HTTP` method on the `addresses` resource is only called before the `DELETE HTTP` method on the `addresses` resource. This is because you are passing the newly created address ID in the `DELETE` call. Normally, tests run in a random order. If the `DELETE` call is made before the `POST` call, then the build will fail, without testing the proper scenarios.

The static `init()` method is annotated with `@BeforeAll` and will be run before all the tests. You are setting up `objectMapper` and the `address` model in this method.

The method's setup would be run before each test is executed because it is marked with the `@BeforeEach` annotation. Here, you are making sure that the login call will only be made if `signedInUser` is `null` or the token has expired.

Let's add an integration test that will verify the `GET /api/v1/addresses` REST endpoint, as shown in the following code:

```
@Test
@DisplayName("returns all addresses")
@Order(6)
public void getAllAddress() throws IOException {
  // given
  MultiValueMap<String, String> headers = new
                              LinkedMultiValueMap<>();
  headers.add(HttpHeaders.CONTENT_TYPE,
      MediaType.APPLICATION_JSON_VALUE);
  headers.add(HttpHeaders.ACCEPT,
      MediaType.APPLICATION_JSON_VALUE);
  headers.add("Authorization", "Bearer " +
            signedInUser.getAccessToken());
  // when
  ResponseEntity<JsonNode> addressResponseEntity =
      restTemplate
      .exchange("/api/v1/addresses", HttpMethod.GET,
          new HttpEntity<>(headers), JsonNode.class);


  // then
  assertThat(addressResponseEntity.getStatusCode())
```

```
        .isEqualTo(HttpStatus.OK);
    JsonNode node = addressResponseEntity.getBody();
    List<Address> addressFromResponse = objectMapper
        .convertValue(node, new
            TypeReference<ArrayList<Address>>(){});
    assertThat(addressFromResponse).hasSizeGreaterThan(0);
    assertThat(addressFromResponse.get(0))
        .hasFieldOrProperty("links");
    assertThat(addressFromResponse.get(0))
        .isInstanceOf(Address.class);
}
```

https://github.com/PacktPublishing/Modern-API-Development-with-Spring-and-Spring-Boot/blob/main/Chapter08/src/integration/java/com/packt/modern/api/controller/AddressControllerIT.java

First, you must set the headers in the given section. Here, you are using the `signedInUser` instance to set the bearer token. Next, you must call the exchange method of `TestRestTemplate`, which takes four arguments – the URI, the HTTP method, `HttpEntity` (that contains the headers and payload if required), and the type of the returned value. You can also use the fifth argument if the template is being used to set `urlVariables`, which expands the template.

Then, you must use the assertions to perform the verification process. Here, you can see that it replicates the actual calls.

Once the tests are run either by running `gradlew clean integrationTest` or `gradlew clean build`, you can find the test report at `Chapter08/build/reports/tests/integrationTest`. The test report should look like this:

Test Summary

12	0	0	8.798s	100%
tests	failures	ignored	duration	successful

Packages   Classes

Package	Tests	Failures	Ignored	Duration	Success rate
com.packt.modern.api.controller	12	0	0	8.798s	100%

Figure 8.3 – Integration test report

You can find all the test address resources in `AddressControllerIT.java`, which contains tests for errors, authentication and authorization, and the create, read, and delete operations.

Similarly, you can add integration tests for other REST resources.

Summary

In this chapter, you explored both manual and automated testing. You learned how to write unit and integration tests using JUnit, the Spring test libraries, AssertJ, and Hamcrest. You also learned how to use the Gherkin *Given > When > Then* language to make tests more readable. You then learned how to separate the unit and integration tests.

Finally, you learned about various test automation skills by automating our unit and integration tests. This would help you to automate your test and catch the bugs and gaps before you deliver the code to quality analysis or customers.

In the next chapter, you will learn how to containerize an application and deploy it in Kubernetes.

Questions

1. What is the difference between unit and integration testing?

2. What is the advantage of having separate unit and integration tests?

3. What is the difference between mocking and spying on an object?

Further reading

- JUnit: `https://junit.org/`

- AssertJ: `https://assertj.github.io/doc/`

- Hamcrest: `http://hamcrest.org/`

- Mockito: `https://site.mockito.org/`

- *API Testing with Postman:* `https://www.packtpub.com/product/api-testing-with-postman-video/9781789616569`

9
Deployment of Web Services

In this chapter, you will learn about the fundamentals of containerization, Docker, and Kubernetes. You will then use these concepts to containerize a sample e-commerce app using Docker. This container will then be deployed as a Kubernetes cluster. You are going to use minikube for Kubernetes, which makes learning and Kubernetes-based development easier.

You'll explore the following topics in this chapter:

- Exploring the fundamentals of containerization
- Building a Docker image
- Deploying an application in Kubernetes

After completing this chapter, you will be able to perform containerization and container deployment in a Kubernetes cluster.

Technical requirements

You will need the following for developing and executing the code in this chapter:

- Docker

- Kubernetes (Minikube)

- Any Java IDE, such as NetBeans, IntelliJ, or Eclipse

- **Java Development Kit (JDK)** 15+

- An internet connection to clone the code (`https://github.com/PacktPublishing/Modern-API-Development-with-Spring-and-Spring-Boot/tree/main/Chapter09`) and download the dependencies and Gradle

- Postman/cURL (for API testing)

So, let's begin!

Exploring the fundamentals of containerization

One problem that's encountered frequently by teams while developing large, complex systems is that the code that works on one machine doesn't work on another. One developer might say it works on their machine, but **Quality Assurance (QA)** claims that it is failing on "their machine" with animated face. The main reason behind these kinds of scenarios is a mismatch of dependencies (such as different versions of Java, a certain web server, or OS), configurations, or files.

Also, setting up a new environment for deploying new products sometimes takes a day or more. This is unacceptable in today's environment and slows down your development turnaround. These kinds of issues can be solved by containerizing the application.

In containerization, an application is bundled, configured, and wrapped with all the required dependencies and files. This bundle can then be run on any machine that support the containerization process. It provides the exact same behavior on all environments. It not only solves bugs related to misconfigurations or dependencies, but also reduces the deployment time to a few minutes or less.

This bundle, which sits on top of a physical machine and its operating system, is called a container. This container shares the kernel, as well as the libraries and binaries of its host operating system, in read-only mode. Therefore, these are lightweight. In this chapter, you are going to use Docker and Kubernetes for containerization and container deployment.

A related concept is virtualization – "the process of creating a virtual environment using the existing hardware system by splitting it into different parts. Each part acts as a separate, distinct, individual system." These distinct individual systems are called **virtual machines** (**VMs**). Each VM runs on its own unique operating system with its own binaries, libraries, and apps. VMs are heavy weighted and can be many gigabytes in size. A hardware system can have VMs with different operating systems such Unix, Linux, and Windows. The following diagram depicts the difference between virtual machines and containers:

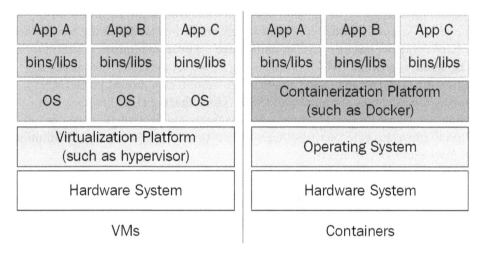

Figure 9.1 – Virtual machines versus containers

Sometimes, people think that virtualization and containerization are the same thing. But they are not. VMs are created on top of the host system, which shares its hardware, whereas containers are executed as isolated processes on top of the hardware and its OS. Containers are lightweight and are only a few MBs, sometimes a GB, whereas VMs are heavyweight and are many GBs in size. Containers run faster than VMs, and they are also more portable.

We'll explore containers in more detail by building a Docker image in the next section.

Building a Docker image

At this point, you know the benefit of containerization and why it is becoming popular: you create an application, product, or service, bundle it using containerization, and give it to the QA team, customer, or DevOps team to run it without any issues.

In this section, you'll learn how to use Docker as a containerization platform. Let's learn about it before creating a Docker image of a sample ecommerce app.

Exploring Docker

Docker is a leading container platform and an open source project. Docker was launched in 2013. 10,000 developers tried it after its interactive tutorial was launched in August 2013. Docker was downloaded 2.75 million times by the time of its 1.0 release in June 2013. Many large corporations have signed a partnership agreement with Docker Inc., including Microsoft, Red Hat, HP, and OpenStack, as well as service providers such as AWS, IBM, and Google.

Docker makes use of Linux kernel features to ensure resource isolation and the packaging of the application, along with its dependencies, such as cgroups and namespaces. Everything in a Docker container executes natively on the host and uses the host kernel directly. Each container has its own user namespace – a **process identifier** (**PID**) for process isolation, a **network** (**NET**) for managing network interfaces, **inter-process communication** (**IPC**) for managing access to IPC resources, a **mount point** (**MNT**) for managing filesystem mount points, and **Unix Time Sharing** (**UTS**) namespaces for isolating kernel and version identifiers. This packaging of dependencies enables an application to run as expected across different Linux operating systems and distributions by supporting a level of portability.

Furthermore, this portability allows developers to develop an application in any language and then easily deploy it from any computer, such as a laptop, to different environments, such as test, stage, or production. Docker runs natively on Linux. However, you can also run Docker on Windows and macOS.

Containers are comprised of just the application and its dependencies, including the basic OS. This makes the application lightweight and efficient in terms of resource utilization. Developers and system administrators are interested in a container's portability and efficient resource utilization.

We'll explore Docker's architecture in the next subsection.

Docker's architecture

As specified in the Docker documentation, Docker uses a client-server architecture. The Docker client (**Docker**) is basically a **command-line interface** (**CLI**) that is used by an end user; clients communicate back and forth with the Docker server (read as a Docker daemon). The Docker daemon does the heavy lifting in that it builds, runs, and distributes your Docker containers. The Docker client and the daemon can run on the same system or on different machines.

The Docker client and daemon communicate via sockets or through a RESTful API. Docker registers are public or private Docker image repositories that you can upload or download images from; for example, Docker Hub (hub.docker.com) is a public Docker registry.

The primary components of Docker are as follows:

- **Docker image**: A Docker image is a read-only template. For example, an image could contain an Ubuntu OS with an Apache web server and your web application installed on it. Docker images are build components of Docker, and images are used to create Docker containers. Docker provides a simple way to build new images or update existing images. You can also use images created by others and/or extend them.

- **Docker container**: A Docker container is created from a Docker image. Docker works so that the container can only see its own processes and has its own filesystem layered on a host filesystem and a networking stack, which pipes to the host-networking stack. Docker containers can be run, started, stopped, moved, or deleted. Docker also provides commands such as `docker stats` and `docker events` for container usage statics, such as CPU and memory usage, and for activities that are performed by the Docker daemons, respectively. These commands help you monitor Docker in a deployed environment.

You also need to be aware of Docker's container life cycle, which is as follows:

1. **Creates a container**: Docker creates a container from the Docker image using the `docker create` command.

2. **Runs the container**: Docker runs the container that was created in *step 1* using the `docker run` command.

3. **Pauses the container** (**optional**): Docker pauses the process running inside the container using the `docker pause` command.

4. **Un pauses the container** (**optional**): Docker un pauses the processes running inside the container using the `docker unpause` command.

5. **Starts the container**: Docker starts the container using the `docker start` command.

6. **Stops the container**: Docker stops the container and processes running inside the container using the `docker stop` command.

7. **Restarts the container**: Docker restarts the container and processes running inside it using the `docker restart` command.

8. **Kills the container**: Docker kills the running container using the `docker kill` command.

9. **Destroys the container**: Docker removes the stopped containers using the `docker rm` command. Therefore, this should only be performed for stopped containers.

At this point, you might be eager to use the Docker container life cycle, but first, you'll need to install Docker by going to `https://docs.docker.com/get-docker/`.

Once you've installed Docker, go to `https://docs.docker.com/get-started/#start-the-tutorial` to execute the first Docker command. You can refer to `https://docs.docker.com/engine/reference/commandline/docker/` to learn more about Docker commands.

For more information, you can look at the overview of Docker that is provided by Docker (`https://docs.docker.com/get-started/overview/`).

Let's make the necessary code changes so that we can create a Docker image for a sample ecommerce app.

Configuring code to build an image

I advise you to go back to *Chapter 8, Testing APIs*, as a base for this chapter's code. You don't need any additional libraries to create a Docker image. However, you do need to add the Spring Boot Actuator dependency, which provides production-ready features for the sample ecommerce app we'll be creating.

We'll add the Actuator dependency and configure it in the next subsection.

Adding Actuator

Actuator's features help you to monitor and manage applications using HTTP REST APIs and **Java Management Extensions** (**JMX**). These endpoints can be found in their respective documentation (`https://docs.spring.io/spring-boot/docs/current/reference/htmlsingle/#production-ready-endpoints`). In this chapter, however, we are only going to use the `/actuator/health` endpoint, which tells us about the application's health status.

You can add Actuator by performing the following steps:

1. Add the Actuator dependency to `build.gradle` (https://github.com/
   PacktPublishing/Modern-API-Development-with-Spring-and-
   Spring-Boot/blob/main/Chapter09/build.gradle):

   ```
   runtimeOnly 'org.springframework.boot:spring-boot-
       starter-actuator'
   ```

2. Next, you need to remove all security from the `/actuator` endpoints. Let's add a
   constant to `Constants.java` (https://github.com/PacktPublishing/
   Modern-API-Development-with-Spring-and-Spring-Boot/
   blob/main/Chapter09/src/main/java/com/packt/modern/api/
   security/Constants.java) for the Actuator URL, as shown here:

   ```
   public static final String ACTUATOR_URL_PREFIX =
       "/actuator/**";
   ```

3. Now, you can update the security configuration in `SecurityConfig.java`, as
   shown here:

   ```
   // rest of the code
   .antMatchers(H2_URL_PREFIX).permitAll()
   .antMatchers(ACTUATOR_URL_PREFIX).permitAll()
   .mvcMatchers(HttpMethod.POST,
       "/api/v1/addresses/**")
   // rest of the code
   ```

**https://github.com/PacktPublishing/Modern-API-Development-with-Spring-and-
Spring-Boot/blob/main/Chapter09/src/main/java/com/packt/modern/api/security/
SecurityConfig.java**

With that, you have added an `antMatcher` with Actuator endpoints. This allows all
Actuator endpoints, both with and without authentication and authorization.

Now, you can configure the Spring Boot plugin's task, called `bootBuildImage`, to
customize the name of the Docker image. We will do this in the next subsection.

Configuring the Spring Boot plugin task

The Spring Boot Gradle plugin already provides a command (`bootBuildImage`) for building Docker images. It becomes available when the `java` plugin is applied in the `plugins` section. However, it is not available when you build the `.war` file. Therefore, you don't need to add any additional plugins as you are going to build a `.jar` file. However, there are certain plugins that are available that you can use if you want.

You can customize the image's name by adding the following code block to the `build.gradle` file:

```
bootBuildImage {
    imageName = "192.168.80.1:5000/${
        project.name}:${project.version}"
}
```

Here, change the IP address and port of the local Docker registry. A Docker image will be built based on your project's name and version. The project version is already defined in the `build.gradle` file's top section. The project name, on the other hand, is picked from the `settings.gradle` file (https://github.com/PacktPublishing/ Modern-API-Development-with-Spring-and-Spring-Boot/blob/main/ Chapter09/settings.gradle). Let's rename it, as shown in the following code snippet:

```
rootProject.name = 'packt-modern-api-development-chapter09'
```

In *Chapter 8, Testing APIs*, the value of `rootProject.name` contains a capital letter, so the Docker image build failed. This is because the plugin has a validation check for capital letters. Therefore, Docker image names should only be in lowercase.

For more information and customization options, please refer to the plugin documentation (https://docs.spring.io/spring-boot/docs/current/ gradle-plugin/reference/htmlsingle/#build-image).

Now that you have configured the code, you can use this to build an image after configure the Docker registry. You will do this in the next subsection.

Configuring the Docker registry

By default, when you build an image (`gradlew bootBuildImage`), it will build an image called `docker.io/library/packt-modern-api-development-chapter09:0.0.1-SNAPSHOT`. You may be wondering why, even though you have given only `name:version`, how come it is prefixing it with `docker.io/library/`. This is because if you don't specify the Docker registry, it takes the `docker.io` registry by default. You need a Docker registry where you can pull and push images from. It is similar to an artifact repository, where you push and pull artifacts such as Spring libraries.

Once the image has been built, you can push it to Docker Hub with by applying your Docker Hub login credentials. Then, you can fetch the image from Docker Hub for deployment in your Kubernetes environment. However, for development purposes, this is not an ideal scenario. The best option is to configure the local Docker registry and then use it for Kubernetes deployment.

> **Using Git Bash on Windows**
>
> You can use Git Bash on Windows to run these commands, which emulates Linux commands.

Let's execute the following commands to check if Docker is up and running:

```
$ docker version
Client:
  Version:           18.06.1-ce
  API version:       1.38
  Go version:        go1.10.3
  Git commit:        e68fc7a
  Built:             Tue Aug 21 17:21:34 2018
  OS/Arch:           windows/amd64
  Experimental:      false
Server: Docker Engine - Community
  Engine:
  Version:              20.10.5
// truncated output for brevity
```

Docker is now up and running. Now, we can pull and start the local Docker registry by using the following command:

```
$ docker run -d -p 5000:5000 -e REGISTRY_STORAGE_DELETE_
ENABLED=true --restart=always --name registry registry:2
```

This command downloads the `register:2` image and then executes the container that was created by it, called `registry`, on port `5000`. There are two port entries: one internal container port and another exposed external port. Both are set to `5000`. The `--restart=always` flag tells Docker to start the `registry` container every time Docker is restarted. The `REGISTRY_STORAGE_DELETE_ENABLED` flag, as its name suggests, it used to remove any images from registry as it is set to true. The default value of this flag is `false`.

Now, let's check the Docker containers:

```
$ docker ps
CONTAINER ID   IMAGE      COMMAND                         CREATED
STATUS         PORTS                     NAMES
10cd2f36af1e   registry:2   "/entrypoint.sh /etc…"   6 days ago
Up 5 seconds   0.0.0.0:5000->5000/tcp   registry
```

This shows that the Docker container registry is up and running and was created using the `registry:2` image.

The host name is import when you're using the containers. Therefore, we'll use the IP number instead of the local hostname for the registry host. This is because the container will refer to its localhost when you use `localhost`, rather than the localhost of your system. In a Kubernetes environment, you need to provide a registry host, so you need to use the IP or proper hostname in place of localhost.

Let's find out what IP we can use by running the following command:

```
$ ipconfig
Windows IP Configuration


Ethernet adapter Ethernet:
   Media State . . . . . . . . . . . : Media disconnected
   Connection-specific DNS Suffix  . :
Ethernet adapter vEthernet (Default Switch):
   Connection-specific DNS Suffix  . :
   Link-local IPv6 Address . . . . . :
ef80::2099:f848:8903:f996%81
   IPv4 Address. . . . . . . . . . . : 192.168.80.1
   Subnet Mask . . . . . . . . . . . : 255.255.240.0
   Default Gateway . . . . . . . . . :
```

You can find your system's IP address in the row highlighted in the preceding output. You can use similar commands on macOS and Linux to find out the IP address of your system.

You haven't configured the **Transport Layer Security** (**TLS**) for your system host, so this registry is an insecure registry. Docker only supports secure registries by default. You must configure Docker so that it can use insecure registries. Please refer to the Docker documentation to learn how to configure an insecure registry (`https://docs.docker.com/registry/insecure/#deploy-a-plain-http-registry`).

Please note that to build and publish the image successfully, the Docker configuration must be performed with a local registry, as explained previously.

> **Note**
> Don't use an insecure registry on any environment other than a local or development environment.

Now, let's create a Docker image for a sample ecommerce app.

Executing a Gradle task to build an image

You need to make a change to the `bootBuildImage` task so that the image's name contains the local Docker registry's prefix. Spring Boot `bootBuildImage` uses Paketo buildpacks to build the docker image. It supports **Long-term Support** (**LTS**) Java releases and only current non-LTS Java release. It means, for non-LTS if Java 16 is released then it would remove support of Java 15. Similarly, when Java 17 gets released, it would remove the Java 16 support. However, it won't remove the Java 17 support when Java 18 would release because Java 17 is LTS release. We can make this change like so:

```
bootBuildImage {
  imageName = "192.168.80.1:5000/${project.name}:${project.version}"
}
// Paketo removes 6 monthly Java release,
// therefore better to use Java 17, which is a LTS
environment = ["BP_JVM_VERSION" : "16"]
```

Here, you have customized the name of the Docker image according to local Docker registry. You should change the IP address and port as per your system and configuration. You have also used the environment property to set the Paketo buildpacks variables. You have set the JVM version to 16, as at the time of writing this chapter, Paketo buildpacks has removed the support of Java 15. It is recommend to use Java 17 (or any future LTS release), which should be release in September 2021. You can find all the supported Paketo buildpacks environment variables at `https://github.com/paketo-buildpacks/bellsoft-liberica#configuration`. Now, you can build the image by executing the following command from your project's home directory:

```
$ gradlew clean build
   build the jar file of app after running the tests


$ gradlew bootBuildImage
> Task :bootBuildImage
Building image '192.168.80.1:5000/packt-modern-api-development-
chapter09:0.0.1-SNAPSHOT'
  > Pulling builder image 'docker.io/paketobuildpacks/
builder:base' ..............................................:....
  ..
  > Pulled builder image 'paketobuildpacks/builder@sha256:e4
6e13f550df3b1fd694000e417d6bed534772716090f11a9876501ddeecb521'
  > Pulling run image 'docker.io/paketobuildpacks/run:base-cnb'
......................................................
  > Pulled run image 'paketobuildpacks/run@
sha256:367a43536f60c21190cea5c06d040d01d29f4102840d6b3e1dcd72
ed2eb71721'
  > Executing lifecycle version v0.10.2
  > Using build cache volume 'pack-cache-8020b69fd072.build'
  // continue...
```

The Spring Boot Gradle plugin uses the *Paketo BellSoft Liberica Buildpack* (`docker.io/paketobuildpacks`) to build an application image. First, it pulls the image from Docker Hub and then runs its container, as shown here:

```
> Running creator
    [creator]      ===> DETECTING
    [creator]      5 of 18 buildpacks participating
    [creator]      paketo-buildpacks/ca-certificates   2.1.0
    [creator]      paketo-buildpacks/bellsoft-liberica 7.0.1
```

```
[creator]         paketo-buildpacks/executable-jar    5.0.0
[creator]         paketo-buildpacks/dist-zip          4.0.0
[creator]         paketo-buildpacks/spring-boot       4.1.0
[creator]         ===> ANALYZING
[creator]         Previous image with name "192.168.80.1:5000/
                  packt-modern-api-development-chapter09:0.0.1-
                  SNAPSHOT" not found
[creator]         // Truncated the output for brevity
[creator]
[creator]         Paketo BellSoft Liberica Buildpack 7.0.1
[creator] https://github.com/paketo-buildpacks/bellsoft-
          liberica
[creator]         // Truncated the output for brevity
[creator] BellSoft Liberica JRE 15.0.2: Contributing to
          layer
[creator]             Downloading from https://github.com/bell-
                      sw/Liberica/releases/download/15.0.2+10/
                      bellsoft-jre15.0.2+10-linux-amd64.tar.gz
[creator]             Verifying checksum
[creator] Expanding to /layers/paketo-buildpacks_bellsoft-
          liberica/jre
[creator]             // Truncated the output for brevity
```

The Spring Boot plugin uses Bellsoft's JRE 15.0.2 with Linux as a base image for building images. It uses the finely grained filesystem layers inside the container to do so:

```
[creator]             Launch Helper: Contributing to layer
[creator]             Creating /layers/paketo-buildpacks_
                      bellsoft-liberica/helper/exec.d/active-
                      processor-count
[creator]             Creating /layers/paketo-buildpacks_
                      bellsoft-liberica/helper/exec.d/java-opts
[creator]             // Truncated the output for brevity
[creator]             JVMKill Agent 1.16.0: Contributing to layer
[creator]             Downloading from https://github.com/
                      cloudfoundry/jvmkill/releases/download/
                      v1.16.0.RELEASE/jvmkill-1.16.0-RELEASE.so
[creator]             Verifying checksum
[creator]         Paketo Executable JAR Buildpack 5.0.0
```

```
[creator]        https://github.com/paketo-buildpacks/
                 executable-jar
[creator]            Class Path: Contributing to layer
[creator]        // Truncated the output for brevity
```

It continues to add the layers and then the labels. At the end, it creates the Docker image:

```
[creator]        Paketo Spring Boot Buildpack 4.1.0
[creator]        // Truncated the output for brevity
[creator]        Adding layer 'paketo-buildpacks/executable-
                 jar:classpath'
[creator]        Adding layer 'paketo-buildpacks/spring-
                 boot:helper'
[creator]        Adding layer 'paketo-buildpacks/spring-
                 boot:spring-cloud-bindings'
[creator]        Adding layer 'paketo-buildpacks/spring-
                 boot:web-application-type'
[creator]        Adding 5/5 app layer(s)
[creator]        Adding layer 'launcher'
[creator]        // Truncated the output for brevity
[creator]        Adding label 'org.springframework.boot.
                 version'
[creator]        Setting default process type 'web'
[creator]        *** Images (672dd1e3fdb4):
[creator]            192.168.80.1:5000/packt-modern-api-
                     development-chapter09:0.0.1-SNAPSHOT
Successfully built image '192.168.80.1:5000/packt-modern-api-
development-chapter09:0.0.1-SNAPSHOT'
BUILD SUCCESSFUL in 58m 19s
```

You can learn more about Spring Boot, Docker, and Kubernetes and their configuration at `https://github.com/dsyer/kubernetes-intro`.

Now that the Docker image has been built, you can use this image to run the sample ecommerce app locally using the following command:

```
$ docker run -p 8,080:8080 192.168.80.1:5000/packt-modern-api-
development-chapter09:0.0.1-SNAPSHOT
```

This command will run the application on port 8080 inside the container. Because it has been exposed on port 8080, you can access the sample ecommerce app on 8080 outside the container too, once the app is up and running. You can test the application by running the following command in a separate Terminal tab/window once the application container is up and running:

```
$ curl localhost:8080/actuator/health
{"status":"UP"}
$ curl localhost:8080/actuator | jq .
{
   "_links": {
     "self": {
       "href": "http://localhost:8080/actuator",
       "templated": false },
     "health-path": {
       "href": "http://localhost:8080/actuator/
                health/{*path}",
       "templated": true   },
     "health": {
       "href": "http://localhost:8080/actuator/health",
       "templated": false },
     "info": {
       "href": "http://localhost:8080/actuator/info",
       "templated": false   }
   }
}
```

The curl localhost:8080/actuator command returns the available Actuator endpoints.

You can also list the containers and their statuses by using the following command:

```
$ docker ps
CONTAINER ID   IMAGE    COMMAND                   CREATED
STATUS                  PORTS                     NAMES
075ee30f733f   192.168.80.1:5000/packt-modern-api-development-
chapter09:0.0.1-SNAPSHOT    "/cnb/process/web"        3 minutes
ago        Up 3 minutes          0.0.0.0:8080->8080/tcp  sharp_
dijkstra
```

```
10cd2f36af1e  registry:2   "/entrypoint.sh /etc…"   6 days ago
Up 3 hours           0.0.0.0:5000->5000/tcp  registry
```

To find out what the available Docker images are, use the following command:

```
$ docker images
REPOSITORY                                                    TAG
IMAGE ID              CREATED              SIZE
paketobuildpacks/run
base-cnb              6281947a9e8d         7 days ago
87.7MB
registry                                                       2
5c4008a25e05          2 weeks ago          26.2MB
paketobuildpacks/builder
<none>                56c025ed0e91         41 years ago
665MB
paketobuildpacks/builder                                      base
a4ec710f3cd8          41 years ago         663MB
192.168.80.1:5000/packt-modern-api-development-chapter09
0.0.1-SNAPSHOT 672dd1e3fdb4 41 years ago      307MB
```

Now, you can tag and push the application image using the following commands:

```
$ docker tag 192.168.80.1:5000/packt-modern-api-development-
chapter09:0.0.1-SNAPSHOT 192.168.80.1:5000/packt-modern-api-
development-chapter09:0.0.1-SNAPSHOT
```

```
$ docker push 192.168.80.1:5000/packt-modern-api-development-
chapter09:0.0.1-SNAPSHOT
```

Similarly, you can also query the local Docker registry container. First, let's run the following command to find all the published images in the registry (the default value is 100):

```
$ curl -X GET http://192.168.80.1:5000/v2/_catalog
{"repositories":["packt-modern-api-development-chapter09"]}
```

Similarly, you can find out what all the available tags are for any specific image by using the following command:

```
>curl -X GET http://192.168.80.1:5000/v2/packt-modern-api-
development-chapter09/tags/list
```

```
{"name":"packt-modern-api-development-
chapter09","tags":["0.0.1-SNAPSHOT"]}
```

For these commands, you can also use localhost instead of the IP, if you are running a local registry container.

We'll deploy this image on Kubernetes in the next section.

Deploying an application in Kubernetes

Docker containers are run in isolation. You need a platform that can execute multiple Docker containers and manage or scale them. Docker Compose does this for us. However, this is where Kubernetes helps. It not only manages the container, but also helps you scale the deployed containers dynamically.

You are going to use Minikube to run Kubernetes locally. You can use it on Linux, macOS, and Windows. It runs a single-node Kubernetes cluster, which is used for learning or development purposes. You can install it by referring to the respective guide (`https://minikube.sigs.k8s.io/docs/start/`).

Once Minikube has been installed, you need to update the local insecure registry in its configuration since, by default, it uses Docker Hub. Adding an image to Docker Hub and then fetching it for local usage is cumbersome for development. You can add a local insecure registry to your Minikube environment by adding your host IP and local Docker registry port to Minikube's config at **HostOptions** > **EngineOptions** > **InsecureRegistry** in `~/.minikube/machines/minikube/config.json` (note: this file is only generated after Minikube has been started once; therefore, start Minikube before modifying `config.json`):

```
$ vi ~/.minikube/machines/minikube/config.json
        "HostOptions": {
35          "Driver": "",
36          "Memory": 0,
37          "Disk": 0,
38          "EngineOptions": {
39              "ArbitraryFlags": null,
40              "Dns": null,
41              "GraphDir": "",
42              "Env": [],
43              "Ipv6": false,
44              "InsecureRegistry": [
```

```
45                    "10.96.0.0/12",
46                    "192.168.80.1:5000"
47            ],
```

Once the insecure registry has been updated, you can start Minikube using the following command:

```
$ minikube start --insecure-registry="192.168.80.1:5000"
 minikube v1.18.1 on Microsoft Windows 10 Pro 10.0.19041 Build
19041
 Automatically selected the docker driver. Other choices:
virtualbox, none
 Starting control plane node minikube in cluster minikube
 Pulling base image ...
 Downloading Kubernetes v1.20.0 preload ...
    > preloaded-images-k8s-v9-v1....: 491.22 MiB / 491.22 MiB
100.00% 25.00 MiB
 Creating docker container (CPUs=2, Memory=4000MB) ...
  Preparing Kubernetes v1.20.2 on Docker 20.10.3 ...
     • Generating certificates and keys ...
     • Booting up control plane ...
     • Configuring RBAC rules ...
 Verifying Kubernetes components...
     • Using image gcr.io/k8s-minikube/storage-provisioner:v4
     • Using image kubernetesui/dashboard:v2.1.0
     • Using image kubernetesui/metrics-scraper:v1.0.4
 Enabled addons: storage-provisioner, dashboard, default-
storageclass
 Done! kubectl is now configured to use "minikube" cluster and
"default" namespace by default
```

Here, we have used the `--insecure-registry` flag while starting Minikube. This is important as it makes the insecure registry work. The Kubernetes cluster uses the `default` namespace by default.

A **namespace** is a Kubernetes special object that allows you to divide the Kubernetes cluster resources among users or projects. However, you can't have nested namespaces. Kubernetes resources can only belong to single namespaces.

You can check whether Kubernetes is working or not by executing the following command once Minikube is up and running:

```
$ kubectl get po -A
NAMESPACE    NAME                                    READY STATUS
RESTARTS    AGE
kube-system coredns-74ff55c5b-6nlbn                 1/1   Running  0
4m10s
kube-system etcd-minikube                           1/1   Running  0
4m32s
kube-system kube-apiserver-minikube                 1/1   Running  0
4m32s
kube-system kube-controller-manager-minikube 1/1   Running  1
4m44s
kube-system kube-proxy-21sz9                        1/1   Running  0
4m10s
kube-system kube-scheduler-minikube                 1/1   Running  0
4m31s
kube-system storage-provisioner                     1/1   Running  1
4m12s
```

The kubectl command is a command-line tool that's used to control a Kubernetes cluster, similar to the docker command for Docker. It is a Kubernetes client that uses Kubernetes REST APIs to perform various Kubernetes operations, such as deploying applications, viewing logs, and inspecting and managing cluster resources.

The get po and get pod parameters allow you to retrieve the pods from your Kubernetes cluster. The -A flag instructs kubectl to retrieve objects from across namespaces. Here, you can see that all the pods are from the kube-system namespace.

These pods are created by Kubernetes and are part of its internal system.

Minikube bundles the Kubernetes dashboard as a user interface for additional insight into your cluster's state. You can start it by using the following command:

```
$ minikube dashboard
  Verifying dashboard health ...
  Launching proxy ...
  Verifying proxy health ...
Opening http://127.0.0.1:12587/api/v1/namespaces/kubernetes-
dashboard/services/http:kubernetes-dashboard:/proxy/ in your
default browser...
```

Running the dashboard will look as follows and allows you to manage the Kubernetes cluster from the UI:

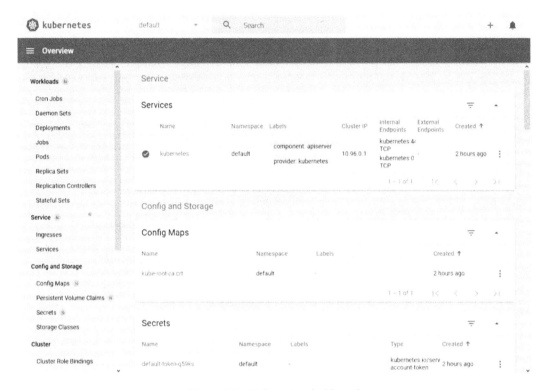

Figure 9.2 – Kubernetes dashboard

Kubernetes uses YAML configuration to create objects. For example, you need a deployment and service object to deploy and access the sample ecommerce application. Deployment will create a pod in the Kubernetes cluster that will run the application container, and the service will allow it to access it. You can create these YAML files either manually or generate them using `kubectl`. You should typically use `kubectl`, which generates the files for you. If you need to, you can modify the content of the file.

Let's create a new directory (`k8s`) in the project's home directory so that we can store the Kubernetes deployment configuration. We can generate the deployment Kubernetes configuration file by using the following commands from newly created `k8s` directory:

```
$ kubectl create deployment chapter09
--image=192.168.80.1:5000/packt-modern-api-development-
chapter09:0.0.1-SNAPSHOT --dry-run=client -o=yaml > deployment.
yaml
$ echo --- >> deployment.yaml
```

```
$ kubectl create service clusterip chapter09 --tcp=8080:8080
--dry-run=client -o=yaml >> deployment.yaml
```

The first command generates the deployment configuration in the `deployment.yaml` file using the `create deployment` command. A **Kubernetes Deployment** defines the scale at which you want to run your application. You can see that the replica is defined as 1. Therefore, Kubernetes will run a single replica of this deployment. Here, you pass the name (`chapter09`) of the deployment, the image name of the application to deploy, the `--dry-run=client` flag to preview the object that will be sent to the cluster, and the `-o=yaml` flag to generate the YAML output.

The second command appends `---` to the end of the `deployment.yaml` file.

Finally, the third command creates the service configuration in `deployment.yaml` with a value of `8080` for both internal and external ports.

Here, you have used the same file for both deployment and service objects. However, you can create two separate files for these – `deployment.yaml` and `service.yaml`. In this case, you need to apply these objects separately in your Kubernetes cluster.

Let's have a look at the content of the `deployment.yaml` file, which was generated by the previous code block:

```
apiVersion: apps/v1
kind: Deployment
metadata:
  creationTimestamp: null
  labels:
    app: chapter09
  name: chapter09
spec:
  replicas: 1
  selector:
    matchLabels:
      app: chapter09
  strategy: {}
  template:
    metadata:
      creationTimestamp: null
      labels:
        app: chapter09
```

```yaml
      spec:
        containers:
        - image: 192.168.80.1:5000/packt-modern-api-
          development-chapter09:0.0.1-SNAPSHOT
          name: packt-modern-api-development-chapter09
          resources: {}
status: {}
---
apiVersion: v1
kind: Service
metadata:
  creationTimestamp: null
  labels:
    app: chapter09
  name: chapter09
spec:
  ports:
  - name: 8080-8080
    port: 8080
    protocol: TCP
    targetPort: 8080
  selector:
    app: chapter09
  type: ClusterIP
status:
  loadBalancer: {}
```

https://github.com/PacktPublishing/Modern-API-Development-with-Spring-and-Spring-Boot/blob/main/Chapter09/k8s/deployment.yaml

Now, you can deploy the sample ecommerce application using the `deployment.yaml` file, as shown in the following code block:

```
$ kubectl apply -f deployment.yaml
deployment.apps/chapter09 created
service/chapter09 created
```

Alternatively, you can perform the following steps to publish the Docker image to Minikube. Start a new Terminal and execute the following commands (the same Terminal window should be used here since the `eval` commands are only valid in an active Terminal):

1. Execute `eval $(minikube docker-env)` to align the Minikube environment with your Docker configuration.

2. Execute `gradle bootBuildImage` to generate an image based on the Minikube environment.

3. Execute the following commands:

    i. `docker tag 192.168.80.1:5000/packt-modern-api-development-chapter09:0.0.1-SNAPSHOT 192.168.80.1:5000/packt-modern-api-development-chapter09:0.0.1-SNAPSHOT`

    ii. `docker push 172.26.208.1192.168.80.1:5000/library/packt-modern-api-development-chapter09:0.0.1-SNAPSHOT`

4. Execute `minikube stop` and `minikube start` to ensure that the new configuration is applied.

5. You can start the Minikube logs by using the following commands:

    i. `minikube -p minikube docker-env`

    ii. `eval $(minikube -p minikube docker-env)`

6. Afterward, deploying with the `kubectl apply -f deploymentTest.yaml` command should work.

This will initiate the application deployment of `chapter09`. You can then either use the Kubernetes dashboard or the `kubectl get all` command to check the status of your pod and service. **Pods** are Kubernetes' smallest and most deployable objects. They contain one or more containers and represent a single instance of a running process in a Kubernetes cluster. A pod's IP address and other configuration details may change because Kubernetes keep track of these and may replace them if a pod goes down. Therefore, **Kubernetes Service** adds an abstraction layer over the pods it exposes the IP addresses of and manages mapping to internal pods.

Let's run the following command to find out the status of the pod and service:

```
$ kubectl get all
NAME                           READY   STATUS    RESTARTS   AGE
pod/chapter09-7788955cf7-rqrn6   1/1    Running   0
```

```
73s
NAME                    TYPE        CLUSTER-IP        EXTERNAL-IP
PORT(S)      AGE
service/chapter09       ClusterIP   10.110.239.183    <none>
8080/TCP     75s
service/kubernetes      ClusterIP   10.96.0.1         <none>
443/TCP      167m
NAME                             READY    UP-TO-DATE   AVAILABLE
AGE
deployment.apps/chapter09        1/1      1            1
79s
NAME                                      DESIRED    CURRENT
READY      AGE
replicaset.apps/chapter09-7788955cf7      1          1          1
77s
```

This returns all the Kubernetes resources in the default namespace. Here, you can see that it returns a running pod, a service, a deployment resource, and a Replica Set for chapter09. You need to run this command multiple times until you find a successful or erroneous response (such as "image is not pullable").

You can't access the application running inside Kubernetes directly. You must either use some kind of proxy or SSH tunneling. Let's quickly create a SSH tunnel using the following command:

```
$ kubectl port-forward service/chapter09 8080:8080
Forwarding from 127.0.0.1:8080 -> 8080
Forwarding from [::1]:8080 -> 8080
```

The application is running on port 8080 internally. It is mapped to the local machine's port, which is 8080. Now, the application can be accessed on port 8080 outside the Kubernetes cluster.

We can access it by using the following command after opening a new Terminal window:

```
$ curl localhost:8080/actuator/health
{"status":"UP","groups":["liveness","readiness"]}
```

With that, the application has been successfully deployed on our Kubernetes cluster. Now, you can use the Postman collection and run all the available REST endpoints.

Summary

In this chapter, you learned about containerization and how it is different from virtualization. You also learned about the Docker containerization platform and how to use the Spring Boot plugin to generate a Docker image for a sample ecommerce app.

Then, you learned about the Docker registry and how to configure a local insecure registry so that you can use it to push and pull images.

You also learned about Kubernetes and its cluster operations by using Minikube. You configured it so that you can pull Docker images from insecure local Docker registries.

Now, you have the necessary skills to build a Docker image of a Spring Boot application and deploy it on a Kubernetes cluster.

In the next chapter, you'll learn about the fundamentals of the gRPC APIs.

Questions

1. What is the difference between virtualization and containerization?
2. What is Kubernetes used for?
3. What is kubectl?

Further reading

- *Kubernetes and Docker – an Enterprise Guide:*

  https://www.packtpub.com/product/kubernetes-and-docker-an-enterprise-guide/9781839213403

- Docker documentation:

  https://docs.docker.com/get-started/overview/

- Minikube documentation:

  https://minikube.sigs.k8s.io/docs/start/

Section 3: gRPC, Logging, and Monitoring

In this section, you will learn about gRPC-based API development. After this section, you will be able to differentiate between REST and reactive APIs with gRPC-based APIs. You will be able to build the server and client using the Protobuf schema. Finally, you will be able to facilitate distributed logging and tracing, and collect the logs as an Elasticsearch index that will be used for debugging and analysis on the Kibana app.

This section comprises the following chapters:

- *Chapter 10, gRPC Fundamentals*
- *Chapter 11, gRPC-based API Development and Testing*
- *Chapter 12, Logging and Tracing*

10
gRPC Fundamentals

gRPC is an open source framework for general-purpose **Remote Procedure Calls** (**RPCs**) across a network. RPC allows a remote procedure (hosted on a different machine) to call as if it is calling a local procedure in connected systems without coding the remote interaction details. RPC has a constant meaning in gRPC abbreviation. It seems logical that the *g* in gRPC would refer to *Google* because it was initially developed there. But the meaning of the *g* has changed with every release. For its first release, version 1.0, the *g* in gRPC stood for gRPC itself. That is, in version 1, it stands for **gRPC Remote Procedure Call**. You are going to use gRPC version 1.37, in which the **g** stands for **gilded**. Therefore, you can refer to gRPC as **gilded Remote Procedure Call** (for version 1.37). You can find out all the meanings of the *g* for different versions at `https://github.com/grpc/grpc/blob/master/doc/g_stands_for.md`.

In this chapter, you'll learn the fundamentals of gRPC such as gRPC architecture, gRPC service definitions, its lifecycle, gRPC server, and client. This chapter will provide you with a foundation that you can use to implement gRPC-based APIs. These fundamentals will help you to implement inter-service communication in a sample e-commerce app.

You will use gRPC-based APIs to develop a basic payment gateway for processing the payments in an e-commerce app in the next chapter.

> **Note**
> gRPC is pronounced as *Jee-Arr-Pee-See*.

You will explore the following topics in this chapter:

- Introduction and gRPC architecture
- Understanding service definitions
- Exploring the RPC life cycle
- Understanding the gRPC server and gRPC stub
- Handling errors

After completing this chapter, you will have learned the gRPC basics that will help you to implement a gRPC-based web service in the next chapter.

Technical requirements

This chapter contains the theory of gRPC. However, you need the following for the development and testing of gRPC-based web services:

- Any Java IDE, such as NetBeans, IntelliJ, or Eclipse
- **Java Development Kit (JDK)** 15
- An internet connection to clone the code and download the dependencies and Gradle
- Postman/cURL (for API testing)

So, let's begin!

Introduction and gRPC architecture

gRPC is an open source framework for general-purpose RPC across a network. gRPC supports full-duplex streaming and is also mostly aligned with HTTP/2 semantics. It supports different media formats, such as Protobuf (default), JSON, XML, and Thrift. The use of **Protocol Buffer** (**Protobuf**) aces the others because of higher performance.

gRPC brings the best of **REST** (**Representational State Transfer**) and RPC to the table and is well suited for distributed network communication through APIs. It offers some prolific features, as follows:

- It is designed for a highly scalable distributed system and offers *low latency*.
- It offers load balancing and failover.

- It can be integrated easily at the application layer for interaction with flow control because of its layered design.

- It supports cascade call cancellation.

- It offers wide communication—mobile app to server, web app to server, and any gRPC client app to the gRPC server app on different machines.

You're already well aware of REST and its implementation. Let's find out the differences between REST and gRPC in the next sub-section, which gives you a different perspective and allows you to choose between REST or gRPC based on requirements and use cases.

REST versus gRPC

gRPC is based on client-server architecture, whereas this is not true for REST.

Both gRPC and REST leverage the HTTP protocol. gRPC supports HTTP/2 specifications and full-duplex streaming communication in contrast to REST, which serves well for various scenarios such as voice or video calls.

You can pass payloads using query parameters, path parameters, and the request body in REST. This means that the request payload/data can be passed using different sources that lead to the parsing of the payload/data from different sources, which adds latency and complexity. On the other hand, gRPC performs better than REST in performance as it uses the static paths and single source of the request payload.

As you know, the REST response error depends on HTTP status codes, whereas gRPC has formalized the set of errors to make it well-aligned with APIs.

The REST API is more flexible in its implementation because it is purely dependent on HTTP. This gives you flexibility, but you need standards and conventions for strict verification and validation. But do you know why you need these strict verifications and validations? It is because you can implement an API in different ways. For example, you can delete a resource using any HTTP method instead of just using the HTTP DELETE method, and this simply sounds horrific.

On top of everything mentioned, gRPC is also built for supporting and handling call cancellations, load balancing, and failovers.

REST is mature and widely adopted, but gRPC brings its advantages. Therefore, you can choose between them based on their pros and cons. (Mind you, we haven't yet discussed GraphQL, which brings its own offerings. You will learn about GraphQL in *Chapter 13, Graph QL Fundamentals,* and *Chapter 14, Graph QL Development and Testing.*)

Let's find out whether we can use gRPC for web communication like REST in the next sub-section.

Can I call the gRPC server from web browsers and mobile apps?

Of course you can. The gRPC framework is designed for communication in distributed systems and is mostly aligned with HTTP/2 semantics. You can call a gRPC API from a mobile app, just like calling any local object. That's the beauty of gRPC! It supports inter-service communication across the intranet and internet and calls from the mobile app and web browser to the gRPC server. Therefore, you can utilize it for all kinds of communications.

gRPC for web (that is, `gRPC-web`) was quite new in 2018, but now (in 2021), it is getting more recognition and is especially being used for **Internet of Things (IoT)** applications. Ideally, you should adopt it first for your internal inter-service communications and then for web/mobile server communication.

Let's find out more about its architecture in the next sub-section.

gRPC architecture overview

gRPC is a general-purpose RPC-based framework. It works very well in the RPC style, which involves the following steps:

1. First of all, you define the service interface, which includes method signatures with their parameters and return types.

2. Then, you implement the defined service interface as a part of the gRPC server. You are now ready to serve the remote calls.

3. Next, you need the stub for clients that you can generate using the service interface. The client application calls the stub, which is a local call. In turn, the stub communicates with the gRPC server and the returned value is passed to the gRPC client. This is shown in the following diagram:

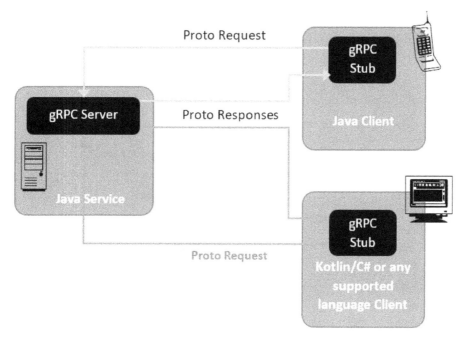

Figure 10.1 – gRPC client-server architecture

For client applications, it is just a local call to stub to get the response. You can have a server on either the same machine or a different machine. This makes it easier for writing distributed services. It is an ideal tool for writing microservices. gRPC is language-independent. You can write servers and clients in different languages. This provides a lot of flexibility for development.

gRPC is a layered architecture that has the following layers to make remote calling possible:

- **Stub**: You know the client calls the server through stubs. A stub is the topmost layer. Stubs are generated from the **Interface Definition Language (IDL)** file that contains service interfaces, methods, and messages. The IDL file will have a `.proto` extension if the interface is defined using Protobuf.

- **Channel**: The stub uses **Application Binary Interfaces (ABIs)** for communicating with the server. The channel is the middle layer that provides these ABIs. In general, the channel provides the connection to the server on a specific host and port. That's the reason the channel has a status such as connected or idle.

- **Transport**: This is the lowest layer and uses HTTP/2 as its protocol. Therefore, gRPC provides full-duplex communication and multiplex parallel calls over the same network connection.

You can develop a gRPC-based service by following these steps:

1. Define the service interface using the `.proto` file (Protobuf).

2. Write the implementation of the service interface defined in *step 1*.

3. Create a gRPC server and register the service with it.

4. Generate the service stub and use it with the gRPC client.

You'll implement the actual gRPC service in the next chapter, *Chapter 11, gRPC-based API Development and Testing*.

> **gRPC stub**
>
> A stub is an object that exposes service interfaces. The gRPC client calls the `stub` method, hooks the call to the server, and gets the response back.

You need to understand Protobuf for defining the service interfaces. Let's explore it in the next sub-section.

Protocol Buffer

Protobuf was created in 2001 and was publicly made available in 2008. It was also used by Google's microservice-based system, Stubby.

gRPC also works well with JSON and other media types. However, you'll define the service interfaces using Protobuf because it is known for its performance. It allows formal contracts, better bandwidth optimization, and code generation. Protobuf is also the default format for gRPC. gRPC makes use of Protobuf not only for data serialization but also for code generation. Protobuf serializes data and unlike JSON, YAML is not human-readable. Let's see how it is built.

Protobuf messages contain a series of key-value pairs. The key specifies the `message` field and its type. Let's examine the following `Employee` message:

```
message Employee {
    int64 id = 1;
    string firstName = 2;
}
```

Let's represent this message using Protobuf (with an id value of 299 and firstName value of Scott), as shown in the following diagram:

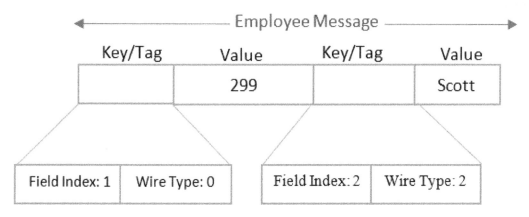

Figure 10.2 – Employee message representation using Protobuf

The Id and firstName fields are tagged with numbers sequenced 1 and 2, respectively, which is required for serialization. The wire type is another aspect that provides information to find the length of the value.

The following table contains the wire types and their respective meanings:

Wire Type	Meaning	Used for
0	Var int	int32, int64, uint32, uint64, sint32, sint64, bool, enum
1	64-bit	fixed64, sfixed64, double
2	Length-delimited	string, bytes, embedded messages, packed repeated fields
3	Start group	groups (deprecated)
4	End group	groups (deprecated)
5	32-bit	fixed32, sfixed32, float

A Protobuf file is created with the .proto extension. You define service interfaces in the form of method signatures and messages (objects) that are referred to in method signatures. These messages can be method parameters or returned types. You can compile a defined service interface with the protoc compiler, which generates the classes for interfaces and given messages. Similarly, you can also generate the stubs for the gRPC client.

Let's have a look at the following sample `.proto` file:

Sample Service Interface of Employee

```
syntax = "proto3";
package com.packtpub;
option java_package = «com.packt.modern.api.proto»;
option java_multiple_files = true;
message Employee {
  int64 id = 1;
  string firstName = 2;
  string lastName = 3;
  int64 deptId = 4;
  double salary = 5;
  message Address {
    string houseNo = 1;
    string street1 = 2;
    string street2 = 3;
    string city = 4;
    string state = 5;
    string country = 6;
    string pincode = 7;
  }
}
message EmployeeCreateResponse {
  int64 id = 1;
}
service EmployeeService {
  rpc Create(Employee) returns (EmployeeCreateResponse);
}
```

Let's understand this code line by line:

1. The first line represents the Protobuf version denoted by the `syntax` keyword. The value of `syntax` (`proto3`) tells the compiler that version 3 of Protobuf is used. The default version is `proto2`. Protobuf version 3 offers more features and simplified syntax and supports more languages. gRPC recommends using Protobuf version 3.

2. Next, you define the `proto` package name using the `package` keyword followed by the package name. It prevents name clashes among message types.

3. Next, you use the `option` keyword for defining the Java package name using the `java_package` parameter.

4. Then, you use the `option` keyword again for generating a separate file for each root-level message type using the `java_multiple_files` parameter.

5. Then, you define the messages using the `messages` keyword, which are nothing but objects. `message` and its fields are defined using the strong types that define the objects with exact specifications. You can define nested messages just like nested classes in Java. The last point contains the table of Protobuf types that you can use for defining the types of message fields.

6. You can use `Employee.Address` for defining the address field in other messages.

7. The tagging of fields marked with a sequence number is required because it is used for serialization and parsing the binary messages.

8. Please note that you cannot change the message structure once it is serialized.

9. Service definitions are defined using the `service` keyword. A service definition contains the methods. You can define methods using the `rpc` keyword. Please refer to the `EmployeeService` service definition for a reference. You'll explore more about service definitions in the next sub-section.

10. Protobuf has predefined types (scalar types). A message field can have one of the Protobuf scalar types. When we compile the `.proto` file, it converts the message field into its respective language type. The following table defines the mapping between Protobuf types and Java types:

Protobuf Types	Java Types	Remarks
double	double	Similar to Java double.
float	float	Similar to Java float.
int32	int	Use sint32 if the field contains negative values because it uses a variable-length encoding, which is inefficient for encoding negative numbers.
int64	long	Use sint64 if a field contains negative values because it uses a variable-length encoding, which is inefficient for encoding negative numbers.
uint32	int	It uses variable-length encoding. Use fixed32 if values are greater than 2^{28}.
uint64	long	It uses variable-length encoding. Use fixed64 if values are greater than 2^{56}.
sint32	int	More efficient for negative numbers encoding because it contains a signed int value. It uses variable-length encoding.
sint64	long	More efficient for negative numbers encoding because it contains a signed int value. It uses variable-length encoding.
fixed32	int	Always 4 bytes.
fixed64	long	Always 8 bytes.
sfixed32	int	Always 4 bytes. More efficient for encoding values that are greater than 2^{28}.
sfixed64	long	Always 8 bytes. More efficient for encoding values that are greater than 2^{56}.
bool	boolean	True or false.
string	String	Contains a UTF-8-encoded string or 7-bit ASCII text, which should not be longer than 2^{32}.
bytes	ByteString	Contains an arbitrary sequence of bytes, which should not be longer than 2^{32}.

Protobuf also allows you to define the enumeration types using the enum keyword and maps using the map<keytype, valuetype> keyword. Please refer to the following code for examples of enumeration and map types:

```
...omitted
message Employee {
  ..omitted
  enum Grade {
    I_GRADE = 1;
    II_GRADE = 2;
    III_GRADE = 3;
    IV_GRADE = 4;
  }
  map<string, int32> nominees = 1;
  ..omitted
}
```

The preceding sample code creates the Employee message, which has a Grade enumeration field with values such as I_GRADE. The nominees field is a map that has a key with type string and a value with type int32.

Let's explore service definitions further in the next section.

Understanding service definitions

You define a service by specifying its methods with respective parameters and return types. These methods are exposed by the server, which can be called remotely. You defined the EmployeeService definition in the previous sub-section, as shown in the next code block:

```
service EmployeeService {
  rpc Create(Employee) returns (EmployeeCreateResponse);
}
```

Here, Create is a method exposed by the EmployeeService service definition. Messages used in the Create service should also be defined as a part of the service definition. The Create service method is a unary service method because the client sends a single request object and receives a single response object in return from the server.

Let's dig further into the types of service methods offered by gRPC:

- **Unary**: We have already discussed the unary service method in the previous example. It would have a one-way response for a single request.

- **Server streaming**: In these types of service methods, the client sends a single object to the server and receives the stream response in return. This stream contains the sequence of messages. The stream is kept open until the client receives all the messages. The message sequence order is guaranteed by gRPC. In the following example, the client would keep receiving the live score messages until the match is over:

```
rpc LiveMatchScore(MatchId) returns (stream
    MatchScore);
```

- **Client streaming**: In these types of service methods, the client sends a sequence of messages to a server and receives a response object in return. The stream is kept open until the client sends all the messages. The message sequence order is guaranteed by gRPC. Once all the messages are sent by the client, it waits for the server response. In the following example, a client sends the data messages to the server until all the data records are sent and it then waits for the report:

```
rpc AnalyzeData(stream DataInput) returns (Report);
```

- **Bidirectional streaming**: This is the simultaneous execution of client and server streaming. It means both the server and client send a sequence of messages using a read-write stream. Here, the order of the sequence is preserved. However, these two streams operate independently. Therefore, each can read and write in whatever order they like. The server can read and reply to the message one by one or at once, or can have any combination. In the following example, processed records can be sent immediately one by one or can be sent later in different batches:

```
rpc BatchProcessing(stream InputRecords) returns
    (stream Response);
```

Now that you have learned about the gRPC service definitions, let's explore the RPC life cycle in the next section.

Exploring the RPC life cycle

In the previous section, you learned about four types of service definitions. Each type of service definition has its own life cycle. Let's find out more about the life cycle of each service definition in this section.

The life cycle of unary RPC

Unary RPC is the simplest form of the service method. Both the client and the server send the single object. Let's find out how it works. Unary RPC is initiated by the client. The client calls a `stub` method. `stub` notifies the server that the RPC call has been invoked. `stub` also provides the server client's metadata, the method name, and the specified deadline, if applicable, with notification.

Metadata is data about the RPC call in the form of key-value pairs such as timeout and authentication details.

Next, in response, the server sends back its initial metadata. Whether the server sends initial metadata immediately or after receiving the client's request message depends on the application. But the server must send it before any response.

The server works on the request and prepares the response after receiving the client's request message. The server sends back the response with the status (code and optional message) and optional trailing metadata for successful calls.

The client receives a response and completes the call (for status `OK`, similar to HTTP status 200).

Next, let's learn about the life cycle of server-streaming RPC.

The life cycle of server-streaming RPC

The life cycle of server-streaming RPC is almost the same as unary RPC. It follows the same steps. The only difference is the way the response is sent because of the stream response. The server sends messages as streams until all the messages are sent. In the end, the server sends back the response with the status (code and optional message) and optional trailing metadata and completes the server-side processing. The client completes the life cycle once it has all the server's messages.

Next, let's learn about the life cycle of client-streaming RPC.

The life cycle of client-streaming RPC

The life cycle of client-streaming RPC is almost the same as unary RPC. It follows the same steps. The only difference is the way request is sent because of stream request. The client sends messages as streams until all the messages are sent to the server. The server sends back the single message response with the status (code and optional message) and optional trailing metadata for successful calls. The server sends the response after receiving all the client's messages in idle scenarios. The client completes the life cycle once it receives the server message.

Next, let's learn about the life cycle of bidirectional streaming RPC.

The life cycle of bidirectional streaming RPC

The first two steps in the life cycle of bidirectional streaming RPC are the same as unary RPC. Streaming processing from both sides is application-specific. Both the server and client can read and write messages in any order because the two streams are independent of each other.

The server can process streams of request messages sent by the client in any order. For example, the server and client can play ping-pong, the client sends the request message and the server process it. Again, the client sends the request message and the server processes it and the process, as you know, goes on. Or, the server waits until it receives all the client's messages before it writes its messages.

The client completes the life cycle once it receives all the server messages.

Events that impact the life cycle

The following events may impact the life cycle of the RPC:

- **Deadlines/timeouts**: gRPC supports deadlines/timeouts. Therefore, a client would wait for the defined deadline/timeout to get the response from the server. If the wait exceeds the defined deadline/timeout, then it throws the DEADLINE_EXCEEDED error. Similarly, the server can query to find out whether a particular RPC has timed out, or how much time is left to complete the RPC.

  Timeout configuration is language-specific. Some language APIs support timeouts (durations of time), and some support deadlines (a fixed point in time). APIs may have a default value of deadline/timeout and some may not.

- **RPC termination**: There are few scenarios where RPC gets terminated because both the client and server make independent and local determinations of the success of the call, and their conclusions may not match. For example, the server may finish its part by sending all its messages but it may fail from the client's side because responses have arrived after the timeout. Another scenario would be when a server decides to complete the RPC before the client sends all messages.

- **Canceling an RPC**: gRPC has a provision to cancel the RPC at any time by either the server or client. This terminates the RPC immediately. However, changes made before the cancellation are not rolled back.

Let's explore the gRPC server and stub a bit more in the next section.

Understanding the gRPC server and gRPC stub

If you closely observe *Figure 10.1*, you'll find that the gRPC server and gRPC stub are core parts of the implementation because gRPC is based on client-server architecture. Once you define the service, you can generate both service interfaces and the stub using the Protobuf compiler, `protoc`, with the gRPC Java plugin. You'll find a practical example in the next chapter, *Chapter 11, gRPC-based API Development and Testing*.

The following types of files are generated by the compiler:

- **Models**: It generates all the messages (that is, models) defined in the service definition file that contains the Protobuf code to serialize, deserialize, and fetch the types of request and response messages.

- **gRPC Java files**: It contains the service base interface and stubs. The base interface is implemented and then used as a part of the gRPC server. Stubs are used by the clients for communication with the server.

First, you need to implement the interface as shown in the following code for `EmployeeService`:

```
public class EmployeeService extends
    EmployeeServiceImplBase {
  // some code


  @Override
  public void create(Employee request,
        io.grpc.stub.StreamObserver<Response>
           responseObserver) {
```

```
      // implementation
   }
}
```

Once you implement the interface, you can run the gRPC server to serve the requests from gRPC clients:

```
public class GrpcServer {
  public static void main(String[] arg) {
    try {
      Server server = ServerBuilder.forPort(8080)
        .addService(new EmployeeService())
        .build();
      System.out.println("Starting gRPC Server Service
                  ...");
      server.start();
      System.out.println("Server has started at port:
                  8080");
      System.out.println("Following services are available:
                  ");
      server.getServices().stream()
        .forEach(
          s -> System.out.println("Service Name: " +
          s.getServiceDescriptor().getName())
        );
      server.awaitTermination();
    } catch (Exception e) {
      // error handling
    }
  }
}
```

For clients, first you need to create the channel using ChannelBuilder, and then you can use the created channels for creating stubs, as shown in the following code:

```
public EmployeeServiceClient(ManagedChannelBuilder<?>
                                    channelBuilder)
```

```
{
    channel = channelBuilder.build();
    blockingStub = EmployeeServiceGrpc.newBlockingStub(
        channel);
    asyncStub = EmployeeServiceGrpc.newStub(channel);
}
```

Here, both blocking and asynchronous stubs have been created using the channel built using the `ManageChannelBuilder` class.

Let's explore error handling in the next section.

Handling errors

Unlike REST, which makes use of the HTTP status codes, gRPC uses a `Status` model that contains its error codes and optional error message (string).

If you remember, you have used the special class called `Error` for containing the error details because HTTP error codes contain limited information. Similarly, the gRPC error `Status` model is limited to code and an optional message (string). You can't have sufficient error details that the client can use to handle the error or retry. You can make use of the richer error model as described at `https://cloud.google.com/apis/design/errors#error_model`, which allows you to pass detailed error information back to the client. You can also find the error models in the next code block for quick reference:

```
package google.rpc;
message Status {
    // actual error code is defined by `google.rpc.Code`.
    int32 code = 1;
    // A developer-facing human-readable error message
    string message = 2;
    // Additional error information that the client code can
    // use
    // to handle the error, such as retry info or a help
    // link.
    repeated google.protobuf.Any details = 3;
}
```

The `details` field contains extra information, and you can use it to pass relevant information, such as `RetryInfo`, `DebugInfo`, `QuotaFailure`, `ErrorInfo`, `PreconditionFailure`, `BadRequest`, `RequestInfo`, `ResourceInfo`, `Help`, and `LocalizedMethod`. All these message types are available at `https://github.com/googleapis/googleapis/blob/master/google/rpc/error_details.proto`.

These richer error models are described using Protobuf. If you would like to use richer error models, you have to make sure that support libraries are aligned with the practical use of APIs as is described for Protobuf.

Error status codes

Similar to REST, errors can be raised by the RPC for various reasons, such as network failure or data validation. Let's have a look at the following REST error codes and their respective gRPC counterparts:

HTTP Status Code	gRPC Status Code	Notes
400	INVALID_ARGUMENT	For invalid arguments.
400	FAILED_PRECONDITION	Action could not be performed due to failed pre-condition.
400	OUT_OF_RANGE	If an invalid range is specified by the client.
401	UNAUTHENTICATED	If the client request is not authenticated, including missing or expired token.
403	PERMISSION_DENIED	Client does not have the sufficient permission.
404	NOT_FOUND	Request resource not found.
409	ABORTED	A conflict for read-write operation or any concurrency conflict.
409	ALREADY_EXISTS	If the request is for creating a new resource that already exists.
429	RESOURCE_EXHAUSTED	If the request reaches the API rate limiting.
499	CANCELLED	If the request is canceled by the client.
500	DATA_LOSS	For unrecoverable data loss or corruption.
500	UNKNOWN	For unknown error at the server side.
500	INTERNAL	For an internal server error.
501	NOT_IMPLEMENTED	API not implemented by the server.
502	N/A	Error due to unreachable network or network misconfiguration.
503	UNAVAILABLE	Server is unavailable due to down or any other reason. The client can perform a retry on such errors.
504	DEADLINE_EXCEEDED	Neither request finishes within the deadline.

gRPC error codes are more readable as you don't need a mapping to understand the number codes.

Summary

In this chapter, you have explored Protobuf, an IDL, and the serialization utility. You have also explored gRPC fundamentals such as service definitions, messages, server interfaces, and methods. You have compared gRPC with REST. I hope this has given you enough perspective to understand gRPC.

You have also learned about the gRPC life cycles, servers, and clients with stubs. You have covered the following in this chapter – Protobuf, gRPC architecture, and gRPC fundamentals – which will allow you to develop gRPC-based APIs and services.

You will make use of the fundamentals you have learned in this chapter in the next chapter for implementing the gRPC server and client.

Questions

1. What is RPC?

2. How is gRPC different in comparison to REST and which one should be used?

3. Which type of service method is useful when you want to view the latest tweets or do similar types of work?

Further reading

- gRPC documentation: `https://grpc.io/`

- Practical gRPC: `https://www.packtpub.com/in/web-development/practical-grpc`

11
gRPC-based API Development and Testing

You will learn how to implement gRPC-based APIs in this chapter. You will learn how to write the gRPC server and client along with writing the APIs based on gRPC. In the later part of this chapter, you will be introduced to microservices and will see how they will help you to design modern scalable architecture. Here, you will go through the implementation of two services – the gRPC server and the gRPC client. gRPC-based APIs are more popular and preferred for inter-service communication in a microservice-based system. Hence, gRPC development skills are an important topic in the API space.

You will explore the following topics in this chapter:

- Writing an API interface
- Developing the gRPC server
- Handling errors
- Developing the gRPC client
- Learning microservice concepts

After completing this chapter, you will have the following skillset – gRPC server and client development, gRPC-based API testing automation, and microservice concepts.

Technical requirements

This chapter concerns the theory of gRPC. However, you need the following for the development and testing of gRPC-based web services:

- Any Java IDE, such as NetBeans, IntelliJ, or Eclipse
- Java Development Kit (JDK) 15
- An internet connection to clone the code and download the dependencies and Gradle
- Postman/cURL (for API testing)

So let's begin!

Please visit the following link to check the code: `https://github.com/PacktPublishing/Modern-API-Development-with-Spring-and-Spring-Boot/tree/main/Chapter11`

Writing an API interface

In this section, we will write the API interface using the protocol buffer (Protobuf) for payment service. If you recall, this is the piece that you haven't yet implemented the sample e-commerce app.

Before writing the API interface, let's set up the Gradle project first.

Setting up the project

The code for this chapter will contain three projects under the `Chapter11` directory – API, server, and client:

- **API**: This is a library project that contains the `.proto` file and its generated Java classes packaged in a jar file. You will publish the `library payment-gateway-api-0.0.1.jar` file locally. This library will then be used in both server and client projects.
- **Server**: This project contains the gRPC server that will implement the gRPC services and serve the gRPC requests.

- **Client**: This project contains the gRPC client that will call the gRPC server. You are going to implement a REST call that will call the gRPC server internally to serve the HTTP request.

Let's first create the server and client project.

Creating the gRPC server and client projects

Either you can use the `Chapter 11` code from a cloned Git repository (`https://github.com/PacktPublishing/Modern-API-Development-with-Spring-and-Spring-Boot`), or you can start by creating the new Spring project from scratch using *Spring Initializer* (`https://start.spring.io/`) for the server and client with the following options (you will create a gRPC `api` library project separately):

- Project → Gradle Project
- Language → Java
- Spring Boot → 2.4.4 (the preferred version should be 2.4+ if not available; later, you can modify it manually in the `build.gradle` file.
- Project metadata:

  a. Group: `com.packt.modern.api`

  b. Artifact: `chapter11`

  c. Name: `chapter11`

  d. Description: `Chapter 11 code of book Modern API Development with Spring and Spring Boot`

  e. Package name: `com.packt.modern.api`
- Packaging → Jar
- Java → 15 (you can change it to another version, such as 16/17, in the `build.gradle` file later, as shown in the following code block):

```
// update following build.gradle file
sourceCompatibility = JavaVersion.VERSION_15
// or for Java 16
// sourceCompatibility = JavaVersion.VERSION_16
// or for Java 17
// sourceCompatibility = JavaVersion.VERSION_17
```

- Dependencies → org.springframework.boot:spring-boot-starter-web

Then, you can click on **GENERATE** and download the project. The downloaded project can be used for creating both the server and client. We have created the separate server and client projects in the Chapter11 directory in the Git repository and copied the extract of the downloaded zipped project.

You can configure the server and client projects later. Let's first create the gRPC API library project.

Creating the gRPC API library project

Create a new directory, api, in the Chapter11 directory. Then, use Gradle to create a new Gradle project using the following command. It will ask for a few options. You should select the options highlighted in the following code block:

```
$ mkdir api
$ cd api
(you can also use gradlew from other chapter's code)
$ ../server/gradlew init
Select type of project to generate:
  1: basic
  2: application
  3: library
  4: Gradle plugin
Enter selection (default: basic) [1..4] 3
Select implementation language:
  1: C++
  2: Groovy
  3: Java
  4: Kotlin
  5: Scala
  6: Swift
Enter selection (default: Java) [1..6] 3
Select build script DSL:
  1: Groovy
  2: Kotlin
Enter selection (default: Groovy) [1..2] 1
```

```
Select test framework:
  1: JUnit 4
  2: TestNG
  3: Spock
  4: JUnit Jupiter
Enter selection (default: JUnit 4) [1..4] 4
Project name (default: demo): api
Source package (default: demo): com.packt.modern.api
BUILD SUCCESSFUL
2 actionable tasks: 2 executed
```

The project is bootstrapped by Gradle. Next, you will configure the `api` project.

Configuring the gRPC API library project

You will configure the plugins section, the Protobuf plugin, and the Maven Publish plugin in this sub-section as key steps. Let's do this as follows:

1. Modify `settings.gradle` in the project's root directory:

    ```
    rootProject.name = 'payment-gateway-api'
    ```

2. Next, you'll modify the `api/lib/build.gradle` file. You'll add the Protobuf and Maven Publish Gradle plugins. You'll also replace the `java-library` plugin with `java`, as shown next:

    ```
    plugins {
        id 'java'
        id 'maven-publish'
        id "com.google.protobuf" version "0.8.15"
    }
    ```

https://github.com/PacktPublishing/Modern-API-Development-with-Spring-and-Spring-Boot/tree/main/Chapter11/api/lib/build.gradle

The Maven Publish plugin will be used to publish the generated `Jar` file to the local Maven repository.

3. Add the group name, version, and source compatibility in build.gradle, as shown in the following code block. The group and version will be used by the Maven Publish plugin to name the published artifact:

```
group = 'com.packt.modern.api'
version = '0.0.1'
sourceCompatibility = JavaVersion.VERSION_15
```

4. Next, add the following dependencies that are required for Protobuf and gRPC (check the highlighted part):

```
def grpcVersion = '1.37.0'
dependencies {
    compile "io.grpc:grpc-protobuf:${grpcVersion}"
    compile "io.grpc:grpc-stub:${grpcVersion}"
    compile "io.grpc:grpc-netty:${grpcVersion}"
    compile 'javax.annotation:javax.annotation-
        api:1.3.2'
    testImplementation 'org.junit.jupiter:junit-
                        jupiter-api:5.6.2'
    testRuntimeOnly 'org.junit.jupiter:junit-jupiter-
                    engine'
}
```

5. Next, let's configure the Protobuf Gradle plugin. This uses the command-line compiler protoc. It searches in the system path by default. However, you can add a Protobuf compiler artifact to the plugin, which will make the build file self-sufficient as far as the gRPC compile task is concerned. Let's configure it as shown in the following code block by adding a protobuf section to the build. gradle file:

```
// protobuf plugin configuration
protobuf {
    generatedFilesBaseDir = "$projectDir/src"
    protoc {
        artifact = "com.google.protobuf:protoc:3.15.8"
    }
    plugins {
        grpc {
```

```
            artifact = "io.grpc:protoc-gen-grpc-
                        java:1.36.1"
        }
    }
    generateProtoTasks {
        all()*.plugins {
            grpc {}
        }
    }
}
```

https://github.com/PacktPublishing/Modern-API-Development-with-Spring-and-Spring-Boot/tree/main/Chapter11/api/lib/build.gradle

Here, generated files will be stored in the src folder according to the value to set for generatedFielsBaseDir. You are also configuring the artifact used by the Protobuf compiler (protoc) and its Java plugin (protoc-gen-grpc-java), which will generate the Java code based on .proto files.

When you run the gradlew build command the first time, Gradle will download the protoc and protoc-gen-grpc-java executables based on the OS, as shown in the following code block:

```
> :generateProto > Resolve dependencies of
:protobufToolsLocator_grpc > protoc-3.15.7-
windows-x86_64.exe > 1.5 MiB/3.5 MiB downloaded
```

```
> :generateProto > Resolve dependencies of
:protobufToolsLocator_grpc > protoc-gen-grpc-java-1.36.1-
windows-x86_64.exe > 662.7 KiB/1.5 MiB downloaded
```

6. The Protobuf Gradle plugin works with the configuration shared hitherto. It works when you run the build command from the command line. However, the IDE may give a compilation error if you don't add the following block to the build.gradle file for adding the generated source files in sourceSets:

```
sourceSets {
    main {
        proto {
            // In addition to the default
            // "src/main/proto"
```

```
            srcDir "src/main/grpc"
            srcDir "src/main/java"
        }
    }
}
```

7. Files generated by the Protobuf compiler and its Java plugin will be stored in the project source directory. Therefore, you need to add the following configuration, which will be doing a clean-up of the generated files when the Gradle clean task is executed:

```
task cleanProtoGen {
    doFirst {
        delete('src/main/grpc')
        delete('src/main/java')
    }
}
clean.dependsOn cleanProtoGen
```

8. Finally, you will add the following block to configure the Maven Publish plugin:

```
publishing {
    publications {
        mavenJava(MavenPublication) {
            artifactId = 'payment-gateway-api'
            from components.java
        }
    }
}
```

You are done with the api project configuration. Before we move on to write the contract of gRPC services in the .proto file, let's learn about the Protobuf Gradle plugin's default locations for the source and generated code for Protobuf files in the next subsection. You can use it for tweaking the location as per your project's requirements.

Location of source and generated files

You are going to use the locations of the source and generated files as per the configuration defined in the previous subsection. However, the plugin allows you to modify these locations, as shown in the following paragraphs.

The default location of sourceSet (the .proto code files) of the Protobuf files is src/$sourceSetName/proto (src/main/proto). You can add other locations to sourceSet for Protobuf files the way you do for Java source files, as shown in the following configuration:

```
sourceSets {
  main {
    java {
      ...
    }
    proto {
      // 'src/main/proto' plus
      // You can add multiple location like next line
      srcDir 'src/main/protobuf'
    }
  }
  test {
    proto {
      // 'src/test/proto' plus
      srcDir 'src/test/protobuf'
    }
  }
}
```

Here, you have added additional source and test code locations of Protobuf files. The Protobuf compiler and Java code generator will look into both of these locations apart from the default location.

Let's find out how you can customize the default location of generated files next.

The default location of **generated Java files** is formed using $generatedFilesBaseDir/$sourceSet/$builtinPluginName. Here, each of the variables contains the following values:

- $generatedFilesBaseDir is set to $buildDir/generated/source/ proto. $buildDir points to the build directory in the root directory of the project. You can change that with the following configuration:

```
protobuf {
  ...
```

```
        generatedFilesBaseDir = "$projectDir/src/generated"
    ...
}
```

- Set the $sourceSet points to main.
- The value of $builtinPluginName is either grpc or java by default. You can change these values by means of the following command:

```
protobuf {
    ...
    generateProtoTasks {
        all()*.plugins {
            grpc {
                // generated files would be copied to
                // "$generatedFilesBaseDir/
                // $sourceSet/grpcjava"
                outputSubDir = 'grpcjava'
            }
        }
    }
}
```

This is how you can change the locations of generated files. However, we'll stick to the locations configured in the previous subsection.

Now, the api project setup is done. We are now good to write the service definitions using the protocol buffer in the next subsection. You haven't yet implemented the payment functionality for our sample e-commerce app. This is because it needs to be integrated with payment gateway services such as Stripe or Paypal. Therefore, you are going to write the sample payment gateway service using gRPC in the next section.

Writing the payment gateway functionalities

Before you write the payment gateway service definition, let's first understand the basic functionality of the payment gateway system in easy terms.

The payment gateway provides a way to capture and transfer a payment from a customer to online sellers and then returns acceptance/decline as a response to the customer. It performs various other stuff during this course of action, such as verification, security, encryption, and communication with all participants.

The following are the actors who participate in this transaction:

- **Payment Gateway**: A web interface that allows the processing of online payments and coordinates with all other actors. This is very similar to physical **point-of-sale** (**POS**) terminals.

- **Merchant**: Merchants are online sellers or service providers, such as Amazon, Uber, and Airbnb.

- **Customer**: This is you, the customer, who performs the buy/pay transaction for products or services and uses credit/debit cards, digital wallets, or online banking.

- **Issuing Bank**: The party that provides the functionality to perform online money transfers, such as Visa/Master/Amex cards, Paypal/Stripe, or traditional banks.

- **Acquirer or Acquiring Bank**: The institution that holds the merchant account of the merchant. It passes the transaction to the issuing bank to receive payment.

  You are going to create two gRPC services – `ChargeService` and `SourceService` as part of the payment gateway service. Don't get confused with the web service, which is an executable/deployable artifact. `ChargeService` and `SourceService` are part of the service component of the protocol buffer's **interface definition language** (**IDL**) file, the same as we learned about in the `EmployeeService` example in the last chapter (the *Protocol buffer* section of *Chapter 10, gRPC Fundamentals*). Both of these services are inspired by Stripe public REST APIs.

Let's understand the transaction flow before we jump into creating the service components of a gRPC-based payment gateway service.

Online payment workflow steps

The following steps are performed when an online transaction takes place:

1. First, the customer should have a payment source (read method) created before initiating the payment. If not, then the customer creates a source, such as card details.

2. Payment is initiated by creating a charge against the payment source (read method).

3. The payment gateway performs all the necessary validation and verification steps and then allows the charge to be captured. These steps trigger the fund transfer from the issuing bank to the merchant account.

You can observe that there are two objects (resources) involved in this workflow (aka source and charge). Therefore, you are going to write two services that function around these two objects. There are various other functionalities performed by the payment gateway, such as disputes, refunds, and payouts. However, you are going to implement only two services, charge and source, in this chapter.

Writing the payment gateway service definitions

Writing a protocol buffer-based IDL is very similar to the way you have defined the OpenAPI specification for REST APIs. In REST, you define the models and API endpoints, whereas in gRPC, you define the messages and RPC procedures wrapped in the service. Let's write our payment gateway service IDL using the following steps:

- First of all, let's create a new file, `PaymentGatewayService.proto`, in the `api/lib/src/main/proto` directory under the root directory of the `api` project.

- After creating a new file, you can add the metadata as shown in the following code block:

```
syntax = "proto3";                          // 1
package com.packtpub.v1;                     // 2
option java_package = "com.packt.modern.api.grpc.v1";
                                             // 3
option java_multiple_files = true;    // 4
```

https://github.com/PacktPublishing/Modern-API-Development-with-Spring-and-Spring-Boot/blob/main/Chapter11/api/lib/src/main/proto/PaymentGatewayService. proto

- Here, the *first line* tells the compiler to use version 3 of the protocol buffer by using the `syntax` specifier. If you don't specify this, then the compiler will use version 2 of the protocol buffer.

- *Line 2* uses the optional `package` specifier to attach the namespace to message types. This prevents name clashes among message types. *We have to postfix it with a package version that allows you to create new versions of APIs with backward compatibility.*

- *Line 3* uses the `java_package` – `option` specifier. Its value will be used as a Java package in generated Java files. If you don't use this `option` specifier and declare the `package` specifier, then the value of `package` will be used as a Java package in generated Java files instead.

- *Line 4* declares the `java_multiple_files` – `option` specifier, which is a Boolean option. It is set to `false` by default. If it is set to `true`, then it generates the separate Java files for each top-level message type, enumeration (`enum`), and services.

Next, let's add the `ChargeService` service, which contains the operations required for charge functionality denoted by `rpc`. Charge objects get created for charging the cards, bank accounts, or digital wallets. Let's add the charge service to the Protobuf (`.proto`) file:

```
service ChargeService {
  rpc Create(CreateChargeReq) returns (
      CreateChargeReq.Response);
  rpc Retrieve(ChargeId) returns (ChargeId.Response);
  rpc Update(UpdateChargeReq) returns (
      UpdateChargeReq.Response);
  rpc Capture(CaptureChargeReq) returns
                    (CaptureChargeReq.Response);
  rpc RetrieveAll(CustomerId) returns (
      CustomerId.Response);
}
```

https://github.com/PacktPublishing/Modern-API-Development-with-Spring-and-Spring-Boot/blob/main/Chapter11/api/lib/src/main/proto/PaymentGatewayService.proto

Each of these procedures in `ChargeService` will perform the following operations:

- **Create**: This procedure creates a new `Charge` object.

- **Retrieve**: This procedure retrieves the `Charge` object based on the given charge ID that has previously been created.

- **Update**: This procedure updates the `Charge` object identified by the given charge ID by setting the values of the parameters passed. Any parameters not provided will be left unchanged.

- **Capture**: This procedure captures the payment of an existing, uncaptured, charge. This is the payment workflow step, where you first create a charge with the capture option set to `false`. Uncaptured payments expire precisely 7 days after they are created. If they are not captured by that point in time, they will be marked as refunded and capture will no longer be allowed.

- **RetrieveAll**: This procedure returns the list of charges that belong to the given customer ID.

> **Empty request or response type**
>
> You can use `google.protobuf.Empty` for void/empty request and response types. This can be used in `.proto` files. You just have to place the following `import` statement before any message/service is defined:
>
> ```
> import "google/protobuf/timestamp.proto";.
> ```
>
> Then, you can use it as shown next:
>
> ```
>  rpc delete(SourceId) returns (google.protobuf.
> Empty);.
> ```

4. The amount is charged on a source that could be a card, bank account, or digital wallet. A variety of payment methods can be used by the customer using a `Source` object. Therefore, you need a service that will allow you to perform operations on the source resource. Let's add the `source` service and its operations to the Protobuf (`.proto`) file:

```
service SourceService {
    rpc Create(CreateSourceReq) returns (
        CreateSourceReq.Response);
    rpc Retrieve(SourceId) returns (SourceId.Response);
    rpc Update(UpdateSourceReq) returns (
        UpdateSourceReq.Response);
    rpc Attach(AttachOrDetachReq) returns
                    (AttachOrDetachReq.Response);
    rpc Detach(AttachOrDetachReq) returns
                    (AttachOrDetachReq.Response);
}
```

https://github.com/PacktPublishing/Modern-API-Development-with-Spring-and-Spring-Boot/blob/main/Chapter11/api/lib/src/main/proto/PaymentGatewayService.proto

Each of these procedures in SourceService will perform the following operations:

- **Create**: This procedure creates a new `Source` object.

- **Retrieve**: This procedure allows you to retrieve the `Source` object based on the given source ID.

- **Update**: This procedure allows you to update the certain fields of the `Source` object passed using the `UpdateSourceReq` object. Any field that is not part of `UpdateSourceReq` will remain unchanged.

- **Attach**: This procedure attaches the `Source` object to the customer. The `AttachOrDetailReq` parameter contains the IDs of both the source and the customer. However, to perform the operation attached, the `Source` object should be in the `CHARGEABLE` or `PENDING` state.

- **Detach**: This procedure will detach the source object from the customer. It will also change the state of the `Source` object to `consumed` and it can no longer be used to create the charge. The `AttachOrDetailReq` parameter contains the IDs of both the source and the customer.

> **The recommended approach for defining the request and response types**
>
> It is recommended to always use the wrapper request and response types. This allows you to add another field to the request or response types.

5. Now the service definitions are done. You can define the given parameters and the returned types of these procedures. Let's first define the parameters and returned types of `ChargeService`. First of all, you will define the `Charge` message type, as shown in the following code block:

```
message Charge {
  string id = 1;
  uint32 amount = 2;
  uint32 amountCaptured = 3;
  uint32 amountRefunded = 4;
  string balanceTransactionId = 5;
  BillingDetails billingDetails = 6;
  string calculatedStatementDescriptor = 7;
  bool captured = 8;
  uint64 created = 9;
  string currency = 10;
  string customerId = 11;
```

```
      string description = 12;
      bool disputed = 13;
      uint32 failureCode = 14;
      string failureMessage = 15;
      string invoiceId = 16;
      string orderId = 17;
      bool paid = 18;
      string paymentMethodId = 19;
      PaymentMethodDetails paymentMethodDetails = 20;
      string receiptEmail = 21;
      string receiptNumber = 22;
      bool refunded = 23;
      repeated Refund refunds = 24;
      string statementDescriptor = 25;
      enum Status {
         SUCCEEDED = 0;
         PENDING = 1;
         FAILED = 2;
      }
      Status status = 26;
      string sourceId = 27;
   }
```

Here, the Charge message contains the following fields:

- id: The unique identifier of the Charge object.

- amount: An amount is a positive number or zero, which is intended to be collected by the payment.

- amountCaptured: This is the captured amount (a positive number or zero). It can be less than the value of the amount field if a partial capture is made.

- amountRefunded: The amount refunded (a positive number or zero). It can be less than the value of the amount field if a partial refund is issued.

- balanceTransactionId: The ID of the balance transaction that describes the impact of this charge on your account balance (not including refunds or disputes).

- billingDetails: The object of the BillingDetails message type that contains billing information associated with the payment method at the time of the transaction.

- `calculatedStatementDescriptor`: The statement description that is passed to card networks, and that is displayed on your customers' credit card and bank statements.

- `captured`: A Boolean field that represents whether a charge has since been captured (it is possible to create a charge without capturing).

- `created`: The timestamp (measured in seconds since the Unix epoch) at which the object was created.

- `currency`: The three-letter ISO currency code.

- `customerId`: The ID of the customer owning the charge.

- `description`: A description of the charge displayed to the user.

- `disputed`: A Boolean field that represents whether the charge has been disputed.

- `failureCode`: The error code of the failure.

- `failureMessage`: A description of the failure. The reason may be stated if this option is available.

- `invoiceId`: The ID of the invoice this charge is for.

- `orderId`: The ID of the order this charge is for.

- `paid`: The Boolean value represents whether the charge succeeded, or was successfully authorized for subsequent capture.

- `paymentMethodId`: The ID of the payment method.

- `paymentMethodDetails`: The object that contains the details of the payment method.

- `receiptEmail`: The email where receipt of the charge will be sent.

- `receiptNumber`: This represents the transaction number in the charge receipt that was sent by email. It should remain null until a charge receipt is sent.

- `refunded`: A Boolean field that represents whether the charge was refunded.

- `refunds`: This contains the list of refunds that have been issued. The keyword `repeated` is used to create a list of `Refund`.

- `statementDescriptor`: The description of a charge for a card.

- `status`: An object of the `Status` enumeration type (`SUCCEEDED`, `PENDING`, `FAILED`) that represents the status of the charge.

- `sourceId`: Id of the `Source` object

A scalar type such as uint32 and string are discussed in the *Protocol buffer* section in the previous chapter (*Chapter 10, gRPC Fundamentals*). You can refer to it for further information.

Predefined well-known types

Apart from scalar types, Protobuf also provides predefined types such as Empty (discussed earlier in *step 4*), Timestamp, and Duration. You can find the complete list at https://developers.google.com/protocol-buffers/docs/reference/google.protobuf.

---

6. Now, you can define the remaining message types of the other parameters (CreateChargeReq, ChargeId, UpdateChargeReq, CaptureChargeReq, and CustomerId) and return the ChargeList type of ChargeService, as shown in the following code block:

```
message CreateChargeReq {
   uint32 amount = 1;
   string currency = 2;
   string customerId = 3;
   string description = 4;
   string receiptEmail = 5;
   Source source Id = 6;
   string statementDescriptor = 7;
   message Response {
      Charge charge = 1;
   }
}
message UpdateChargeReq {
   string sourceId = 1;
   string customerId = 2;
   string description = 3;
   string receiptEmail = 4;
   message Response {
      Charge charge = 1;
   }
}
message CaptureChargeReq {
   string sourceId = 1;
```

```
      uint32 amount = 2;
      string receiptEmail = 3;
      string statementDescriptor = 4;
      message Response {
        Charge charge = 1;
      }
    }
  message ChargeId {
    string id = 1;
    message Response {
      Charge charge = 1;
    }
  }
  message CustomerId {
    string id = 1;
    message Response {
      repeated Charge charge = 1;
    }
  }
```

Here, the `CreateChargeReq` type contains the required attribute's charge amount (`amount`) and `currency`. It also contains a number of optional attributes – `customerId`, `receiptEmail`, `source`, and `statementDescriptor`.

`UpdateChargeReq` contains all the optional attributes – `customerId`, `description`, and `receiptEmail`.

`CaptureChargeReq` contains all the optional attributes – `amount`, `receiptEmail`, and `statementDescriptor`.

Less well-known Google common types

`Money` and `Date` (not `Timestamp`) are less known common types that can be used. However, you have to copy the definitions instead of importing them (the way you can do for `Empty` and `Timestamp`). You can copy it from Google's API common types: Money from `https://github.com/googleapis/googleapis/blob/master/google/type/money.proto`, and Date from `https://github.com/googleapis/googleapis/blob/master/google/type/date.proto`. Other common types are also available in the repository that you can use.

---

7. Now, you can define the parameters and return the SourceService types. First of all, let's define the Source message type, as shown in the following code.

    The source uses the flow value that could be one of REDIRECT, RECEIVER, CODE VERIFICATION, and NONE. Similarly, the usage value could be REUSABLE or SINGLEUSE. Therefore, let's first create the Flow and Usage enumerations using enum:

    ```
    enum Flow {
        REDIRECT = 0;
        RECEIVER = 1;
        CODEVERIFICATION = 2;
        NONE = 3;
    }
    enum Usage {
        REUSABLE = 0;
        SINGLEUSE = 1;
    }
    ```

    Now, you can use this Flow enum in the Source message:

    ```
    message Source {
        string id = 1;
        uint32 amount = 2;
        string clientSecret = 3;
        uint64 created = 4;
        string currency = 5;
        Flow flow = 6;
        Owner owner = 7;
        Receiver receiver = 8;
        string statementDescriptor = 9;
        enum Status {
            CANCELLED = 0;
            CHARGEABLE = 1;
            CONSUMNED = 2;
            FAILED = 3;
            PENDING = 4;
        }
        Status status = 10;
    ```

```
    string type = 11;
    Usage usage = 12;
}
```

8. Now, you can define the remaining messages types of the other parameters
   of SourceService – CreateSourceReq, UpdateSourceReq,
   AttachOrDetachReq, and SourceId, as shown in the following code block:

```
message CreateSourceReq {
    string type = 1;
    uint32 amount = 2;
    string currency = 3;
    Owner owner = 4;
    string statementDescriptor = 5;
    Flow flow = 6;
    Receiver receiver = 7;
    Usage usage = 8;
    message Response {
        Source source = 1;
    }
}
message UpdateSourceReq {
    string sourceId = 1;
    uint32 amount = 2;
    Owner owner = 3;
    message Response {
        Source source = 1;
    }
}
message SourceId {
    string id = 1;
    message Response {
        Source source = 1;
    }
}
message AttachOrDetachReq {
    string sourceId = 1;
```

```
    string customerId = 2;
    message Response {
        Source source = 1;
    }
}
```

The other message types used in these messages can be referred to in the payment gateway definition file located at `https://github.com/PacktPublishing/Modern-API-Development-with-Spring-and-Spring-Boot/blob/main/Chapter11/api/lib/src/main/proto/PaymentGatewayService.proto`.

> **Multiple proto files**
>
> You can also create a separate definition file for each service, such as `ChargeService.proto` and `SourceService.proto` for modularity. You can then import these files into another Protobuf file using `import "SourceService.proto";`.
>
> You can find more information about importing at `https://developers.google.com/protocol-buffers/docs/proto#importing_definitions`.

You are now done with the payment gateway service definitions in the `Protobuf` file. Now, you can use this file to generate the gRPC server interface and stubs for the gRPC client.

Next, you will publish the Java classes generated from the `Protobuf` file packaged in the `Jar` file.

Publishing the payment gateway service gRPC server, stubs, and models

You can use the following command, which should be executed from the `api` project's root directory:

```
$ gradlew clean publishToMavenLocal
```

This command will first remove the existing files. Then it will generate the Java files (the `generateProto` Gradle task) from the Protobuf file, build it (the `build` Gradle task), and then publish the artifact to your local Maven repository (the `publishToMavenLocal` Gradle task).

The generateProto Gradle task will generate the two types of Java classes in two directories, as shown next:

- **Models**: This Protobuf Gradle plugin generates messages (aka models) and classes in separate Java files in the /api/lib/src/main/java directory, such as Card. java or Address.java. This directory will also contain the Java files of request and response objects used in operation contracts, such as CreateChargeReq, CreateSourceReq, Charge.java, and Source.java.

- **gRPC classes**: This Protobuf Gradle plugin generates the service definitions of both services (ChargeServiceGrpc.java and SourceServiceGrpc.java) in the /api/lib/src/main/grpc directory. Each of these gRPC Java files contains a base class, stub classes, and methods for each operation defined in the service descriptor for the Charge and Source services.

  The following key *static* classes are defined in ChargeServiceGrpc:

  i. ChargeServiceImplBase (abstract base class)

  ii. Stubs: ChargeServiceStub, ChargeServiceBlockingStub, and ChargeServiceFutureStub

  Similarly, the following key *static* classes are defined in SourceServiceGrpc:

  i. SourceServiceImplBase (abstract base class)

  ii. Stubs: SourceServiceStub, SourceServiceBlockingStub, and SourceServiceFutureStub

The abstract base classes described earlier contain the operations defined in the service block in the Protobuf file. You can use these base classes to implement the business logic for operations offered by these services, just like you had implemented the REST endpoints from the swagger-generated API interfaces.

These abstract classes should be implemented to provide the business logic implementations to the services offered by the gRPC server. Let's develop the gRPC server next.

Developing the gRPC server

You need to configure the Server project before implementing these abstract classes. Let's configure the server project first.

The server project directory structure will look like the following. The project root directory contains the build.gradle and settings.gradle files:

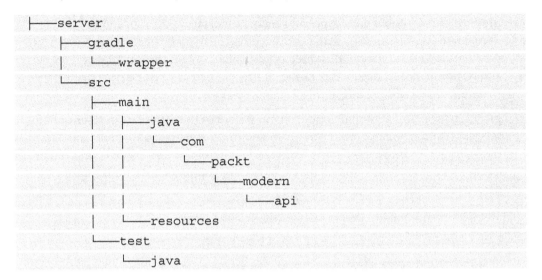

```
├──server
    ├──gradle
    │    └──wrapper
    └──src
        ├──main
        │    ├──java
        │    │    └──com
        │    │         └──packt
        │    │              └──modern
        │    │                   └──api
        │    └──resources
        └──test
             └──java
```

The resources directory will contain the application.properties file.

> **Using the gRPC libraries for dependencies**
>
> There are two Spring Boot starter projects that you can use. However, we'll stick to libraries provided by gRPC for a simplified solution and to aid understanding of the gRPC concepts. These libraries are available at the following links:
>
> https://github.com/LogNet/grpc-spring-boot-starter.
>
> https://github.com/yidongnan/grpc-spring-boot-starter.

Let's perform the following steps to configure the project:

1. First, you need to modify the project name in the Chapter11/server/ settings.gradle file to represent the server, as shown here:

    ```
    rootProject.name = 'chapter11-server'
    ```

2. Next, you can add the dependencies required for `Server` projects to the
   `Chapter11/server/build.gradle` file:

```
def grpcVersion = '1.37.0'
dependencies {
  implementation 'com.packt.modern.api:payment-gateway-
                 api:0.0.1'

  // gRPC dependencies
  implementation "io.grpc:grpc-protobuf:${grpcVersion}"
  implementation "io.grpc:grpc-stub:${grpcVersion}"
  implementation "io.grpc:grpc-netty:${grpcVersion}"
  implementation 'com.google.api.grpc:googleapis-
                 common-protos:0.0.3'

  implementation 'org.springframework.boot:spring-boot-
starter-web'
  testImplementation 'org.springframework.boot:spring-
                      boot-starter-test'
  testImplementation "io.grpc:grpc-
                      testing:${grpcVersion}"
}
```

https://github.com/PacktPublishing/Modern-API-Development-with-Spring-and-Spring-Boot/tree/main/Chapter11/server/build.gradle

3. The `payment-gateway-api` dependency is published in the local Maven repository. Therefore, you need to add the local Maven repository to the repository section, as shown in the following code block:

```
repositories {
    mavenCentral()
    mavenLocal()
}
```

You are done with the Gradle configuration! Now, you can write the gRPC server. However, before writing the server, you need to implement the base abstract classes generated by the Protobuf. Once the source and charge services (using base classes) are implemented, you can write the gRPC server code.

Implementation of the gRPC base classes

You are going to use an in-memory database (`ConcurrentHashMap`) for storing and retrieving the data. If you want, you can use the external database the way it is used in REST web services. This is done to keep the focus on gRPC server implementation.

First, create the in-memory database for both the charge and source data stores. Create a new file, `server/src/main/java/com/packt/modern/api/server/repository/DbStore.java`, and add code, as shown in the following code block:

```java
@Component
public class DbStore {
  private static final Map<String, Source> sourceEntities =
    new ConcurrentHashMap<>();
  private static final Map<String, Charge> chargeEntities =
    new ConcurrentHashMap<>();
  public DbStore() {
    // Seed Source for testing
    Source source = Source.newBuilder()
      .setId(RandomHolder.randomKey()).setType(
          "card").setAmount(100)
      .setOwner(createOwner()).setReceiver(createReceiver())
      .setCurrency("USD").setStatementDescriptor(
          "Statement Desc")
      .setFlow(Flow.RECEIVER).setUsage(Usage.REUSABLE)
      .setCreated(Instant.now().getEpochSecond()).build();
    sourceEntities.put(source.getId(), source);
    // Seed Charge for testing
    Charge charge = Charge.newBuilder()
      .setId(RandomHolder.randomKey()).setAmount(1000)
      .setCurrency("USD").setCustomerId("ab1ab2ab3ab4ab5")
      .setDescription("Charge Description")
      .setReceiptEmail("receipt@email.com")
      .setStatementDescriptor("Statement Descriptor")
      .setSourceId(source.getId())
      .setCreated(Instant.now().getEpochSecond()).build();
    chargeEntities.put(charge.getId(), charge);
  }
```

Here, you create two `ConcurrentHashMap` objects for storing the charge and store objects, respectively. You create two seed objects of each of these in the constructor using the builder and store them to their respective hash maps.

According to the operations defined in the service contract, you create the methods in the database store to perform the operations. These operations are implemented with basic business logic to keep the flow and logic concise and to the point.

Let's now add the `createSource()` method to implement the `create()` contract of `SourceService` defined in the `Protobuf` file, as shown in the following code block:

```
public CreateSourceReq.Response createSource(
        CreateSourceReq req) {
    // validate request object
    // Owner and receiver should be taken from req. in the
    // form of ID
    Source source = Source.newBuilder()
        .setId(
            RandomHolder.randomKey()).setType(req.getType())
        .setAmount(req.getAmount()).setOwner(createOwner())
        .setReceiver(createReceiver()).setCurrency(
            req.getCurrency())
        .setStatementDescriptor(req.getStatementDescriptor())
        .setFlow(req.getFlow()).setUsage(req.getUsage())
        .setCreated(Instant.now().getEpochSecond()).build();
    sourceEntities.put(source.getId(), source);
    return CreateSourceReq.Response.newBuilder().setSource(
            source)
        .build();
}
```

This method creates a source object from the values received from the request object (`CreateSourceReq`). This newly created `Source` object is then saved in a hash map `sourceEntities` and returned to the caller.

Similarly, other contract methods for source- and charge-related data storage logic can be implemented here. Find the full source of this class at https://github.com/ PacktPublishing/Modern-API-Development-with-Spring-and-Spring- Boot/blob/main/Chapter11/server/src/main/java/com/packt/ modern/api/server/repository/DbStore.java.

Now, you have the in-memory DbStore. Next, let's use its repository classes.

Writing repository classes

The in-memory DbStore can be used in the ChargeRepositoryImpl repository class, as shown here:

```
@Repository
public class ChargeRepositoryImpl implements
      ChargeRepository {
  private DbStore dbStore;
  public ChargeRepositoryImpl(DbStore dbStore) {
    this.dbStore = dbStore;
  }
  @Override
  public CreateChargeReq.Response create(
      CreateChargeReq req) {
    return dbStore.createCharge(req);
  }
  // Other methods removed for brevity
```

https://github.com/PacktPublishing/Modern-API-Development-with-Spring-and- Spring-Boot/blob/main/Chapter11/server/src/main/java/com/packt/modern/api/ server/repository/ChargeRepositoryImpl.java

First, an interface is declared that contains the public methods mapped to the abstract base class of the charge service, which is available at https://github.com/ PacktPublishing/Modern-API-Development-with-Spring-and-Spring- Boot/blob/main/Chapter11/server/src/main/java/com/packt/ modern/api/server/repository/ChargeRepository.java.

Then, ChargeRepositoryImpl implements the ChargeRepository interface and makes use of DbStore to perform the operations.

Similarly, you can create the `SourceRepositoryImpl` class, which implements `SourceRespository`, as shown here:

```
@Repository
public class SourceRepositoryImpl implements
      SourceRepository {
  private DbStore dbStore;
  public SourceRepositoryImpl(DbStore dbStore) {
    this.dbStore = dbStore;
  }
  @Override
  public UpdateSourceReq.Response update(UpdateSourceReq
      req) {
    return dbStore.updateSource(req);
  }
  // Other methods removed for brevity
```

https://github.com/PacktPublishing/Modern-API-Development-with-Spring-and-Spring-Boot/blob/main/Chapter11/server/src/main/java/com/packt/modern/api/server/repository/SourceRepositoryImpl.java

`SourceRepository` contains the methods mapped to the `SourceService` abstract base class. You can refer to it at `https://github.com/PacktPublishing/ Modern-API-Development-with-Spring-and-Spring-Boot/blob/main/ Chapter11/server/src/main/java/com/packt/modern/api/server/ repository/SourceRepository.java`.

These repository classes can be used from the service classes, which implement the abstract generated base classes. Let's write these next.

Implementing service classes

Now you have the underlying implementation ready in the form of repository and `DB store` classes that can be used to implement the gRPC service's base classes.

Let's implement the `Source` service first, as shown next:

1. Create a new file, `SourceService.java`, in the `server/src/main/com/packt/modern/api/server/service` directory.

2. Add the implementations to operations defined in the `SourceService` abstract base class, as shown next:

```
@Service
public class SourceService extends
    SourceServiceImplBase {
  private SourceRepository repository;
  public SourceService(SourceRepository repository) {
    this.repository = repository;
  }
  @Override
  public void create(CreateSourceReq req,
      StreamObserver<CreateSourceReq.Response>
        resObserver) {
    CreateSourceReq.Response resp =
        repository.create(req);
    resObserver.onNext(resp);
    resObserver.onCompleted();
  }
  // Other methods removed for brevity
```

https://github.com/PacktPublishing/Modern-API-Development-with-Spring-and-Spring-Boot/blob/main/Chapter11/server/src/main/java/com/packt/modern/api/server/service/SourceService.java

Here, the `SourceServiceImplBase` abstract class is autogenerated by the `Protobuf` plugin that contains the contract methods of the `Source` service. A very unique part of the method signature generated is the second argument, `StreamObserver`. `StreamObserver` receives notifications from observable streams. It is being used here for service implementation. Similarly, it is also used in the client stubs. The gRPC library provides the `StreamObserver` argument for outgoing messages. However, you also have to implement it for incoming messages.

`StreamObserver` arguments are not thread-safe, so you have to take care of multithreading and should use synchronized calls.

3. There are three primary methods of `StreamObserver`:

- `onNext()`: This method receives the value from the stream. It can be called multiple times. However, it should not be called after `onCompleted()` or `onError()`. Multiple `onNext()` calls are required for streams when multiple data is sent to clients.

- `onCompleted()`: This marks the completion of the stream and no further method calls are allowed after that. It can only be called once.

- `onError()`: This method receives the termination error from the stream. Like `onCompleted()`, it can only be called once and no further method calls are allowed.

4. Similarly, you can implement the other methods of an abstract class.

Next, in the same way that you have implemented the `Source` service, similarly, you can implement the `Charge` service.

Let's implement the `Source` service first, as shown next:

1. Create a new file, `ChargeService.java`, in the `server/src/main/com/packt/modern/api/server/service` directory.

2. Add the implementations to operations defined in the `SourceService` abstract base class, as shown here:

```
@Service
public class ChargeService extends
        ChargeServiceImplBase {
  private ChargeRepository repository;
  public ChargeService(ChargeRepository repository) {
    this.repository = repository;
  }
  @Override
  public void create(CreateChargeReq req,
        StreamObserver<CreateChargeReq.Response>
        resObserver) {
    CreateSourceReq.Response resp =
        repository.create(req);
    resObserver.onNext(resp);
    resObserver.onCompleted();
```

```
    }
    // Other methods removed for brevity
```

https://github.com/PacktPublishing/Modern-API-Development-with-Spring-and-Spring-Boot/blob/main/Chapter11/server/src/main/java/com/packt/modern/api/server/service/ChargeService.java

This is along the same lines as the manner in which the `SourceService create` method was implemented.

3. Similarly, you can implement the other methods of the abstract class. Please refer to the source link shown in the previous code block for complete code implementation.

Now, you have gRPC service implementation ready. Let's implement the server next.

Implementation of the gRPC server

Spring Boot application runs on its own server. However, we want to run the gRPC server, which internally uses the Netty web server. Therefore, we first need to modify the Spring Boot configuration to stop running its web server. You can do that by modifying the `application.properties` file, as shown in the following code block:

```
spring.main.web-application-type=none
grpc.port=8080
```

https://github.com/PacktPublishing/Modern-API-Development-with-Spring-and-Spring-Boot/blob/main/Chapter11/server/src/main/resources/application.properties

Next, let's create the gRPC server. It will have three methods – `start()`, `stop()`, and `block()` for starting up the server, stopping the server, and serving requests until a termination request is received, respectively.

Create a new file, `GrpcServer.java`, in the `server/src/main/com/packt/modern/api/server` directory and the code, as shown in the following code block:

```
@Component
public class GrpcServer {
    private final Logger LOG =
        LoggerFactory.getLogger(getClass());
    @Value("${grpc.port:8080}")
    private int port;
```

```java
private Server server;
private ChargeService chargeService;
private SourceService sourceService;
private ExceptionInterceptor exceptionInterceptor;
public GrpcServer(…) {
  // code removed for brevity
}
public void start() throws IOException,
  InterruptedException {
  server = ServerBuilder.forPort(port)
      .addService(sourceService).addService(chargeService)
      .intercept(exceptionInterceptor).build().start();
  server.getServices().stream().forEach(s -> LOG.info(
      "Service Name: {}", s.getServiceDescriptor().
        getName()));
  Runtime.getRuntime().addShutdownHook(new Thread(() -> {
   GrpcServer.this.stop();
  }));
}
private void stop() {
  if (server != null) {
    server.shutdown();
  }
}
public void block() throws InterruptedException {
  if (server != null) {
    // received the request until application is
    // terminated
    server.awaitTermination();
  }
}
}
```

https://github.com/PacktPublishing/Modern-API-Development-with-Spring-and-Spring-Boot/blob/main/Chapter11/server/src/main/java/com/packt/modern/api/server/GrpcServer.java

The server library of gRPC provides the server builder for building the server. You can see that both the services are added to the server. The builder also allows you to add interceptors that can intercept the incoming request and response. We are going to use the interceptor in the *Handling errors* section.

The GrpcServer start() method also added a shutdown hook that calls the stop() method, which internally calls the server.shutdown() method.

The server code is ready. Now, you need an interface to start the server. You are going to use the CommandLineRunner function interface to run the server.

Create a new file, GrpcServerRunner.java in the same directory where you created the GrpcServer.java file and add the following code:

```
@Profile("!test")
@Component
public class GrpcServerRunner implements CommandLineRunner {
    private GrpcServer grpcServer;
    public GrpcServerRunner(GrpcServer grpcServer) {
        this.grpcServer = grpcServer;
    }
    @Override
    public void run(String... args) throws Exception {
        grpcServer.start();
        grpcServer.block();
    }
}
```

https://github.com/PacktPublishing/Modern-API-Development-with-Spring-and-Spring-Boot/blob/main/Chapter11/server/src/main/java/com/packt/modern/api/server/GrpcServerRunner.java

Here, you override the CommandLineRunner run() method and call the start and block methods. Therefore, when you execute the jar file, GrpcServerRunner will be executed using its run() method and will start the gRPC server.

Another thing to remember is that you have marked the GrpcServerRunner class with the @Profile annotation with the "!test" value, which means that when the test profile is active, this class won't be loaded, and hence not executed.

You are now done with both service and server implementation, so let's test it in the next subsection.

Testing the gRPC server

First of all, you need to set the active profile to test in your test classes because it will disable the GrpcServerRunner. Let's do this and test it, as shown in the following code block:

```
@ActiveProfiles("test")
@SpringBootTest
@TestMethodOrder(OrderAnnotation.class)
class ServerAppTests {
  @Autowired
  private ApplicationContext context;
  @Test
  @Order(1)
  void beanGrpcServerRunnerTest() {
    assertNotNull(context.getBean(GrpcServer.class));
    assertThrows(NoSuchBeanDefinitionException.class,
        () -> context.getBean(GrpcServerRunner.class),
      "GrpcServerRunner should not be loaded during
         test");
  }
  // continue…
```

https://github.com/PacktPublishing/Modern-API-Development-with-Spring-and-Spring-Boot/blob/main/Chapter11/server/src/test/java/com/packt/modern/api/ServerAppTests.java

The beanGrpcServerRunnerTest() method tests the loading of the GrpcServer class and GrpcServerRunner and the test should be passed if the profile is set correctly.

Now, let's move on to test the gRPC services.

The gRPC test library provides a special class, GrpcCleanupRule, that manages the shutdown of registered servers and channels gracefully. You need to annotate with the JUnit @Rule to make it effective. The gRPC test library also provides the InProcessServerBuilder builder class, which allows you to build the server, and the InProcessChannelBuilder builder class, which allows you to build the channel. These three classes are all you need to build and manage the server and channel.

Therefore, you first need to declare the required instances and then set up the method so that the execution environment is available before you fire the requests to the gRPC Source service.

Let's add the required class instances and test `setup()` method in the following code:

```
@Rule
private final static GrpcCleanupRule grpcCleanup = new
    GrpcCleanupRule();
private static SourceServiceGrpc.
    SourceServiceBlockingStub blockingStub;
private static String newlyCreatedSourceId = null;

@BeforeAll
public static void setup(@Autowired SourceService
    sourceService,
    @Autowired ChargeService chargeService,
    @Autowired ExceptionInterceptor exceptionInterceptor)
    throws IOException {
  String serverName =
            InProcessServerBuilder.generateName(); // 1
  grpcCleanup.register(InProcessServerBuilder
      .forName(serverName)
      .directExecutor().addService(sourceService)
      .intercept(exceptionInterceptor).build().start());
                                                    // 2
  blockingStub =
      SourceServiceGrpc.newBlockingStub(grpcCleanup
      .register(InProcessChannelBuilder.forName(serverName)
      .directExecutor().build()));
                                                    // 3

}
```

Here, the setup method creates the server and channel with the Source service. Let's understand each of the lines mentioned in the setup() method:

- *Line 1* generates the unique name of the server.

- *Line 2* registers the newly created server and adds the Source service and server interceptor to it. We'll discuss ExceptionInterceptor in the *Handling errors* section. Then it starts the server for serving requests.

- *Line 3* creates the blocking stub that will be used as a client for making the calls to the server. Here again, GrpcCleanUpRule is used to create the client channel.

Once the setup is executed, it provides us with the environment to carry out the tests. Let's test our first request, as shown in the following code block:

```
@Test
@Order(2)
@DisplayName("Creates the source object using create RPC
    call")
public void SourceService_Create() {
  CreateSourceReq.Response response =
      blockingStub.create(
              CreateSourceReq.newBuilder().setAmount(100)
              .setCurrency("USD").build());
  assertNotNull(response);
  assertNotNull(response.getSource());
  newlyCreatedSourceId = response.getSource().getId();
  assertEquals(100, response.getSource().getAmount());
  assertEquals("USD", response.getSource().getCurrency());
}
```

All the complex aspects of the setup() method are complete. These tests now look pretty simple. You just use the blocking stub to make a call. You create the request object and use the stub to call the server. Finally, validate the server responses.

Similarly, you can test the validation error, as shown in the following code block:

```
@Test
@Order(3)
@DisplayName("Throws the exception when invalid source id
              is passed to retrieve RPC call")
public void SourceService_RetrieveForInvalidId() {
```

```
        Throwable throwable = assertThrows(
            StatusRuntimeException.class, () -> blockingStub
                .retrieve(
                    SourceId.newBuilder().setId("").build()));
        assertEquals("INVALID_ARGUMENT: Invalid Source ID is
                passed.", throwable.getMessage());
    }
```

You can also test for the valid response for source retrieval, as shown in the following code block:

```
@Test
@Order(4)
@DisplayName("Retrieves source obj created using create
    RPC call")
public void SourceService_Retrieve() {
  SourceId.Response response =
      blockingStub.retrieve(SourceId
              .newBuilder().setId(newlyCreatedSourceId).
              build());
  assertNotNull(response);
  assertNotNull(response.getSource());
  assertEquals(100, response.getSource().getAmount());
  assertEquals("USD", response.getSource().getCurrency());
}
```

This is the way in which you can write the test for the gRPC server and test the exposed RPC calls. You can use the same approach for writing the rest of the test cases.

After writing the test, you may have an idea of how the client is going to send the request to the server.

We have not yet discussed the exception interceptor that we have used in both the server code and test. Let's discuss this in the next section.

Handling errors

You may have already gone through the *Handling errors* section in *Chapter 10, gRPC Fundamentals*, where google.rpc.Status and gRPC status codes are discussed. You may like to revisit this section before going through this section.

io.grpc.ServerInterceptor is a thread-safe interface for intercepting the incoming calls that can be used for cross-cutting calls, such as authentication and authorization, logging, and monitoring. Let's use it to write the ExceptionInterceptor, as shown in the following code block:

```
@Component
public class ExceptionInterceptor implements ServerInterceptor
{

  @Override
  public <RQT, RST> ServerCall.Listener<RQT> interceptCall(
          ServerCall<RQT, RST> serverCall, Metadata metadata,
          ServerCallHandler<RQT, RST> serverCallHandler) {
      ServerCall.Listener<RQT> listener = serverCallHandler
                              .startCall(serverCall, metadata);
      return new ExceptionHandlingServerCallListener
          <>(listener, serverCall, metadata);
  }
  // continue…
```

https://github.com/PacktPublishing/Modern-API-Development-with-Spring-and-Spring-Boot/blob/main/Chapter11/server/src/main/java/com/packt/modern/api/server/interceptor/ExceptionInterceptor.java

Here, RQT represents the request type, and RST represents the response type.

We are going to use it for exception intercepting. An interceptor will pass the call to the server listener (ExceptionHandlingServerCallListener). ExceptionHandlingServerCallListener is a private class in ExceptionInterceptor that extends the ForwardingServerCallListener. SimpleForwardingServerCallListener abstract class.

It (the private listener class) has overridden events, onHalfClose() and onReady(), that will catch the exception and pass the call to the handleException() method. The handleException() method will use the ExceptionUtils method for tracing the actual exception and respond with error details. It returns StatusRuntimeException, which is used for closing the server call with an error status.

Let's see how this flow looks in code in the next code block:

```
private class ExceptionHandlingServerCallListener<RQT, RST>
      extends ForwardingServerCallListener
              .SimpleForwardingServerCallListener<RQT> {
```

```
private final ServerCall<RQT, RST> serverCall;
private final Metadata metadata;
ExceptionHandlingServerCallListener(
    ServerCall.Listener<RQT>
    listener,ServerCall<RQT, RST> serverCall, Metadata
        metadata) {
  super(listener);
  this.serverCall = serverCall;
  this.metadata = metadata;
}
@Override
public void onHalfClose() {
  try { super.onHalfClose();}
  catch (RuntimeException e) {
    handleException(e, serverCall, metadata);
    throw e;
  }
}
@Override
public void onReady() {
  try { super.onReady();}
  catch (RuntimeException e) {
    handleException(e, serverCall, metadata);
    throw e;
  }
}
private void handleException(RuntimeException e,
          ServerCall<RQT, RST> serverCall, Metadata
          metadata) {
  StatusRuntimeException status =
      ExceptionUtils.traceException(e);
  serverCall.close(status.getStatus(), metadata);
}
}
```

Let's write the `ExceptionUtils` class next to complete the exception handling core components. Then, you can use these components in a service implementation to raise the exceptions.

The `ExceptionUtils` class will have two types of overloaded methods:

- `observerError()`: This method will use `StreamObserver` to raise the `onError()` event.

- `traceException()`: This method will trace the error from `Throwable` and return the `StatusRuntimeException` instance.

You can use the following code to write the `ExceptionUtils` class:

```
@Component
public class ExceptionUtils {
  public static StatusRuntimeException
      traceException(Throwable e) {
    return traceException(e, null);
  }
  public static <T extends GeneratedMessageV3> void
      observeError(
        StreamObserver<T> responseObserver, Throwable e) {
    responseObserver.onError(traceException(e));
  }
  public static <T extends GeneratedMessageV3> void
      observeError(
          StreamObserver<T> responseObserver, Exception e,
                                      T defaultInstance) {
    responseObserver.onError(traceException(e,
        defaultInstance));
  }
  // continue...
```

You can see that the observerError() method is also calling traceException() internally for onError events. Let's write the last overloaded method, traceException(), next:

```java
public static <T extends com.google.protobuf.
        GeneratedMessageV3>
        StatusRuntimeException traceException(Throwable e,
        T defaultInstance) {
    com.google.rpc.Status status;
    StatusRuntimeException statusRuntimeException;
    if (e instanceof StatusRuntimeException) {
      statusRuntimeException = (StatusRuntimeException) e;
    } else {
      Throwable cause = e;
      if (cause != null && cause.getCause() != null
                              && cause.getCause() !=
                              cause) {
        cause = cause.getCause();
      }
      if (cause instanceof SocketException) {
        String errorMessage = "Sample exception message";
        status = com.google.rpc.Status.newBuilder()
            .setCode(com.google.rpc.Code.UNAVAILABLE_VALUE)
            .setMessage(errorMessage + cause.getMessage())
            .addDetails(Any.pack(defaultInstance)).build();
      } else {
        status = com.google.rpc.Status.newBuilder()
            .setCode(com.google.rpc.Code.INTERNAL_VALUE)
            .setMessage("Internal server error")
            .addDetails(Any.pack(defaultInstance)).build();
      }
      statusRuntimeException =
                      StatusProto.toStatusRuntimeException(
                          status);
    }
    return statusRuntimeException;
}
```

SocketException is shown by way of an example. You can add a check for another kind of exception here. You may notice that here we are using com.google.rpc.Status to build the status. Then, this instance of status is passed to toStatusRuntimeException() of StatusProto, which converts the status to StatusRuntimeException.

Let's add the validation error in the DbStore class to make use of these exception handling components, as shown in the following code block:

```
public SourceId.Response retrieveSource(String sourceId) {
  if (Strings.isBlank(sourceId)) {
    com.google.rpc.Status status =
        com.google.rpc.Status.newBuilder()
      .setCode(Code.INVALID_ARGUMENT.getNumber())
      .setMessage("Invalid Source ID is passed.")
      .addDetails(Any.pack(
          SourceId.Response.getDefaultInstance())).build();
    throw StatusProto.toStatusRuntimeException(status);
  }
  Source source = sourceEntities.get(sourceId);
  if (Objects.isNull(source)) {
    com.google.rpc.Status status =
        com.google.rpc.Status.newBuilder()
      .setCode(Code.INVALID_ARGUMENT.getNumber())
      .setMessage("Requested source is not
                available").build();
    throw StatusProto.toStatusRuntimeException(status);
  }
  return SourceId.Response.newBuilder().setSource(source)
    .build();
}
```

https://github.com/PacktPublishing/Modern-API-Development-with-Spring-and-Spring-Boot/blob/main/Chapter11/server/src/main/java/com/packt/modern/api/server/repository/DbStore.java

You can similarly raise `StatusRuntimeException` in any part of the service implementation. You can also use the `addDetails()` method of `com.google.rpc.Status` to add more details to the error status, as shown in the `traceException(Throwable e, T defaultInstance)` code.

Finally, you can capture the error raised by the `retrieve()` method of `SourceService` in the Service implementation class, as shown next:

```
@Override
public void retrieve(SourceId sourceId,
                    StreamObserver<SourceId.Response>
resObserver) {
  try {
    SourceId.Response resp =
        repository.retrieve(sourceId.getId());
    resObserver.onNext(resp);
    resObserver.onCompleted();
  } catch (Exception e) {
    ExceptionUtils.observeError(resObserver, e,
                          SourceId.Response.
                            getDefaultInstance());
  }
}
```

https://github.com/PacktPublishing/Modern-API-Development-with-Spring-and-Spring-Boot/blob/main/Chapter11/server/src/main/java/com/packt/modern/api/server/service/SourceService.java

Exception handling is explained simply and constructively in this chapter. You can enhance it more as per your application requirements.

Now, let's write the gRPC client in the next section.

Developing the gRPC client

A client project's directory structure will look as follows. The project root directory contains the `build.gradle` and `settings.gradle` files:

Project directory structure of Client project

```
├──client
    ├──gradle
    │    └──wrapper
    └──src
        ├──main
        │    ├──java
        │    │    └──com
        │    │        └──packt
        │    │            └──modern
        │    │                └──api
        │    └──resources
        └──test
            └──java
```

The resources directory will contain the `application.properties` file.

Let's perform the following steps to configure the project:

1. First, you need to modify the project name in the `Chapter11/client/settings.gradle` file to represent the server, as shown here:

   ```
   rootProject.name = 'chapter11-client'
   ```

2. Next, you can add the dependencies required for client projects in the `Chapter11/client/build.gradle` file. The `grpc-stub` library provides the stubs-related APIs, and `protobuf-java-util` provides the utility methods for Protobuf and JSON conversions:

   ```
   def grpcVersion = '1.37.0'
   dependencies {
     implementation 'com.packt.modern.api:payment-gateway-
                 api:0.0.1'

     // gRPC dependencies
   ```

```
implementation "io.grpc:grpc-stub:${grpcVersion}"
implementation "com.google.protobuf:protobuf-java-
           util:3.15.8"
implementation 'org.springframework.boot:spring-
           bootstarter-web'
testImplementation 'org.springframework.boot:spring-
           boot-starter-test'
}
```

https://github.com/PacktPublishing/Modern-API-Development-with-Spring-and-Spring-Boot/tree/main/Chapter11/client/build.gradle

3. The `payment-gateway-api` dependency is published in the local Maven repository. Therefore, you need to add the local Maven repository to the repository section, as shown in the following code block:

```
repositories {
    mavenCentral()
    mavenLocal()
}
```

You are done with the Gradle configuration. Now, you can write the gRPC client.

Implementing the gRPC client

As you know, the Spring Boot application runs on its own server. Therefore, the client's application port should be different to the gRPC server port. Also, we need to provide the gRPC server host and port. These can be configured in `application.properties`:

```
server.port=8081

grpc.server.host=localhost
grpc.server.port=8080
```

https://github.com/PacktPublishing/Modern-API-Development-with-Spring-and-Spring-Boot/blob/main/Chapter11/client/src/main/resources/application.properties

Next, let's create the gRPC client. This client will be used for configuring the gRPC service stubs with the channel. The channel is responsible for providing the virtual connection to a conceptual endpoint in order to perform gRPC calls.

Create a new file, GrpcClient.java, in the client/src/main/com/packt/modern/api/client directory and the code, as shown in the following code block:

```java
@Component
public class GrpcClient {
  @Value("${grpc.server.host:localhost}")
  private String host;
  @Value("${grpc.server.port:8080}")
  private int port;
  private ManagedChannel channel;
  private SourceServiceBlockingStub sourceServiceStub;
  private ChargeServiceBlockingStub chargeServiceStub;
  public void start() {
    channel = ManagedChannelBuilder.forAddress(host, port)
                                        .usePlaintext().
                                    build();
    sourceServiceStub = SourceServiceGrpc.newBlockingStub(
        channel);
    chargeServiceStub = ChargeServiceGrpc.newBlockingStub(
        channel);
  }
  public void shutdown() throws InterruptedException {
    channel.shutdown().awaitTermination(1,
        TimeUnit.SECONDS);
  }
  public SourceServiceBlockingStub getSourceServiceStub() {
    return this.sourceServiceStub;
  }
  public ChargeServiceBlockingStub getChargeServiceStub() {
    return this.chargeServiceStub;
  }
}
```

https://github.com/PacktPublishing/Modern-API-Development-with-Spring-and-Spring-Boot/blob/main/Chapter11/client/src/main/java/com/packt/modern/api/client/GrpcClient.java

Here, start () is the key that initialized the Source and Charge service stubs. ManagedChannelBuilder is used for building the ManagedChannel. ManagedChannel is a channel that also provides life cycle management. This managed channel is passed to stubs.

You are using plain-text communication. However, it also provides encrypted communication.

We are now done with the client's code. Now, we need to call the start () method. You are going to implement CommandLineRunner the way it was implemented for the GrpcServerRunner class.

It can be implemented as shown here:

```java
@Profile("!test")
@Component
public class GrpcClientRunner implements CommandLineRunner {
  @Autowired
  GrpcClient client;
  @Override
  public void run(String... args) {
    client.start();
    Runtime.getRuntime().addShutdownHook(new Thread(() -> {
      try { client.shutdown();}
      catch (InterruptedException e) {
        // logging and handling
      }
    }));
  }
}
```

https://github.com/PacktPublishing/Modern-API-Development-with-Spring-and-Spring-Boot/blob/main/Chapter11/client/src/main/java/com/packt/modern/api/client/GrpcClientRunner.java

This will initiate the stub instantiation following the start of the application. You can then call the stub methods.

Now, to call the stub methods, let's add a simple REST endpoint. This will demonstrate how to use the charge service stub to call its retrieve method.

You can create a new `ChargeController.java` file for the REST controller in the `src/main/java/com/packts/modern/api/controller` directory and add the code as shown here:

```java
@RestController
public class ChargeController {
  private GrpcClient client;
  public ChargeController(GrpcClient client) {
    this.client = client;
  }
  @GetMapping("/charges")
  public String getSources(@RequestParam(
          defaultValue = "ab1ab2ab3ab4ab5") String
              customerId)
            throws InvalidProtocolBufferException {
    var req = CustomerId.newBuilder().setId(
        customerId).build();
    CustomerId.Response resp = client
                        .getChargeServiceStub().
                          retrieveAll(req);
    var printer = JsonFormat
                        .printer().
                        includingDefaultValueFields();
    return printer.print(resp);
  }
}
```

https://github.com/PacktPublishing/Modern-API-Development-with-Spring-and-Spring-Boot/blob/main/Chapter11/client/src/main/java/com/packt/modern/api/controller/ChargeController.java

Here, we have created a REST endpoint, `/charges`. This uses the `GrpcClient` instance to call the `retrieveAll()` RPC method of the Charge gRPC service using `ChargeServiceStub`.

Then, the response is converted to a JSON-formatted string using the `JsonFormat` class from the `protobuf-java-util` library and returned as a response. Generated JSON-formatted strings will also contain the fields with default values.

We are done with our development. Let's now test the complete flow in the next subsection.

Testing the gRPC client

Make sure that your gRPC server is up and running before testing the client. It is assumed that the gRPC `api` project has been built and that its latest artifacts have been published to the local Maven repository:

1. First, start the server using the following command. Run it from the server project's root directory:

```
// Server project root directory
$ gradlew clean build
$ java -jar build\libs\chapter11-server-0.0.1-SNAPSHOT.
jar
```

```
  INFO …GrpcServer : gRPC server started and listening on
port: 8080
 .INFO …GrpcServer : Following service are available:
  INFO …GrpcServer : Service Name: com.packtpub.
v1.SourceService
  INFO …GrpcServer : Service Name: com.packtpub.
v1.ChargeService
```

2. Next, start the client using the following command in a new terminal window. Run it from the client project's root directory:

```
// Client project root directory
$ gradlew clean build
$ java -jar build\libs\chapter11-client-0.0.1-SNAPSHOT.
jar
```

```
  INFO …GrpcServer : Tomcat started on port(s): 8081
(http) …
  INFO …GrpcServer : …
  INFO …GrpcServer : gRPC client connected to
  localhost:8080
```

3. Open a new terminal window and execute the following command (the output is truncated):

```
$ curl http://localhost:8081/charges
{
    "charge": [{
        "id": "3igf4pd8vi1e1oadflov5gcnce64h5i3",
        "amount": 1000,
        "amountCaptured": 0,
        ...

        ...
        "created": "1619461208",
        "currency": "USD",
        "customerId": "ab1ab2ab3ab4ab5",
        "description": "Charge Description",
        ...

        ...
        "receiptEmail": "receipt@email.com",
        "statementDescriptor": "Statement Descriptor",
        "status": "SUCCEEDED",
        "sourceId": "cb863hsulr0u0deh5et1hfatqapng2jj"
    }]
}
```

A REST endpoint is used for demonstration purposes only. Similarly, you can use the gRPC client for calling other services and their methods. gRPC is often used for inter-service communication, which is essential for microservice-based applications. However, it can also be used for web-based communication.

Let's learn a bit about microservices in the next section.

Learning microservice concepts

Microservices are self-contained lightweight processes that communicate over a network. Microservices provides narrowly focused APIs to their consumers. These APIs can be implemented using REST, gRPC, or events.

Microservices are not new—they have been around for many years. For example, Stubby, a general-purpose infrastructure based on **Remote Procedure Call** (**RPC**), was used in Google data centers in the early 2000s to connect several services with and across data centers.

Its recent rise is due to its popularity and visibility. Before microservices became popular, monolithic architectures were mainly being used for developing on-premises and cloud-based applications.

A monolithic architecture allows the development of different components, such as presentation, application logic, business logic, and **Data Access Objects** (**DAOs**), and then you either bundle them together in an **Enterprise Archive** (**EAR**) or a **Web Archive** (**WAR**) or store them in a single directory hierarchy (such as Rails or Node.js).

Many famous applications, such as Netflix, have been developed using a microservices architecture. Moreover, eBay, Amazon, and Groupon have evolved from monolithic architectures to microservices architectures. Nowadays, microservices-based application development is very common. The gRPC server that we have developed in this chapter could be called a microservice (obviously, if you keep the scope of the server to either the Source service or Charge server).

Let's have a look at simple monolithic and microservice application designs in the next subsection.

Design differences in monolithic and microservice-based systems

Here, you'll have a look at the different system designs, which are designed using a monolithic design, an SOA monolithic design, and a microservices design. Let's discuss each of these in turn.

Traditional monolithic design

The following diagram depicts the traditional monolithic application design. This design was widely used before SOA became popular:

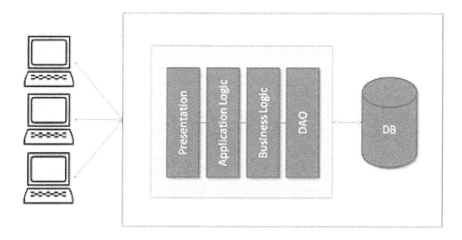

Figure 11.1 – Traditional monolithic application design

In a traditional monolithic design, everything is bundled in the same archive (all the presentation code is bundled in with the presentation archive, the application logic goes into the application logic archive, and so on), regardless of how it all interacts with the database files or other sources.

Monolithic design with services

After SOA, applications started being developed based on services, where each component provides services to other components or external entities. The following diagram depicts a monolithic application with different services; here, services are being used with a presentation component. All services, the presentation component, or any other components are bundled together:

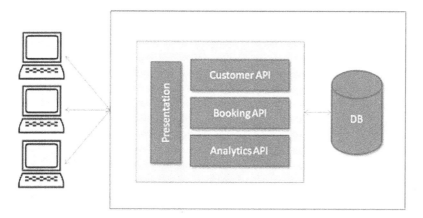

Figure 11.2 – Monolithic design with services

So, everything is bundled together in the form of EAR with a modules approach. However, few SOA services may be deployed separately, but overall, it will be monolithic, too. However, the database is shared across.

Microservices design

The following diagram depicts the microservices design. Here, each component is autonomous. Each component could be developed, built, tested, and deployed independently. Here, even the application's **User Interface** (**UI**) component could also be a client and consume the microservices. For our example, the layer designed is used within the microservice.

The API gateway provides an interface where different clients can access the individual services and solve various problems, such as what to do when you want to send different responses to different clients for the same service. For example, a booking service could send different responses to a mobile client (minimal information) and a desktop client (detailed information), providing different details to each, before providing something different again to a third-party client.

A response may require the fetching of information from two or more services:

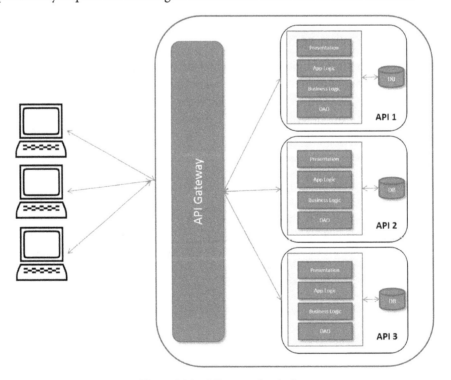

Figure 11.3 – Microservice design

Each API service will be developed and deployed as a separate process and communication among these will happen based on exposed APIs.

For a sample e-commerce app, you can divide the application based on domains and bounded context and then develop a separate microservice for each of the domains. A quick look provides the following microservice:

- Customers

- Orders

- Billing

- Shipping

- Invoicing

- Inventory

- Payment collection and so on

You can develop each of these separately and use inter-process (inter-service) communication to stitch the solution together.

Summary

In this chapter, you have explored the protocol buffer, aka Protobuf, and gRPC-based service implementation. You have developed the gRPC server and then consumed its services by developing a gRPC client. You have learned about unit testing the gRPC server and handing the exception for gRPC-based services, and you have also learned about the basic concepts of microservices.

You now have the skills to develop the gRPC-based services (servers) and clients by defining the services using the Protobuf.

In the next chapter, you will learn about distributed logging and tracing in web services.

Questions

1. Why should you use gRPC for binary large object transfers via HTTP/2?

2. You have implemented exception handling using `com.google.rpc.Status`. Can you do so without using this?

3. What is the difference between `com.google.rpc.Status` and `io.grpc.Status`?

Further reading

- Protocol Buffer (Protobuf) version 3 documentation: `https://developers.google.com/protocol-buffers/docs/proto3`

- Protobuf's well-known types: `https://developers.google.com/protocol-buffers/docs/reference/google.protobuf`

- Practical gRPC: `https://www.packtpub.com/in/web-development/practical-grpc`

12
Logging and Tracing

In this chapter, you will explore the logging and monitoring tools, the **ELK (Elastic Search, Logstash, Kibana)** stack and Zipkin. This tool will then be used to implement the distributed logging and tracing of the request/response of API calls. Spring Sleuth will be used to inject the tracing information into API calls. You will learn how to publish and analyze the logging and tracing of different requests and logs related to responses.

These aggregated logs will help you to troubleshoot web services. You will call one service (such as gRPC client), which will then call another service (such as gRPC server), and link them with a trace identifier. Then, using this trace identifier, you can search the centralized logs and debug the request flows. In this chapter, we will use this sample flow. However, the same can be used where service calls require more internal calls. You will also use Zipkin to ascertain the performance of each API call.

You will explore the following topics in this chapter:

- Introducing logging and tracing
- Understanding the ELK stack
- Installing the ELK stack
- Implementing logging and tracing
- Distributed tracing with Zipkin

After completing this chapter, you will have an understanding of distributed logging and monitoring and the ELK stack.

Technical requirements

This chapter contains the theory of gRPC. However, the following is required for the development and testing of gRPC-based web services:

- Any Java IDE, such as NetBeans, IntelliJ, or Eclipse
- **Java Development Kit (JDK)** 15
- An internet connection to clone the code and download the dependencies and Gradle
- Postman/cURL (for API testing)
- Docker and Docker Compose

So let's begin!

Please visit the following link to check the code: `https://github.com/PacktPublishing/Modern-API-Development-with-Spring-and-Spring-Boot/tree/main/Chapter12`

Introducing logging and tracing

Today, products and services are divided into multiple small parts and executed as separate processes or deployed as separate services, unlike a monolithic system. An API call may make several other internal API calls. Therefore, you need distributed and centralized logging to trace the request that spans multiple web services. This tracing can be done using the trace identifier (`traceId`), which can also be referred to as a correlation identifier (`correlationId`). This identifier is a collection of characters that form a unique string, which is populated and assigned to an API call that requires multiple inter-service calls. Then, the same trace identifier is propagated to subsequent API calls for tracking purposes.

Errors and issues are imminent in the production system. You need debugging to ascertain the root cause. One of the key tools associated with debugging is logs. Logs can also give you warnings relating to the system if the system is designed to cater to warnings. Logs also offer throughput, capacity, and monitoring of the health of the system. Therefore, you need an excellent logging platform and strategy that enables effective debugging.

There are different open source and enterprise tools available in the market for logging, including Splunk, Graylog, and the ELK stack. The ELK stack is the most popular of these and allows you to use it for free unless you won't offer it as SaaS. You are going to use the ELK stack for logging throughout this chapter.

Let's understand the ELK stack in the next subsection.

Understanding the ELK stack

The **ELK stack** comprises three components – Elasticsearch, Logstash, and Kibana. All three products are part of Elasticsearch B.V. (`https://www.elastic.co/`). The ELK stack performs the aggregation, analysis, visualization, and monitoring of logs. The ELK stack provides a complete blogging platform that allows you to analyze, visualize, and monitor all types of logs, including product and system logs.

You are going to use the following workflow for publishing logs, as shown in the following diagram:

Figure 12.1 – Log flows in the ELK stack

Let's understand the diagram:

- Services/system logs are pushed to Logstash on the TCP port.

- Logstash pushes the logs to Elasticsearch for indexing.

- Kibana then uses the Elasticsearch index to query and visualize the logs.

In an ideal production system, you should use one more layer. A broker layer such as Redis/Kafka/RabbitMQ should be placed between the service logs and Logstash. This prevents data loss and can handle the sudden spike in input load.

> **Tips for the ELK stack configuration**
>
> The ELK stack is fully customizable and comes with a default configuration. However, if you are using the Elasticsearch cluster (more than one instance of Elasticsearch is deployed), it is better to use an odd number of Elasticsearch nodes (instances) to avoid the split-brain problem.
>
> It is recommended to use the appropriate data type for all the fields (in log JSON input). This will allow you to perform the logical check and comparators while querying the log data. For example, you can't perform the `http_status < 400` check if the `http_status` field type is a string.

If you are already familiar with the ELK stack, you can skip and move on to the next section. Here, you'll find a brief introduction to each of the tools in the ELK stack.

Elasticsearch

Elasticsearch is one of the most popular enterprise full-text search engines that is based on Apache Lucene and developed using Java. Elasticsearch is also a high-performance, full-featured text search engine library. It is restricted open source software as per recent changes in the licensing terms that prevent you from offering Elasticsearch or the ELK stack as an SaaS. It is distributable and supports multi-tenancy. A single Elasticsearch server stores multiple indexes (each index represents a database), and a single query can search the data of multiple indexes. It is a distributed search engine and supports clustering.

It is readily scalable and can provide near-real-time searches with a latency of 1 second. Elasticsearch APIs are extensive and are very elaborative. Elasticsearch provides JSON-based schema-less storage and represents data models in JSON. Elasticsearch APIs use JSON documents for HTTP requests and responses.

Logstash

Logstash is an open source data collection engine with real-time pipeline capabilities. It performs three major operations – it collects the data, filters the information, and outputs the processed information to data storage, in the same way as Elasticsearch does. It allows you to process any event data, such as logs from a variety of systems, because of data pipeline capabilities.

Logstash runs as an agent that collects the data, parses it, filters it, and sends the output to a designated data store, such as Elasticsearch, or as simple standard output on a console.

On top of that, it has a rich set of plugins.

Kibana

Kibana is an open source web application that is used for visualizing and performing information analytics. It interacts with Elasticsearch and provides easy integration with it. You can perform searching, display, and interact with the information stored in Elasticsearch indices.

It is a browser-based web application that lets you perform advanced data analysis and visualize your data in a variety of charts, tables, and maps. Moreover, it is a zero-configuration application. Therefore, it does not require any coding or additional infrastructure following installation.

Next, let's learn how to install the ELK stack

Installing the ELK stack

You can use various methods to install the ELK stack, such as installing individual components as per the operating system, or downloading the Docker images and running them individually, or executing the Docker images using Docker Compose/Docker Swarm/Kubernetes. You are going to use Docker Compose in this chapter.

Let's understand the grammar of the Docker Compose file before we create the ELK stack Docker Compose file. The Docker Compose file is defined using YAML. The Docker Compose file contains four important top-level keys:

- `version`: This denotes the version of the Docker Compose file format. You can use the appropriate version based on installing the Docker Engine. You can check the link at `https://docs.docker.com/compose/compose-file/` to ascertain the mapping between the Docker Compose file version and the Docker Engine version.

- `services`: This contains one or more service definitions. The service definition represents the service executed by the container that contains the container name (`container_name`), Docker image (`image`), environment variables (`environment`), external and internal ports (`port`), the command to be executed when running the container (`command`), the network to be used for communicating with other services (`networks`), mapping of the host filesystem with a running container (`volume`), and the container to be executed once the dependent service has started (`depends_on`).

- `networks`: This represents the (top-level) named network that needs to be created to establish a communication channel among the defined services. Then, this network is used by the service to communicate based on the `networks` key of the defined service. The top-level network key contains the driver field that can be `bridge` for a single host and `overlay` when used in Docker Swarm. You are going to use `bridge`.

- `volumes`: A top-level volumes key is used to create the named volume that mounts the host path. Make sure to use it only if required by the multiple services, otherwise, you can use the `volumes` key inside the service definition, which would be service-specific.

Now, let's create the Docker Compose file, `docker-compose.yaml`, in the `Chapter12` directory for defining the ELK stack. Then, you can add the following code to this file:

```
version: "3.2"
services:
  elasticsearch:
```

```
container_name: es-container
image: docker.elastic.co/elasticsearch/elasticsearch:7.12.1
environment:
  - xpack.security.enabled=false
  - "discovery.type=single-node"
networks:
  - elk-net
ports:
  - 19200:9200
```

https://github.com/PacktPublishing/Modern-API-Development-with-Spring-and-Spring-Boot/blob/main/Chapter12/docker-compose.yaml

Here, you have a defined version of the Docker Compose file. Then, create the `services` key section that contains the `elasticsearch` service. The service contains the container name, Docker image, environment variables, and network (because you want ELK components to communicate with one another). Finally, ports are defined in `external:internal` format. You are going to use port `19200` from the browser to access it. However, other services will use port `9200` for communicating with Elasticsearch.

Similarly, you can define the `logstash` service next, as shown in the following code block:

```
logstash:
  container_name: ls-container
  image: docker.elastic.co/logstash/logstash:7.12.1
  environment:
    - xpack.security.enabled=false
  command: logstash -e 'input { tcp { port => 5001 codec =>
          "json" }} output { elasticsearch { hosts =>
          "elasticsearch:9200" index => "modern-api" }}'
  networks:
    - elk-net
  depends_on:
    - elasticsearch
  ports:
    - 5002:5001
```

The Logstash configuration contains two extra service keys:

- First, a command key that contains the `logstash` command with a given configuration (using `-e`). The Logstash configuration normally contains three important parts:

  i. `input`: Logstash input channels, such as `tcp` or `File`. You are going to use a TCP input channel. This means that gRPC Server and gRPC client services will push the logs in JSON format (a JSON-coded plugin is used) to `logstash` on port `5001`.

  ii. `filter`: The filter key contains various filter expressions using different means, such as `grok`. You don't want to filter anything from logs, so you should opt out of this key.

  iii. `output`: Where to send the input data after filtering out the information. Here, you are using Elasticsearch. Logstash pushes the received log information to Elasticsearch on port `9200` and uses the `modern-api` Elasticsearch index. This index is then used on Kibana for querying, analyzing, and visualizing the logs.

- A second key, `depends_on`, tells Docker Compose to start Elasticsearch before executing the `logstash` service.

Next, let's add the final service, `kibana`, as shown in the following code block:

```
kibana:
    container_name: kb-container
    image: docker.elastic.co/kibana/kibana:7.12.1
    environment:
      - ELASTICSEARCH_HOSTS=http://es-container:9200
    networks:
      - elk-net
    depends_on:
      - elasticsearch
    ports:
      - 5600:5601
networks:
  elk-net:
    driver: bridge
```

The service's `kibana` definition is in line with other defined services. It uses the environment variable `ELASTICSEARCH_HOSTS` to connect to Elasticsearch.

At the end of the Docker Compose file, you define the `elk-net` network, which uses the `bridge` driver.

You are done with configuring the ELK stack Docker Compose file. Let's now start Docker Compose using the following command (you can use the `-f` flag if you use other filenames apart from `docker-compose.yaml` or `docker-compose.yml`):

```
$ docker-compose up -d
Creating network "chapter12_elk-net" with driver "bridge"
Creating es-container ... done
Creating ls-container ... done
Creating kb-container ... done
```

Here, the `-d` option is used, which will start Docker Compose in the background. It starts `es-container` first based on dependencies (the `depends_on` key).

> **Note**
>
> Elasticsearch uses 2GB of heap size by default. Docker also uses 2GB memory by default in some systems such as Mac.
>
> This may causes errors such as error-137. Therefore, you should increase the default Docker memory to 4GB+ to avoid such issues.
>
> Please refer link `https://docs.docker.com/config/ containers/resource_constraints/#memory` for Docker memory configurations.

You can see that first. It creates the network, and then creates the service containers and starts them. Once all the containers are up, you can hit the URL, `http:// localhost:19200/` (contains the external port defined for the `Elasticsearch service`), in the browser to check whether the `Elasticsearch` instance is up or not.

It may respond as per the following code block if the Elasticsearch service is up:

```
{
    "name" : "e0c115e8a785",
    "cluster_name" : "docker-cluster",
    "cluster_uuid" : "VKBFi9-sQoqg8xzAM1Ep-g",
    "version" : {
      "number" : "7.12.1",
      "build_flavor" : "default",
      "build_type" : "docker",
```

```
    "build_hash" : "3186837139b9c6b6d23c3200870651f10d3343b7",
    "build_snapshot" : false,
    "lucene_version" : "8.8.0",
    "minimum_wire_compatibility_version" : "6.8.0",
    "minimum_index_compatibility_version" : "6.0.0-beta1"
  },
  "tagline" : "You Know, for Search"
}
```

Next, let's check the Kibana dashboard by hitting the URL http://localhost:5600 (which contains the external port defined for the kibana service) in the browser. This should load the home page of Kibana, as shown in the following screenshot:

Figure 12.2 – Kibana home page

You may be wondering how the logs can be viewed since you have used the -d option. You can use the docker-compose logs [service name] command. If you don't provide the service name, then it would show the logs of all the services. You can use the --tail flag to filter the number of lines. The --tail="all" flag would show all the lines:

```
// Don't use flag -t as it is a switch that turns on the
timestamp
docker-compose logs --tail="10" elasticsearch
docker-compose logs --tail="10" kibana
```

You can use the following command to stop the Docker Compose file:

```
$ docker-compose down
Stopping kb-container ... done
Stopping ls-container ... done
Stopping es-container ... done
Removing kb-container ... done
Removing ls-container ... done
Removing es-container ... done
Removing network chapter12_elk-net
```

It stops the containers first based on dependencies and then removes them. Finally, it removes the network.

Next, let's make the code changes to integrate the application with the ELK stack.

Implementing logging and tracing

Logging and tracing go hand-in-hand. Logging in application code is already taken care of by default. You use Logback for logging. Logs are either configured to display on the console or pushed to the filesystem. However, you also need to push the logs to the ELK stack for indexing and analysis. For this purpose, you make certain changes to the Logback configuration file, `logback-spring.xml`, to push the logs to Logstash. On top of that, these logs should also contain tracking information.

Correlation/trace identifiers should be populated and propagated in distributed transactions for tracing purposes. A distributed transaction refers to the main API call that internally calls other services to serve the request. Spring provides a **Spring Cloud Sleuth** library that takes care of distributing tracing. It generates the trace ID along with the span identifier. The trace ID gets propagated to all the participant services during the distributed transaction. The span ID also participates in a distributed transaction. However, the scope of the span identifier belongs to its service (the one it populates).

You can copy and enhance the code from *Chapter 11, gRPC-based API Development and Testing*, found at `https://github.com/PacktPublishing/Modern-API-Development-with-Spring-and-Spring-Boot/tree/main/Chapter11`, to implement logging and tracing, or refer to this chapter's code at `https://github.com/PacktPublishing/Modern-API-Development-with-Spring-and-Spring-Boot/tree/main/Chapter12` for changes.

First of all, you'll make the changes to the gRPC server code in the following subsection.

Changing the gRPC server code

To enable tracing and the publishing of logs to the ELK stack, you need to make the following code changes, as demonstrated in the following steps:

1. Add the following dependencies to the `build.gradle` file:

    ```
    // logging
    implementation 'net.logstash.logback:logstash-logback-
    encoder:6.6'

    implementation 'org.springframework.cloud:spring-cloud-
    starter-sleuth:3.0.2'

    implementation 'io.zipkin.brave:brave-instrumentation-
    grpc:5.13.2'
    ```

    https://github.com/PacktPublishing/Modern-API-Development-with-Spring-and-Spring-Boot/blob/main/Chapter12/server/build.gradle

    You are going to add these three dependencies:

    a. **Logstash-Logback Encoder**: This library provides the Logback encoder that publishes the logs to Logstash. This will be configured in the `spring-logback.xml` file.

    b. **Spring Cloud Starter Sleuth**: This is the one that takes care of managing the trace and span IDs in `log` statements.

    c. **Brave Instrumentation for gRPC**: This library is only required for gRPC-based code because the gRPC server is statically generated and not managed by the Spring Context. You don't need this dependency for RESTful web services. This library provides a server interceptor for tracing.

2. Add/modify the `spring-logback.xml` file with the following content:

    ```xml
    <?xml version="1.0" encoding="UTF-8"?>
    <configuration>
      <springProperty scope="context" name="applicationName"
                                        source="spring.
      application.name"/>
      <springProperty scope="context" name="logstashDestination"
                                        source="logstash.
      destination" />
      <property name="LOG_PATTERN" value="%d{yyyy-MM-dd
    ```

```
HH:mm:ss.SSS}
    %5p [${applicationName},%X{traceId:-},%X{spanId:-}]
${PID:-}
        --- [%15.15t] %-40.40logger{39} : %msg%n"/>
    <property name="LOG_FILE" value="${chapter12-grpc-
        server.service.logging.file:-chapter12-grpc-
server-logs}"/>
    <property name="LOG_DIR" value="${chapter12-grpc-
        server.service.logging.path:-chapter12-grpc-
server-logs}"/>
    <property name="SERVICE_ENV" value="${service.env:-
dev}"/>
    <property name="LOG_BASE_PATH" value="${LOG_
DIR}/${SERVICE_ENV}"/>
    <property name="MAX_FILE_SIZE"
    value="${chapter12.service.logging.rolling.
maxFileSize:-100MB}"/>
    <!-- other configuration has been remove for brevity
-->
```

https://github.com/PacktPublishing/Modern-API-Development-with-Spring-and-Spring-Boot/blob/main/Chapter12/server/src/main/resources/logback-spring.xml

Here, you define the properties. The values of two of them are taken from the Spring configuration file (`application.properties` or `application.yaml`). Let's now add the Logstash encoder, as shown in the following code block:

```
<appender name="STASH"
    class="net.logstash.logback.appender.
LogstashTcpSocketAppender">
    <destination>${logstashDestination}</destination>
    <encoder class="net.logstash.logback.encoder.
LogstashEncoder" />
</appender>
<!-- other configuration has been remove for brevity  -->
```

Here, the STASH appender is defined, and uses the TCP socket for pushing the logs to Logstash. It contains the destination element that is used for assigning Logstash's <HOST>:<TCP Port> value. Another element encoder contains the fully qualified class name, LogstashEncoder.

Finally, you add the STASH appender to the root element, as shown next:

```
<!-- other configuration has been remove for brevity  -->
  <root level="INFO">
    <appender-ref ref="STDOUT"/>
    <appender-ref ref="STASH"/>
    <appender-ref ref="FILE"/>
  </root>
<!-- other configuration has been remove for brevity  -->
```

The root level is set as INFO.

3. Next, let's add the Spring properties used in this logback-spring.xml file to application.properties, as shown in the following code block:

```
spring.application.name=grpc-server
spring.main.web-application-type=none
grpc.port=8080
logstash.destination=localhost:5002
```

https://github.com/PacktPublishing/Modern-API-Development-with-Spring-and-Spring-Boot/blob/main/Chapter12/server/src/main/resources/application.properties

Here, the Logstash destination host is set to localhost. If you are running on any remote machine, change the host accordingly. The Logstash TCP port is set as per the Logstash external port set in the Docker Composer file.

4. The required dependencies and configurations are now set. You can add the tracing server interceptor to the gRPC server (you don't need a tracing interceptor if you are using the RESTful web service as Spring's autoconfiguration mechanism takes care of this)

First of all, let's define a new bean in a configuration file, as shown here:

```
@Configuration
public class Config {
  @Bean
  public GrpcTracing grpcTracing(RpcTracing rpcTracing) {
    return GrpcTracing.create(rpcTracing);
  }
}
```

https://github.com/PacktPublishing/Modern-API-Development-with-Spring-and-Spring-Boot/blob/main/Chapter12/server/src/main/java/com/packt/modern/api/server/Config.java

The `RpcTracing` bean is already available as a result of autoconfiguration, which is used to create `GrpcTracing`. Now, you have the bean available for `GrpcTracing`, which can be used to create the tracing server interceptor.

5. Let's modify the gRPC server Java file to add the tracing server interceptor to the gRPC server, as shown in the following code block:

```
@Component
public class GrpcServer {
    // code truncated for brevity
    private GrpcTracing grpcTracing;

    public GrpcServer(SourceService sourceService,
ChargeService
        chargeService, ExceptionInterceptor
exceptionInterceptor,
        GrpcTracing grpcTracing) {
    // code truncated for brevity
    this.grpcTracing = grpcTracing;
    }
    public void start() throws IOException,
InterruptedException {
        server = ServerBuilder.forPort(port)
            .addService(sourceService).
addService(chargeService)
            .intercept(exceptionInterceptor)
            .intercept(grpcTracing.newServerInterceptor())
            .build().start();
    // code truncated for brevity
    }
// code truncated for brevity
```

https://github.com/PacktPublishing/Modern-API-Development-with-Spring-and-Spring-Boot/blob/main/Chapter12/server/src/main/java/com/packt/modern/api/server/GrpcServer.java

Here, you can see that the bean created in the configuration file in the previous step is injected using the constructor. Later, the `grpcTracing` bean was used to create the server interceptor.

The changes required to enable log publishing to the ELK stack and tracing are made to the gRPC server. You can rebuild the gRPC server and run its jar to see the effective changes. Check the following code block by way of reference:

```
$ java -jar build\libs\chapter12-server-0.0.1-SNAPSHOT.jar

  .   ____          _            __ _ _
 /\\ / ___'_ __ _ _(_)_ __  __ _ \ \ \ \
( ( )\___ | '_ | '_| | '_ \/ _` | \ \ \ \
 \\/  ___)| |_)| | | | | || (_| |  ) ) ) )
  '  |____| .__|_| |_|_| |_\__, | / / / /
 =========|_|==============|___/=/_/_/_/
 :: Spring Boot ::                (v2.4.4)
// Logs truncated for brevity
2021-04-30 09:28:17.999         INFO [grpc-server,,]       14580 ---
[             main] com.packt.modern.api.server.GrpcServer    :
gRPC server started and listening on port: 8080.
```

You can see that the logs are following the pattern configured in `logback-spring.xml`. The log block printed after `INFO` contains the application/service name, as well as the trace and span IDs. The highlighted line is showing a blank trace ID and span ID because no external call is made that involves the distributed transaction. The Trace and span IDs only get added to logs if the distributed transaction (inter-service communication) is called.

Now, similarly, you can add the logging and tracing implementation in the gRPC client next.

Changing the gRPC client code

To enable tracing and the publishing of logs to the ELK stack, you need to make the code changes in gRPC client as well, which are very similar to the changes implemented in the gRPC code. Refer to the following steps for more information:

1. Add the following dependencies to the `build.gradle` file:

    ```
    // logging
    implementation 'net.logstash.logback:logstash-logback-
    encoder:6.6'
    ```

```
implementation 'org.springframework.cloud:spring-cloud-
starter-sleuth:3.0.2'
implementation 'io.zipkin.brave:brave-instrumentation-
grpc:5.13.2'
```

https://github.com/PacktPublishing/Modern-API-Development-with-Spring-and-Spring-Boot/blob/main/Chapter12/client/build.gradle

These are the same dependencies as the ones you added to the gRPC server code.

2. Add/modify the `spring-logback.xml` file with the following content:

```
<?xml version="1.0" encoding="UTF-8"?>
<configuration>
  <springProperty scope="context" name="applicationName"
                  source="spring.application.name"/>
  <springProperty scope="context" name="logstash.
      destination" source="logstashDestination" />
  <property name="LOG_PATTERN" value="%d{yyyy-MM-dd
      HH:mm:ss.SSS}
    %5p [${applicationName},%X{traceId:-},%X{spanId:-}]
        ${PID:-}
      --- [%15.15t] %-40.40logger{39} : %msg%n"/>
  <property name="LOG_FILE" value="${chapter12-grpc-
        client.service.logging.file:-chapter12-grpc-
        client-logs}"/>
  <property name="LOG_DIR" value="${chapter12-grpc-
        client.service.logging.path:-chapter12-grpc-
        client-logs}"/>
  <property name="SERVICE_ENV" value="${service.env:-
        dev}"/>
  <property name="LOG_BASE_PATH" value="${LOG_
        DIR}/${SERVICE_ENV}"/>
  <property name="MAX_FILE_SIZE"
   value="${chapter12.service.logging.rolling.
        maxFileSize:-100MB}"/>
  <!-- other configuration has been remove for brevity
 -->
```

https://github.com/PacktPublishing/Modern-API-Development-with-Spring-and-Spring-Boot/blob/main/Chapter12/client/src/main/resources/logback-spring.xml

Here, you define the properties. It is similar to the gRPC server. Only the gRPC client-related values have changed. Let's now add the Logstash encoder, as shown in the following code block:

```
<appender name="STASH"
    class="net.logstash.logback.appender.
LogstashTcpSocketAppender">
  <destination>${logstashDestination}</destination>
  <encoder class="net.logstash.logback.encoder.
LogstashEncoder" />
</appender>
<!-- other configuration has been remove for brevity  -->
```

This is exactly the same as the gRPC server. Finally, you add the STASH appender to the root element, as shown next:

```
<!-- other configuration has been remove for brevity  -->
  <root level="INFO">
    <appender-ref ref="STDOUT"/>
    <appender-ref ref="STASH"/>
    <appender-ref ref="FILE"/>
  </root>
<!-- other configuration has been remove for brevity  -->
```

3. Next, let's add the spring properties used in this logback-spring.xml file to application.properties, as shown in the following code block:

```
spring.application.name=grpc-client
server.port=8081
grpc.server.host=localhost
grpc.server.port=8080
logstash.destination=localhost:5002
```

https://github.com/PacktPublishing/Modern-API-Development-with-Spring-and-Spring-Boot/blob/main/Chapter12/client/src/main/resources/application.properties

4. The required dependencies and configurations are now set. Now, you can add the tracing to the gRPC client. `ManagedChannelBuilder`, provided by the gRPC library, supports channel creation statically. Therefore, you would use `SpringAwareManagedChannelBuilder` to create a channel for enabling tracing. `SpringAwareManagedChannelBuilder` is a wrapper around the gRPC's managed channel class. It is provided by the `brave-instrumention-grpc` library (you don't need additional tracing changes if you are using the RESTful web service; Spring autoconfiguration takes care of this). First of all, let's define a new bean, `SpringAwareManagedChannelBuilder`, and its dependent beans in a configuration file, as shown next:

```java
@Configuration
public class Config {
  @Bean
  public SpringAwareManagedChannelBuilder
managedChannelBuilder(
    Optional<List<GrpcManagedChannelBuilderCustomizer>>
customizers) {
      return new
SpringAwareManagedChannelBuilder(customizers);
  }
  @Bean
  GrpcManagedChannelBuilderCustomizer
    tracingManagedChannelBuilderCustomizer(GrpcTracing
grpcTracing) {
      return new
TracingManagedChannelBuilderCustomizer(grpcTracing);
  }
  @Bean
  public GrpcTracing grpcTracing(RpcTracing rpcTracing) {
    return GrpcTracing.create(rpcTracing);
  }
}
```

https://github.com/PacktPublishing/Modern-API-Development-with-Spring-and-Spring-Boot/blob/main/Chapter12/client/src/main/java/com/packt/modern/api/client/Config.java

The RpcTracing bean is already available by means of autoconfiguration, which is used to create GrpcTracing. It is then used to create the bean of GrpcManagedChannelBuilderCustomizer, which is used to create the Spring-aware managed channel builder.

5. Let's now modify the gRPC client Java file to use the Spring-aware managed channel builder for creating the channel:

```
@Component
public class GrpcClient {
  @Autowired
  private SpringAwareManagedChannelBuilder builder;
  // code truncated for brevity
  public void start() {
    channel = builder.forAddress(host, port).
        usePlaintext().build();
    sourceServiceStub = SourceServiceGrpc.
        newBlockingStub(channel);
    chargeServiceStub = ChargeServiceGrpc.
        newBlockingStub(channel);
  }
  // code truncated for brevity
```

https://github.com/PacktPublishing/Modern-API-Development-with-Spring-and-Spring-Boot/blob/main/Chapter12/client/src/main/java/com/packt/modern/api/client/Config.java

Here, you can see that the bean created in the configuration file in the previous step is autowired. Later, the SpringAwareManagedChannelBuilder bean is used to build the channel.

6. Now, you need to make the final change – adding a REST endpoint that thereby allows the client to call the server. This is created for testing purposes so that you can initiate the distributed transaction that would use the gRPC client to call the gRPC server API. Let's add the REST endpoint, as shown in the following code block:

```
@RestController
public class ChargeController {
  private Logger LOG = LoggerFactory.
      getLogger(getClass());
  private GrpcClient client;
```

```java
    public ChargeController(GrpcClient client) {
      this.client = client;
    }
    @GetMapping("/charges")
    public String getSources(@RequestParam(defaultValue =
        "ab1ab2ab3ab4ab5") String customerId)
      throws InvalidProtocolBufferException {
    LOG.info("CustomerId : {}", customerId);
    var req = CustomerId.newBuilder().setId(customerId).
        build();
    CustomerId.Response resp = client.
        getChargeServiceStub().retrieveAll(req);
    var printer = JsonFormat.printer().
        includingDefaultValueFields();
    LOG.info("Server response received in Json Format:
        {}", resp);
    return printer.print(resp);
  }
}
```

https://github.com/PacktPublishing/Modern-API-Development-with-Spring-and-Spring-Boot/blob/main/Chapter12/client/src/main/java/com/packt/modern/api/controller/ChargeController.java

The changes required to facilitate log publishing to the ELK stack and tracing are also done for the gRPC client. You can rebuild the gRPC client and run its `jar` to see the effective changes. Check the following block for reference:

```
$ java -jar build\libs\chapter12-client-0.0.1-SNAPSHOT.jar

  .   ____          _            __ _ _
 /\\ / ___'_ __ _ _(_)_ __  __ _ \ \ \ \
( ( )\___ | '_ | '_| | '_ \/ _` | \ \ \ \
 \\/  ___)| |_)| | | | | || (_| |  ) ) ) )
  '  |____| .__|_| |_|_| |_\__, | / / / /
 =========|_|==============|___/=/_/_/_/
 :: Spring Boot ::            (v2.4.4)
// Logs truncated for brevity
2021-04-30 09:35:34.122       INFO [grpc-client,,]      6416 ---
```

```
[              main] com.packt.modern.api.ClientApp        :
Started ClientApp in 60.985 seconds (JVM running for 72.022)
2021-04-30 09:35:35.002      INFO [grpc-client,,]      6416 ---
[              main] com.packt.modern.api.client.GrpcClient  :
gRPC client connected to localhost:8080
```

You can see that the logs follow the pattern configured in logback-spring.xml. The log block printed after INFO contains the application/service name, trace ID, and span ID. Trace and span IDs are blank because they only get added to logs if distributed transactions (inter-service communication) are called.

The changes required to enable log aggregation and distributed tracing in both the gRPC server and client services are now complete.

Next, you'll make the changes and view the logs in Kibana.

Testing the logging and tracing changes

Before beginning testing, make sure that the ELK stack is up and running. Also, make sure that you first start the gRPC server and then the gRPC client service.

You can add the appropriate log statements to your services for verbose logs.

Let's run the following command in the new terminal window. This will call the newly created REST endpoint/charges in the gRPC client service:

```
$ curl http://localhost:8081/charges
```

It should respond with the following JSON output:

```
charge {
  "id": "kopek7qkec3fkfve5s2aqu2gu03srhnb"
  "amount": 1000
  "created": 1619755094
  "currency": "USD"
  "customerId": "ab1ab2ab3ab4ab5"
  "description": "Charge Description"
  "receiptEmail": "receipt@email.com"
  "statementDescriptor": "Statement Descriptor"
  "sourceId": "7e1ra7pc67120gu0ns5htukgko9q561s"
}
```

This should generate logs similar to the following logs in the gRPC client:

```
2021-04-30 10:06:39.620          INFO
[grpc-client,1053588e335284a4,1053588e335284a4] 6416 --- [nio-
8081-exec-2] c.p.m.api.controller.ChargeController : CustomerId
: ab1ab2ab3ab4ab5

2021-04-30 10:06:41.367          INFO
[grpc-client,1053588e335284a4,1053588e335284a4] 6416 --- [nio-
8081-exec-2] c.p.m.api.controller.ChargeController : Server
response received in Json Format: charge {
  id: "kopek7qkec3fkfve5s2aqu2gu03srhnb"
  amount: 1000
  created: 1619755094
  currency: "USD"
  customerId: "ab1ab2ab3ab4ab5"
  description: "Charge Description"
  receiptEmail: receipt@email.com
  statementDescriptor: "Statement Descriptor"
  sourceId: "7e1ra7pc67120gu0ns5htukgko9q561s"
}
```

Here, the blocks highlighted first show the application name (grpc-client), trace ID (1053588e335284a4), and span ID (1053588e335284a4). Both trace and span IDs are the same since the API request is initiated here.

It should also generate logs similar to the following logs in the gRPC server:

```
2021-04-30 10:06:41.358          INFO
[grpc-server,1053588e335284a4,298c90606395413d]        14580
--- [ault-executor-2] c.p.m.api.server.repository.
DbStore       : Request for retrieving charges with customer ID
: ab1ab2ab3ab4ab5
```

Here, the blocks highlighted first shows the application name (grpc-server), trace ID (1053588e335284a4), and span ID (298c90606395413d). The trace ID is the same as what is displayed in the gRPC client logs. The span IDs are different from the gRPC client service because span IDs belong to respective individual services. This is how the trace/correlational ID helps you trace the requests call across different services because it will be propagated to all the services it involves.

Tracing this request was simple as logs contain just a few lines and are scattered across only two services. What if you have a few gigabytes of logs scattered across various services? Then, you can make use of the ELK stack to search the log index using different query criteria. However, we are going to use the trace ID for this purpose.

First, open the Kibana home page in the browser. Then, click on the hamburger menu in the top-left corner, as shown in the following screenshot. Then, click on the **Discover** option in the menu that appears:

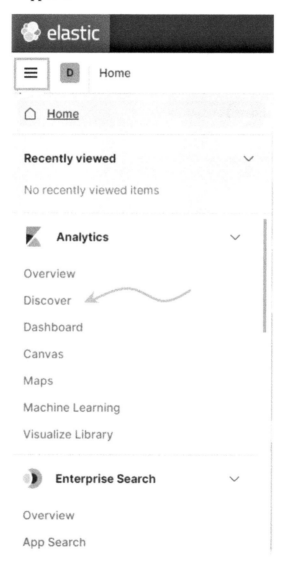

Figure 12.3 – Kibana hamburger menu

This should open the **Discover** page, as shown on the following page. However, the first time you have to create the index pattern, that will filter out the indexes available in Elasticsearch:

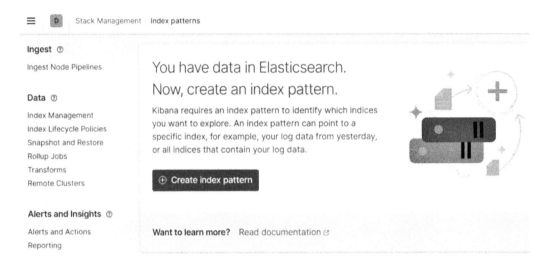

Figure 12.4 – Kibana's Discover page

Next, click on the **Create index pattern** button, which opens the following page to define the index pattern as *step 1*. Here, you should enter the index name (`modern-api`) given in the Logstash configuration in the ELK stack's Docker Compose file:

Create index pattern

An index pattern can match a single source, for example, `filebeat-4-3-22` , or **multiple** data sources, `filebeat-*` . Read documentation

Step 1 of 2: Define an index pattern

Index pattern name

modern-api*   Next step >

Use an asterisk (*) to match multiple indices. Spaces and the characters \, /, ?, ", <, >, | are not allowed.

◯ ✕ Include system and hidden indices

✓ Your index pattern matches 1 source.

modern-api   Index

Rows per page: 10 ∨

Figure 12.5 – Kibana's Create index pattern page – step 1

In *step 2*, you have to select the time field (timestamp) from the pop-up menu:

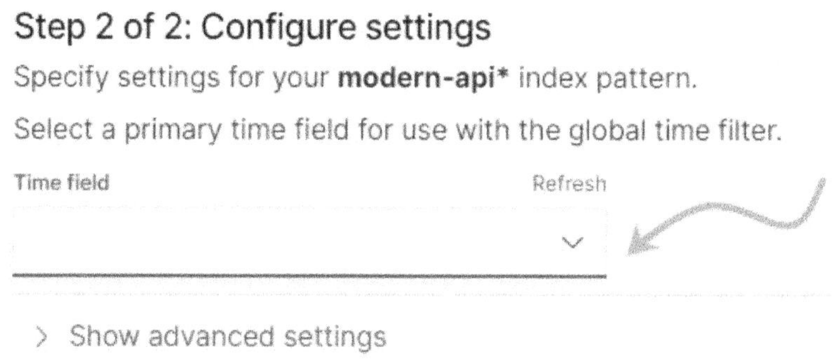

Step 2 of 2: Configure settings

Specify settings for your **modern-api*** index pattern.

Select a primary time field for use with the global time filter.

Time field   Refresh

> Show advanced settings

Figure 12.6 – Kibana's Create index pattern page – step 2

Then, click on **Next** and select all available fields from the Elasticsearch index. Then, go to the **Discover** page again to execute the filter query as shown in the following screenshot.

You can add the filter query to the **Search** textbox and the **Date/Duration** menu at the top of the **Discover** page. Query criteria can be input using the **KQL** (**Kibana Query Language**) language, which allows you to add different comparator and logical operators. For more information, refer to https://www.elastic.co/guide/en/kibana/master/kuery-query.html.

We have entered criteria (traceId: 1053588e335284a4) and kept the **Duration** field as its default (the last 15 minutes). The left-hand side also shows you how to select the Elasticsearch index and all the fields available in it.

Once you press the **Enter** key or click on the refresh button after entering the criteria, the search displays the available logs from all the services. The searched values are highlighted in yellow.

You can observe in the following screenshot that the searched trace ID shows logs from both server and client services:

Figure 12.7 – Kibana's Discover page following index creation

The searched **Discovery** page also shows the graph that shows the number of calls made during a particular period. You can generate more logs and with some errors, and then you can use different criteria to filter the results and explore more.

You can also save the searches and perform more operations, such as customizing the dashboard. Please refer to `https://www.elastic.co/guide/en/kibana/master/index.html` for more information.

Distributed tracing with Zipkin

The ELK stack is good for log aggregation, filtering, and debugging using the trace ID and other fields. However, it can't check the performance of API calls – the time taken by the call. It is especially important when you have a microservice-based application.

This is where Zipkin (OpenZipkin), along with Spring Cloud Sleuth, not only helps you to trace transactions across multiple service invocations, but also to capture the response time taken by each service involved in the distributed transaction. Zipkin also shows this information using nice graphs. It helps you to locate the performance bottlenecks and drill down the specific API call that creates the latency issue. You can find out the total time taken by the main API call as well as its internal API call time.

Services developed with Spring Boot facilitate their integration with Zipkin. You just have to make two code changes – the addition of the Sleuth-Zipkin dependency and the addition of the Zipkin URL property.

You can make these two changes to both the gRPC server and client, as shown next:

1. First, add the highlighted dependency to `build.gradle` (both the gRPC server and client projects):

```
implementation 'net.logstash.logback:logstash-logback-encoder:6.6'

implementation 'org.springframework.cloud:spring-cloud-starter-sleuth:3.0.2'

implementation 'io.zipkin.brave:brave-instrumentation-grpc:5.13.2'

implementation 'org.springframework.cloud:spring-cloud-sleuth-zipkin:3.0.2'
```

2. Next, add the following properties to the `application.properties` file (for both the gRPC server and client):

```
zipkin.baseUrl: localhost:9411
spring.sleuth.sampler.probability: 1.0
```

The Zipkin `baseUrl` property points to the Zipkin application. The `spring.sleuth. sampler.probability` property refers to the probability of requests that should be sampled. 1 represents 100% of the requests that should be sampled. For more information about the available configuration, you can refer to `https://docs.spring.io/ spring-cloud-sleuth/docs/3.0.2/reference/html/appendix. html#appendix`. This link refers to the documentation of the 3.0.2 version. Please use the appropriate version if you change the dependency version.

You are done with the changes required in the code for publishing the tracing information to Zipkin. Rebuild both the server and client services after making these changes

Let's start Zipkin in the next subsection.

Executing Zipkin

There are various ways to install and run Zipkin. Please refer to `https://zipkin.io/pages/quickstart` to find out about these options. For development purposes, you should fetch the latest release as a self-contained executable jar from `https://search.maven.org/remote_content?g=io.zipkin&a=zipkin-server&v=LATEST&c=exec` and then start it using the following command (make sure to change the version in the jar file based on the downloaded file):

```
java -jar zipkin-server-2.23.2-exec.jarThis will start Zipkin
with an in-memory database. For production purposes, it is
recommended to use a persistence store such as Elasticsearch.
```

It should start with the default port `9411` at `http://127.0.0.1:9411/` if executed at the localhost.

Once the Zipkin server and the ELK stack are up and running, you can start both gRPC server and client services and execute the following command:

```
$ curl http://localhost:8081/charges
```

This command should print a log similar to the following log statement in the gRPC client service:

```
2021-05-03 01:15:15.817          INFO
[grpc-client,79a4fd9c639d2e5b,79a4fd9c639d2e5b]          10212 ---
[nio-8081-exec-1] c.p.m.api.controller.ChargeController     :
CustomerId : ab1ab2ab3ab4ab5
```

Keep the trace ID handy as you are going to use it in the Zipkin UI. Open the Zipkin home page by accessing `http://localhost:9411`. It will look as follows:

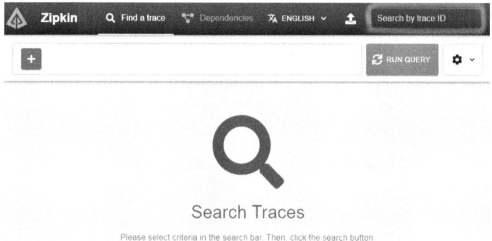

Figure 12.8 – Zipkin home page

You can observe that Zipkin also allows you to run queries. However, we'll make use of the trace ID. Paste the copied trace ID in the **Search by trace ID** textbox in the top-right corner and then press **Enter**:

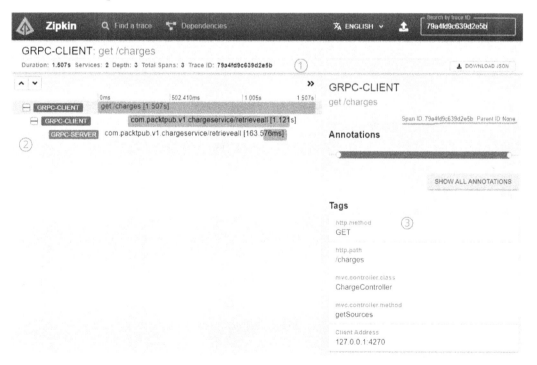

Figure 12.9 – Zipkin search result page

The Zipkin trace ID shows complete API call information at the top marked with point 1 if the trace ID is available. In the left-hand side section, it shows all the corresponding API calls with a hierarchy, which shows the individual call times in a graphical way. These API call rows are selectable. Selected call details are displayed on the right-hand side. The **Show All Annotations** button on the right-hand side displays the individual call start and finish times, as shown in the following screenshot for the grpc-server call:

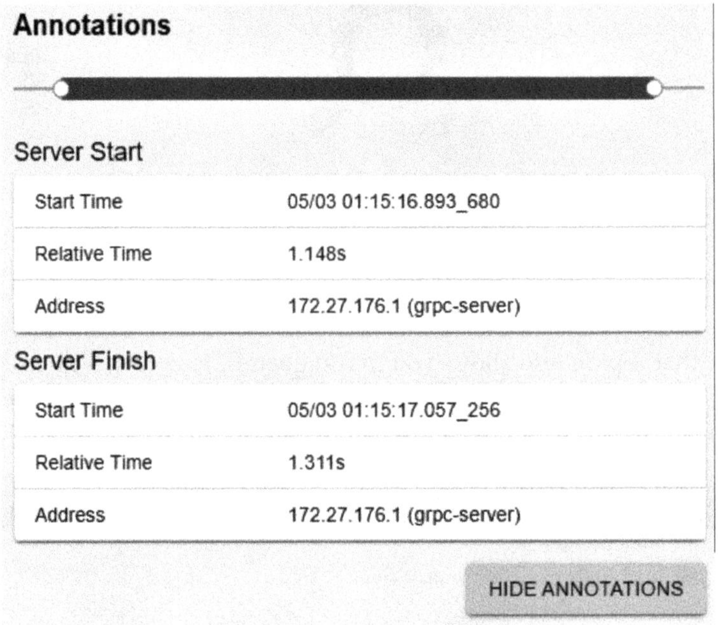

Annotations

Server Start

Start Time	05/03 01:15:16.893_680
Relative Time	1.148s
Address	172.27.176.1 (grpc-server)

Server Finish

Start Time	05/03 01:15:17.057_256
Relative Time	1.311s
Address	172.27.176.1 (grpc-server)

HIDE ANNOTATIONS

Figure 12.10 – Annotation details

Time tracking at a granular level for each distributed API call allows you to identify the latency issues and relative time tracking for performance tuning.

Summary

You have learned how the trace/correlation ID is important and how it can be set up using Spring Cloud Sleuth. You can use these generated IDs to find the relevant logs and API call durations. You have integrated the Spring Boot services with the ELK stack and Zipkin.

You have also implemented extra code and configurations, which are required for enabling distributed tracing for gRPC-based services.

You have acquired the log aggregation and distributed tracing skills using Spring Cloud Sleuth, the ELK stack, and Zipkin in this chapter.

In the next chapter, you are going to learn about the fundamentals of the GraphQL API.

Questions

1. What is the difference between the trace ID and span ID?

2. Should you use the broker between services that generate the logs and the ELK stack? If yes, why?

3. How does Zipkin work?

Further reading

- Elasticsearch documentation: `https://www.elastic.co/guide/en/elasticsearch/reference/current/index.html`

- Kibana documentation: `https://www.elastic.co/guide/en/kibana/master/index.html`

- Kibana Query Language: `https://www.elastic.co/guide/en/kibana/master/kuery-query.html`

- Logstash documentation: `https://www.elastic.co/guide/en/logstash/master/index.html`

- Advanced Elasticsearch 7.0: `https://www.packtpub.com/product/advanced-elasticsearch-7-0/9781789957754`

- Zipkin documentation: `https://zipkin.io/pages/quickstart`

Section 4: GraphQL

In this section, you will learn about GraphQL-based API development. After completing this section, you will be able to differentiate between REST and reactive APIs, as well as gRPC APIs, with GraphQL. You will then be able to write the GraphQL schema, which will be used to generate Java code. Then, you will be able to write the data fetchers and loaders for resolving the query fields.

This section comprises the following chapters:

- *Chapter 13, GraphQL Fundamentals*
- *Chapter 14, GraphQL API Development and Testing*

13
GraphQL Fundamentals

In this chapter, you will learn about the fundamentals of GraphQL, including its **Schema Definition Language** (**SDL**), queries, mutations, and subscriptions. This knowledge will help you in the next chapter, when you will implement an API based on GraphQL.

We will cover the following topics in this chapter:

- Introducing GraphQL
- Learning about the fundamentals of GraphQL
- Designing a GraphQL schema
- Testing GraphQL query and mutation
- Solving the N+1 problem

After completing this chapter, you will know about the basics of GraphQL, including its semantics, schema design, and everything you need to develop a GraphQL-based API using Spring and Spring Boot.

Technical requirements

This chapter covers the theory surrounding GraphQL. However, you need the following for developing and testing the GraphQL-based service code presented in the next chapter:

- Any Java IDE, such as NetBeans, IntelliJ, or Eclipse
- **Java Development Kit (JDK)** 15
- An internet connection to clone the code and download the dependencies and the Gradle software

So, let's begin!

Introducing GraphQL

You might have heard of or be aware of **GraphQL**, which has become more popular and is the preferred way of implementing APIs for handheld devices and the web.

GraphQL is a declarative query and manipulation language, and a server-side runtime for APIs. GraphQL empowers the client to query exactly the data they want – no more, no less.

We'll discuss its brief history in the next subsection.

Brief history of GraphQL

In 2011, Facebook was facing challenges in terms of improving the performance of its website on mobile browsers. They started building their own mobile app with mobile-native technologies. However, APIs were not up to the mark because of hierarchical and recursive data. They wanted to optimize their network calls. Note that in those days, mobile network speed was in Kb/s in some parts of the world. Having a high quality and fast mobile app was going to be the key to their success since their consumers had started shifting to mobile devices.

In 2012, a few engineers on Facebook – Lee Byron, Dan Schafer, and Nick Schrock – teamed up to create GraphQL. Initially, it was used to design and develop Facebook's newsfeed feature, but later on, it was used across its infrastructure. It was being used internally in Facebook before it was open sourced in 2015. This is when they published the GraphQL specification and its JavaScript implementation to the public. Soon, other language implementations of the GraphQL specification started rolling out, including Java.

I think you would enjoy watching this GraphQL documentary at `https://www.youtube.com/watch?v=783ccP__No8`, which walks through GraphQL's journey from being an internal Facebook tool to its current success.

Did you know?

Netflix and Coursera were also working on a similar idea to build efficient and performant APIs. Coursera didn't take it forward, but Netflix open sourced Falcor.

Comparing GraphQL with REST

You developed APIs using REST in the *Section 1* of this book. In fact, a sample e-commerce UI app also consumed REST APIs to implement its e-commerce functionality. We are going to keep referring to REST in this chapter so that we can understand the necessary GraphQL concepts wherever they're applicable. This correlation should help you grasp the GraphQL concept easily.

GraphQL is more powerful, flexible, and efficient than REST. Let's understand why.

When a user logs into the sample e-commerce UI app and automatically navigates to the product listing page, the UI app consumes three different endpoints, as follows:

- The user endpoint, to fetch the user's information
- The product endpoint, to fetch the product list
- The cart endpoint, to fetch the cart items from the user's cart

So, basically, you have to make three calls to fetch the required information in a fixed structure (you can't change the fields that are sent in response) from the backend.

On the other hand, GraphQL can fetch a user's information, a user's cart data, and the product list in a single call. This It reduces the network call from three to one. GraphQL just exposes a single endpoint, unlike REST, where you have to define an endpoint for each use case. You might say that you can write a new REST endpoint that does that. Yes, that may solve this specific use case, but it is not flexible; it won't allow for quick change iteration.

Moreover, GraphQL lets you describe the fields you want to fetch from the backend in a request. The server provides a response according to the requested fields only – no more, no less.

You don't have to create a new REST endpoint if you need a new set of fields. For example, you may want to add user reviews to products. For this, you just need to add the `reviews` field to the GraphQL query. Similarly, you don't need to consume extra fields. You just add those fields that you need in the GraphQL query. Instead, REST's response contains predefined fields, regardless of whether you need certain fields in the response object or not. Then, you have to filter the required fields at the client end. Therefore, you can say that GraphQL uses network bandwidth effectively by avoiding over/underfetching problems.

GraphQL APIs don't need constant changes like REST does, where you may need to change the API or add new APIs for a requirement change. This improves the development speed and iteration. You can easily add new fields or mark existing deprecated ones (fields not being used by a client anymore). Therefore, you can make the changes in the client without impacting the backend. In short, you can write evolving APIs without any versioning and breaking changes.

REST offers caching using built-in HTTP specifications. However, GraphQL does not follow the HTTP specifications; instead, it makes use of libraries such as Apollo/Relay for caching. However, REST is based on HTTP and does not follow any specification for implementation, which may lead to inconsistent implementations, as we discussed while comparing REST with gRPC. You can use the HTTP GET method to delete a resource.

GraphQL is superior to REST APIs in terms of its usage in mobile clients. The capabilities of GraphQL APIs are also defined using strong types. These types are part of the schema that contains the API definitions. These types are written in the schema using **SDL**.

GraphQL acts as a contract between the server and the client. You can correlate the GraphQL schema with the gRPC **Interface Definition Language** (**IDL**) file and OpenAPI specification file.

We'll discuss the fundamentals of GraphQL in the next section.

Learning about the fundamentals of GraphQL

GraphQL APIs contain three important **root types** – **query**, **mutation**, and **subscription**. These are all defined in the GraphQL schema using special SDL syntax.

GraphQL provides a single endpoint that returns the JSON response based on the request, which can be a query, a mutation, or a subscription.

First, let's understand queries.

Exploring the Query type

The Query type is used for reading operations that fetch information from the server. A single Query type can contain many queries. Let's write a query using SDL to retrieve the logged-in user, as shown in the following GraphQL schema:

```
type Query {
    me: LogginInUser
    # You can add other queries here
}
type LoggedInUser {
    id: ID
    accessToken: String
    refreshToken: String
    username: String
}
```

Here, you have done two things:

1. You have defined the query root of the GraphQL interface, which contains the query you can run. It just contains a single query type, me, that returns an instance of the LoggedInUser type.

2. You have defined the user-defined LoggedInUser object type, which contains four fields. These fields are followed by their types. Here, you have used GraphQL's built-in *scalar types*, called ID and String, to define the types of the fields. We'll discuss these types later in this chapter, when we discuss built-in scalar types in detail.

Once you have this schema implementation on the server and fired the following GraphQL query, you will only get the fields you requested, along with their values, as a JSON object in response.

You can find the me query and its JSON response in the following code block:

```
# Request input
{
    me {
        id
        username
    }
}
```

```
}
#JSON response
{
   "data": {
      "me": {
         "id": "asdf90asdkqwe09kl",
         "username": "scott"
      }
   }
}
```

Here, GraphQL's request input does not start with a query because it is the default. *This is called an anonymous query.* However, if you want, you can also prefix the request input with query, as shown here:

```
query {
   me {
      id
      username
   }
}
```

As you can see, you can only query those fields that you need. Here, only the id and username fields were requested from the LoggedInUser type, and the server responded with only these two fields. The request payload is enclosed in curly braces { }. You can use # for commenting in the schema.

Now, you know how to define Query and object types in a GraphQL schema. You also learned how to form a GraphQL request payload according to its query type and the expected JSON response.

We'll learn about GraphQL mutations in the next subsection.

Exploring the Mutation type

The Mutation type is used in GraphQL requests for all the add, update, or delete operations that get performed on the server. A single Mutation type can contain many mutations. Let's define an addItemInCart mutation that adds a new item to the cart:

```
type Mutation {
   addItemInCart(productId: ID, qty: Int): [Item]
```

```
    # You can add other mutation here
}
type Item {
    id: ID!
    productId: ID
    qty: Int
}
```

Here, you have defined the `Mutation` type and a new `object` type called `Item`. The mutation is added and called `addItemInCart`. The `Query`, `Mutation`, and `Subscription` types can pass arguments. To define the necessary parameters, you can enclose the named arguments with `()` brackets; the arguments are divided by commas. The signature of `addItemInCart` contains two arguments and returns a list of cart items. *A list is marked using* `[]` *brackets.*

> **Optional and required arguments**
>
> Let's say you declare an argument with a default value, such as the following mutation:
>
> ```
> pay(amount: Float, currency: String = "USD"):
> Payment
> ```
>
> Here, the currency is an optional argument. It contains the default value, whereas the amount is a required field because does not contain any default value.

Please note that `Int` is a built-in scalar type for signed 32-bit integers. Default values are null in GraphQL. If you want to force a non-nullable value for any field, then its type should be marked with an exclamation mark (`!`). Once it has been applied to any field in the schema, the GraphQL server will always provide a value, instead of a null for that field when it is placed in the request payload by the client. You can also declare a list with exclamation marks; for example, `items: [Item]!` and `items: [Item!]!`. Both declarations would provide zero or more items in a list. However, the latter would provide valid `Item` object.

Once you have this schema implementation on the server, you can use the following GraphQL query. You will get only the fields you requested, along with their values, as a JSON object:

```
# Request input
mutation {
    addItemInCart(productId: "qwer90asdkqwe09kl", qty: 2) {
```

```
    id
    productId
  }
}
```

You can see that this time, the GraphQL request input starts with the `mutation` keyword. If you don't start a mutation with the `mutation` keyword, then you might get an error, with a message similar to `Field ' addItemInCart' doesn't exist on type 'Query'`. This is because the server treats the request payload as a query.

Here, you must add the required arguments to the `addItemInCart` mutation and then add the fields (`id`, `productId`) you want to retrieve in response. Once the request has been processed successfully, you will get a JSON output similar to the following:

```
#JSON response
{
  "data": {
    addItemInCart: [
      {
        "id": "zxcv90asdkqwe09kl",
        "productId": "qwer90asdkqwe09kl"
      }
    ]
  }
}
```

Here, the value of the `id` field is generated by the server. Similarly, you can write other mutations, such as delete and update, in the schema. Then, you can use the payload in the GraphQL request to process the mutation accordingly.

We'll explore the GraphQL `Subscription` type in the next subsection.

Exploring the Subscription type

The concept of subscriptions will be new to you if you are only familiar with REST. In absence of GraphQL, you might use polling or WebSockets to implement a similar functionality. There are many use cases where you will need the subscription feature, including the following:

- Live score updates or election results
- Batch processing updates

There are many such cases where you will need to immediately update events. GraphQL provides a subscription feature for this use case. In such cases, the client subscribes to the event by initiating and holding a steady connection. When the subscribed event occurs, the server pushes the resultant event data to the client. This resultant data is sent as a stream through an initiated connection, rather than through a request/response kind of communication (which happened in the case of query/mutation).

> **Recommended approach**
>
> It is recommended that a subscription should only be used when a small update occurs for a large object (such as batch processing) or there are live updates with low latency, such as a live score update. Otherwise, you should use polling (execute a query periodically at a specified interval).

Let's create a subscription in a schema, as shown here:

```
type Subscription {
    orderShipped(customerID: ID!): Order
    # You can add other subscription here
}
# Order type contains order information and another object
# Shipping
# Shipping contains id and estDeliveryDate and carrier
# fields
type Order {
    # other fields omitted for brevity
    shipping: Shipping
}
type Shipping {
    Id: ID!
    estDeliveryDate: String
    carrier: String
}
```

Here, we have defined an `orderShipped` subscription that accepts the customer ID as an argument and returns `Order`. When clients subscribe to this event, whenever an order is shipped for the given `customerId`, the server will push the requested order details to the client using a stream.

You can use the following GraphQL request to subscribe to the GraphQL subscription:

```
# Request Input
subscription {
  orderShipped(customerID: "customer90asdkqwe09kl") {
    shipping {
      estDeliveryDate
      trackingId
    }
  }
}
# JSON Output
{
  "data": {
    "orderShipped": {
      "estDeliveryDate": "13-Aug-2022",
      "trackingId": "tracking90asdkqwe09kl"
    }
  }
}
```

The client will request a JSON response whenever any order belonging to a given customer is shipped. The server pushes these updates to all the clients who subscribed to this GraphQL subscription.

In this section, you learned how to declare the Query, Mutation, and Subscription types in a GraphQL schema.

You have defined scalar types and the user-defined object types in a schema. You also explored how to write a GraphQL request input for a query/mutation or subscription.

Now, you know how to define the operation parameters in root types and pass arguments while sending GraphQL requests. Note that the non-nullable field in the schema can be marked by an exclamation mark (!). For arrays or lists of objects, you must use square brackets ([]).

In the next section, we'll deep dive into GraphQL schema.

Designing a GraphQL schema

A schema is a GraphQL file that is written using DSL syntax. Primarily, it contains root types (query, mutation, and subscription) and the respective types that are used in root types, such as object types, scalar types, interfaces, union types, input types, and fragments.

First, let's discuss these types. You learned about root types (query, mutation, and subscription) and object types in the previous section. Now, let's learn more about scalar types.

Understanding scalar types

Scalar types resolve concrete data. There are three kinds of scalar types – built-in scalar types, custom scalar types, and enumeration types. Let's discuss built-in scalar types first. GraphQL provides the following five kinds of built-in scalar types:

- `Int`: This stores integers and is represented by a signed 32-bit integer.
- `Float`: This stores a signed, double-precision, floating-point value.
- `String`: This stores a sequence of UTF-8 characters.
- `Boolean`: This stores a Boolean value – true or false.
- `ID`: This is used to define the object identifier string. This can only be serialized as a string only, and is not human-readable.
- You can also define your own scalar types, which are known as *custom scalar types*. This includes types such such as `Date`.
- The `Date` custom scalar type can be defined like so:

```
scalar Date
```

You need to write an implementation that determines the serialization, deserialization, and validation of these custom scalar types. For example, the date can be treated as a Unix timestamp or a string with a particular data format in a custom scalar `Date` type case.

Another special scalar type is the enumeration type (`enum`), which is used to define a particular set of allowed values. Let's define the order status enumeration, as shown here:

```
enum OrderStatus {
    CREATED
    CONFIRMED
    SHIPPED
```

```
    DELIVERED
    CANCELLED
}
```

Here, the `OrderStatus` enumeration type represents the order status at a given point in time. We'll understand GraphQL fragments in the next subsection before exploring other types.

Understanding fragments

You may encounter conflicting scenarios while querying on the client side. You may have two or more queries that return the same result (the same object or set of fields). To avoid this conflict, you can give the query result a name. This name is known as an **alias**.

Let's use an alias in the following query:

```
query HomeAndBillingAddress {
    home: getAddress(type: "home") {
        number
        residency
        street
        city
        pincode
    }
    billing: getAddress(type: "home") {
        number
        residency
        street
        city
        pincode
    }
}
```

Here, `HomeAndBillingAddress` is a named query that contains the `getAddress` query operation. `getAddress` is being used twice, which results in it returning the same set of fields. Therefore, the `home` and `billing` aliases are being used to differentiate the result object.

The `getAddress` query may return the `Address` object. The `Address` object may have additional fields, such as `type`, `state`, `country`, and `contactNo`. So, when you have queries that may use the same set of fields, you can create a **fragment** and use it in queries.

Let's create a fragment and replace the common fields in the previous code block:

```
query HomeAndBillingAddress {
  home: getAddress(type: "home") {
    ...addressFragment
  }
  billing: getAddress(type: "home") {
    ...addressFragment
  }
}
fragment addressFragment on Address {
    number
    residency
    street
    city
    pincode
}
```

Here, the `addressFragment` fragment has been created and used in the query.

You can also create an *inline fragment* in the query. Inline fragments can be used when a querying field returns an `Interface` or `Union` type. We will explore inline fragments in more detail later.

We'll look at GraphQL interfaces in the next subsection.

Understanding interfaces

GraphQL interfaces are abstract. You may have a few fields that are common across multiple objects. You can create an interface type for such a common set of fields. For example, a product may have some common attributes, such as ID, name, and description. The product can also have other attributes based on its type. For example, a book may have several pages, an author, and a publisher, while a bookcase may have material, width, height, and depth attributes.

Let's define these three objects (`Product`, `Book`, and `Bookcase`) using interfaces:

```
interface Product {
  id: ID!
  name: String!
  description: string
}
type Book implements Product {
  id: ID!
  name: String!
  description: string
  author: String!
  publisher: String
  noOfPages: Int
}
type Bookcase implements Product {
  id: ID!
  name: String!
  description: string
  material: [String!]!
  width: Int
  height: Int
  depth: Int
}
```

Here, an abstract type called `Product` has been created using the `interface` keyword. This interface can be implemented when we wish to create new the object types – `Book` and `Bookcase`.

Now, you can simply write the following query, which returns all the products (books and bookcases):

```
type query {
  allProducts: [Product]
}
```

Now, you can use the following query on the client side to retrieve all the products:

```
query getProducts {
  allProducts {
    id
    name
    description
  }
}
```

You might have noticed that the preceding code only contains attributes from the Product interface. If you want to retrieve attributes from Book and Bookcase, then you have to use **inline fragments**, as shown here:

```
query getProducts {
  allProducts {
    id
    name
    description
    ... on Book {
      author
      publisher
    }
    ... on BookCase {
      material
      height
    }
  }
}
```

Here, an operation (...) is being used to create the inline fragments. This way, you can fetch the fields from the type that implements the interface.

We'll understand Union types in the next subsection.

Understanding Union types

Let's say there are two object types – `Book` and `Author`. Here, you want to write a GraphQL query that can return both books and authors. Note that the interface is not there; so, how can we combine both objects in the query result? In such cases, you can use a `Union` type, which is a combination of two or more objects.

Consider the following before creating a `Union` type:

- You don't need to have a common field.

- Union members should be of a concrete type. Therefore, you can't use union, interface, input, or scalar types.

Let's create a `Union` type that can return any object included in the `union` type – books and bookcases – as shown in the following code block:

```
union SearchResult = Book | Author

type Book {
    id: ID!
    name: String!
    publisher: String
}
type Author {
    id: ID!
    name: String!
}

type Query {
    search(text: String): [SearchResult]
}
```

Here, the union keyword is being used to create a `union` type for the `Book` and `Author` objects. A pipe symbol (|) is being used to separate the included objects. At the end, a query has been defined, which returns the collection of books or authors that contains the given text.

Now, let's write this query for the client, as shown here:

```
# Request Input
{
```

```
    search(text: "Malcolm Gladwell") {
      __typename
      ... on Book {
        name
        publisher
      }
      ... on Author {
        name
      }
    }
  }
```

Response JSON

```
{
  "data": {
    "search": [
      {
        "__typename": "Book",
        "name": "Blink",
        "publisher": "Back Bay Books"
      },
      {
        "__typename": "Author",
        "name": " Malcolm Gladwell ",
      }
    ]
  }
}
```

As you can see, an inline fragment is being used in the query. Another important point is the extra field, called __typename, which refers to the object it belongs to and helps you differentiate between different objects in the client.

We'll look at input types in the next subsection.

Understanding input types

So far, you have used scalar types as arguments. GraphQL also allows you to pass object types as arguments in mutations. The only difference is that you have to declare them with `input` instead of using the `type` keyword.

Let's create a mutation that accepts an input type as an argument:

```
type Mutation {
    addProduct(prodInput: ProductInput): Product
}
input ProductInput {
    name: String!
    description: String
    price: Float!
    # other fields...
}
type Product {
    # Product Input fields. Truncated for brevity.
}
```

Here, the `addProduct` mutation accepts `ProductInput` as an argument and returns a `Product`.

Now, let's use the GraphQL request to add a product to the client, as shown here:

```
# Request Input
mutation AddProduct ($input: ProductInput) {
    addProduct(prodInput: $input) {
        name
    }
}


#---- Variable Section ----
{
    "input": {
        name: "Blink",
        description: "a book",
        "price": 10.00
    }
```

```
}

# JSON Output
{
  "data": {
    addProduct {
      "name": "Blink"
    }
  }
}
```

Here, you are running a mutation that uses an input type. You might have observed that `Variable` is being used here to pass `ProductInput`. The named mutation is being used for the variable. If variables are defined in the mutation, along with their types, then they should be used in the mutation.

Variable values should be assigned in the variable section (or beforehand in the client). The value of a variable's input is assigned using a JSON object that should map to `ProductInput`.

We'll look at the tools we can use while designing a GraphQL schema in the next subsection.

Tools that help with designing a schema

You can use the following tools for design and work with GraphQL. Each has its own offerings:

- **GraphiQL**: It is pronounced *graphical*. It is an official GraphQL Foundation project that provides the web-based GraphQL **Integrated Development Environment (IDE)**. It makes use of **Language Server Protocol (LSP)**, which uses the JSON-RPC-based protocol between the source code editor and the IDE. It is available at `https://github.com/graphql/graphiql`.

- **GraphQL Playground**: This is also a GraphQL IDE that provides better features than GraphiQL. It is available at `https://github.com/graphql/graphql-playground`.

- **GraphQL Faker**: This provides the mock data for your GraphQL APIs. It is available at `https://github.com/APIs-guru/graphql-faker`.

- **GraphQL Editor**: This allows you to design your schema visually and then transform it into code. It is available at `https://github.com/graphql-editor/graphql-editor`.

- **GraphQL Voyager**: This converts your schema into interactive graphs, such as entity diagrams and all its relationships. It is available at `https://github.com/APIs-guru/graphql-voyager`.

In the next section, you'll test the knowledge that you have acquired throughout this chapter.

Testing GraphQL queries and mutations

Let's write queries and mutations in a real GraphQL schema to test the skill you have learned throughout this chapter.

You are going to use GitHub's GraphQL API explorer in this section. Let's perform the following steps:

1. First, go to `https://docs.github.com/en/graphql/overview/explorer`.

2. You might have to authorize it using your GitHub account, so that you can execute GraphQL queries.

3. GitHub Explorer is based on GraphiQL. It is divided into three vertical sections (from left to right):

   a. There are two two subsections – an upper section for writing a query and a bottom section for defining variables.

   b. The middle vertical section shows the response.

   c. Normally, the rightmost section is hidden. Click on the **Docs** link to display it. It shows the respective documentation and schema, along with the root types that you can explore.

4. Let's fire this query to find out the ID of the repository you wish to mark as star:

```
query {
  repository (name:
      "Modern-API-Development-with-Spring-and-Spring-
  Boot",
      owner: "PacktPublishing") {
    id
```

```
      owner {
        id
        login
      }
      name
      description
      viewerHasStarred
      stargazerCount
    }
  }
```

Here, you are querying this book's repository by providing two arguments – the repository's name and its owner. You are fetching a few of the fields from here. One of the most important ones is `stargazerCount` because we are going to perform an `addStar` mutation. This count will tell us whether the mutation was successful or not.

5. Click on the **Execute Query** button on the top bar, or press *Ctrl + Enter* to execute the query. You might get the following output once this query executes successfully:

```
{
  "data": {
    "repository": {
      "id": "MDEwOlJlcG9zaXRvcnkyOTMyOTU5NDA=",
      "owner": {
        "id": "MDEyOk9yZ2FuaXphdGlvbjEwOTc0OTA2",
        "login": "PacktPublishing"
      },
      "name": "Modern-API-Development-with-Spring-and-
                Spring-Boot",
      "description": "Modern API Development with
        Spring and Spring Boot, published by Packt",
      "viewerHasStarred": false,
      "stargazerCount": 1
    }
  }
}
```

Here, you need to copy the value of id (highlighted) from the response because you need it to mark the start.

6. Execute the following query to perform the addStar mutation:

```
mutation {
  addStar(input: {
    starrableId: "MDEwOlJlcG9zaXRvcnkyOTMyOTU5NDA="
  }) {
    clientMutationId
  }
}
```

This performs the addStar mutation for the given repository ID.

7. Once the previous query has executed successfully, you must reexecute the query from *step 4* to find out about the change. If you get an access issue, then you can choose your own GitHub repository to perform these steps.

You can also explore other queries and mutations to deep dive into GraphQL.

Finally, let's understand the N+1 problem in GraphQL queries before we jump into the implementation in the next chapter.

Solving the N+1 problem

The N+1 problem is not new to Java developers. You might have encountered this problem in hibernation, which occurs if you don't optimize your queries or write entities properly.

Let's understand what the N+1 problem is.

Understanding the N+1 problem

The N+1 problem normally occurs when associations are involved. There are one-to-many relationships between the customer and the order. One customer can have many orders. If you need to find all the customers and their orders, you may do the following:

1. Find all the users.

2. Find all the user's orders based on the user's ID, which was received in the first step by setting the relation.

So, here, you fire two queries. If you optimize the implementation any further, you can place a joint between these two entities and receive all the records in a single query.

If this is so simple, then why does GraphQL encounter the N+1 problem? You need to understand the `resolver` function to answer this question.

If you go by the database schema we created in *Chapter 4, Writing Business Logic for APIs,* you can say that the `getUsersOrders` query will lead to the following SQL statements being executed:

```
SELECT * FROM ecomm.user;
SELECT * FROM ecomm.orders WHERE customer_id in (1);
SELECT * FROM ecomm.orders WHERE customer_id in (2);
...

...

SELECT * FROM ecomm.orders WHERE customer_id in (n);
```

Here, it's executing a query on the user to fetch all the users. Then, it executes N queries on orders. This is why it is called the N+1 problem. This is not efficient because ideally, it should execute a single query or in the worst case, two queries.

GraphQL can only respond with the values of fields that have been requested in the query due to resolvers. Each field has its own resolver function in the GraphQL server implementation that fetches the data for its corresponding field. Let's assume we have the following schema:

```
type Mutation {
  getUsersOrders: [User]
}
type User {
  name: String
  orders: [Order]
}
type Order {
  id: Int
  status: Status
}
```

Here, we have a mutation that returns a collection of users. Each user may have a collection of orders. You might use the following query in the client:

```
{
  getUsersOrders {
    name
```

```
    orders {
        id
        status
    }
  }
}
```

Let's understand how this query will be processed by the server. In the server, each field will have its own resolver function that fetches the corresponding data.

The first resolver will be for the user and will fetch all the users from the data store. Next, the resolver will be ordered for each user. It will fetch the orders from the data store based on the given user ID. Therefore, the `orders` resolver would execute n times, where n is the number of users that have been fetched from the data store.

We'll learn how to resolve the N+1 problem in the next subsection.

Solution for the N+1 problem

You need to have a solution that waits until all the orders have been loaded. Once all the user IDs have been retrieved, a database call should be made to fetch all the orders in a single data store call. You can use the batch if the size of the database is huge. Then, it can resolve the individual order resolvers. However, this is easier said than done. GraphQL provides a library called *DataLoader* (`https://github.com/graphql/dataloader`) that does this job for you.

Java provides a similar library called *java-dataloader* (`https://github.com/graphql-java/java-dataloader`) that can help you solve this problem. You can find out more about it at `https://www.graphql-java.com/documentation/v16/batching/`.

Summary

In this chapter, you learned about GraphQL, its advantages, and how it compares to REST. You learned how GraphQL solves overfetching and underfetching problems. You then learned about GraphQL's root types – queries, mutations, and subscriptions – and how different blocks can help you design the GraphQL schema. Finally, you understood how resolvers work, how they can lead to the N+1 problem, and the solution to this problem.

Now that you know about the fundamentals of GraphQL, you can start designing GraphQL schemas. You also learned about GraphQL's client-side queries and how to make use of aliases, fragments, and variables to resolve common problems.

In the next chapter, you will use the GraphQL skills you acquired in this chapter to implement a GraphQL server.

Questions

1. Is GraphQL better than REST? If yes, then in what way?

2. When should you use fragments?

3. How can you use variables in a GraphQL query?

Further reading

- GraphQL specifications: `https://spec.graphql.org/`

- GraphQL documentation: `https://graphql.org/learn/`

- *GraphQL and Apollo with Android from Novice to Expert* (video): `https://www.packtpub.com/product/graphql-and-apollo-with-android-from-novice-to-expert-video/9781800564626`

14

GraphQL API Development and Testing

You will learn about GraphQL-based API development and its testing in this chapter. You will implement GraphQL-based APIs for a sample application in this chapter. GraphQL server implementation will be developed based on a **design-first** approach, the way you defined the OpenAPI specification in *Chapter 3, API Specifications and Implementation*, and designed the schema in *Chapter 11, gRPC-based API Development and Testing*.

The following topics will be covered in this chapter:

- Workflow and tooling
- Implementing the GraphQL server
- Documenting APIs
- Test automation

After completing this chapter, you will have learned how to practically implement the GraphQL concepts learned in the previous chapter and the implementation of the GraphQL server using Java and Spring and its testing.

Technical requirements

You need the following for developing and testing the GraphQL-based service code presented in this chapter:

- Any Java IDE such as NetBeans, IntelliJ, or Eclipse

- **Java Development Kit (JDK)** 15

- An internet connection to clone the code and download the dependencies and Gradle (version 7+)

So, let's begin!

Please visit the following link to check the code files: `https://github.com/PacktPublishing/Modern-API-Development-with-Spring-and-Spring-Boot/tree/main/Chapter14`

Workflow and tooling for GraphQL

As a per-data graph way of thinking in GraphQL, data is exposed using an API consisting of graphs of objects. These objects are connected using relations. GraphQL only exposes a single API endpoint. Clients query this endpoint that uses a *single data graph*. On top of that, the data graph may resolve data from a single source, or multiple sources, by following the **OneGraph principle** of GraphQL. These sources can be a database, legacy system, or services that expose data using REST/gRPC/SOAP.

The GraphQL server can be implemented in the following two ways:

- **Standalone GraphQL service**: A standalone GraphQL service contains a single data graph. It could be a monolithic app or microservice architecture that fetches the data from single or multiple sources (having no GraphQL API).

- **Federated GraphQL services**: It's very easy to query a single data graph for comprehensive data fetching. However, enterprise applications are made using multiple services and hence you can't have a single data graph unless you build a monolithic system. If you don't build a monolithic system, then you would have multiple service-specific data graphs.

This is where you make use of federated GraphQL services. A federated GraphQL service contains a *single distributed graph* exposed using a gateway. Clients would call the gateway, which is an entry point to the system. The data graph would be distributed among multiple services and each service can maintain its own development and release cycle independently. Having said that, federated GraphQL services would still follow the OneGraph principle. Therefore, the client would query the single endpoint for fetching any part of the graph.

Let's assume that a sample e-commerce app is developed using GraphQL federated services. It would have products, orders, shipping, inventory, customers, and other services that would expose their domain-specific data graphs using the GraphQL API.

Let's draw a high-level diagram of GraphQL federated e-commerce services as follows:

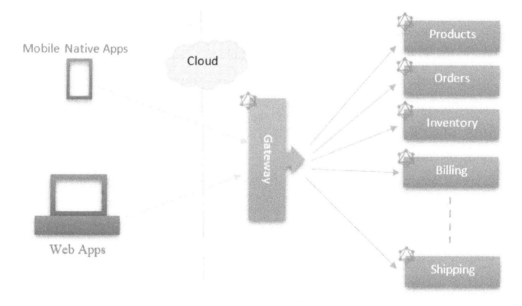

Figure 14.1 – Federated GraphQL services

Let's say the GraphQL client queries for a list of most ordered products with the least inventory by calling the **Gateway** endpoint. This query may have fields from **Orders**, **Products**, and **Inventory**. Each service is responsible for resolving only the respective part of a data graph. **Orders** would resolve order-related data, **Products** would resolve product-related data, **Inventory** would resolve inventory-related data, and so on. **Gateway** then consolidates the graph data and sends it back to the clients.

The `graphql-java` library (`https://www.graphql-java.com/`) provides the Java implementation of the GraphQL specification. Its source code is available at `https://github.com/graphql-java/graphql-java`.

There are many Spring Boot Starter projects for GraphQL, for example, at `https://github.com/graphql-java-kickstart/graphql-spring-boot`. However, we are going to use Netflix's **Domain Graph Service** (**DGS**) framework (`https://netflix.github.io/dgs/`). Netflix's DGS provides not only the GraphQL Spring Boot Starter but also the full set of tools and libraries that you need to develop production-ready GraphQL services. It is built on top of Spring Boot and uses the `graphql-java` library.

Netflix open-sourced the DGS framework after using it in production in February 2021. It is continuously being enhanced and supported by the community. Netflix uses the same open-sourced code based on their production, which gives the assurance of the code's quality and future maintenance.

It provides the following features:

- Provides a Spring Boot Starter and integration with Spring Security
- Gradle plugin for code generation from a GraphQL schema
- Support interfaces and union types, plus provides custom scalar types
- Supports GraphQL subscriptions using WebSocket and server-sent events
- Error handling
- Pluggable instrumentation
- GraphQL federated services by easy integration with GraphQL Federation
- File upload
- GraphQL Java client
- GraphQL test framework

Full WebFlux support could be available in the future. The release candidate build was available at the time of writing this chapter.

Let's write a GraphQL server using Netflix's DGS framework in the next section.

Implementation of the GraphQL server

You are going to develop a standalone GraphQL server in this chapter. The knowledge you acquire while developing the standalone GraphQL server can be used to implement federated GraphQL services.

Let's create the Gradle project first in the next subsection.

Creating the gRPC server project

Either you can use the `Chapter 14` code from a cloned Git repository (`https://github.com/PacktPublishing/Modern-API-Development-with-Spring-and-Spring-Boot`), or you can start by creating a new Spring project from scratch using *Spring Initializr* (`https://start.spring.io/`) for the server and client with the following options (you will create the gRPC `api` library project separately):

- **Project**: Gradle Project

- **Language**: Java

- **Spring Boot**: `2.4.4` (the preferred version is 2.4+; if not available, you can later modify it manually in the `build.gradle` file)

- **Project Metadata**:

  Group: `com.packt.modern.api`

  Artifact: `chapter14`

  Name: `chapter14`

  Description: `Chapter 14 code of book Modern API Development with Spring and Spring Boot`

  Package name: `com.packt.modern.api`

- **Packaging**: `JAR`

- **Java**: `11` (you can change it to another version such as 15/16/17 in the `build.gradle` file later as shown in the following code block):

  ```
  // update following build.gradle file
  sourceCompatibility = JavaVersion.VERSION_15
  // or for Java 16
  // sourceCompatibility = JavaVersion.VERSION_16
  // or for Java 17
  // sourceCompatibility = JavaVersion.VERSION_17
  ```

- **Dependencies**: `org.springframework.boot:spring-boot-starter-web`

Then, you can click on the **GENERATE** button and download the project. The downloaded project will be used for creating the GraphQL server.

Next, let's add the GraphQL DGS dependencies to the newly created project.

Adding the GraphQL DGS dependencies

Once the Gradle project is available, you can modify the build.gradle file to include the GDS dependencies and plugin as shown in the following code:

```
plugins {
    id 'org.springframework.boot' version '2.4.4'
    id 'io.spring.dependency-management' version
      '1.0.11.RELEASE'
    id 'java'
    id 'com.netflix.dgs.codegen' version '4.6.4'
}
// other part removed from brevity
def dgsVersion = '3.12.1'
dependencies {
    implementation platform("com.netflix.graphql.dgs:
        graphql-dgs-platform-dependencies:${dgsVersion}")
    implementation 'com.netflix.graphql.dgs:graphql-dgs-
        spring-boot-starter'
    implementation 'com.netflix.graphql.dgs:graphql-dgs-
        extended-scalars'
    implementation 'org.springframework.boot:spring-boot-
        starter-web'
    testImplementation 'org.springframework.boot:spring-boot-
        starter-test'
    implementation 'com.github.javafaker:javafaker:1.0.2'
}
```

https://github.com/PacktPublishing/Modern-API-Development-with-Spring-and-Spring-Boot/blob/main/Chapter14/build.gradle

Here, the DGS Codegen plugin is added, which will generate the code from the GraphQL schema file. Next, the following three dependencies have been added:

- `graphql-dgs- platform-dependencies`: The DGS platform dependencies for the DGS **bill of material (BOM)**

- `graphql-dgs-spring-boot-starter`: The DGS Spring Boot Starter library for DGS Spring support

- `graphql-dgs-extended-scalars`: The DGS extended scalars library for custom scalar types

Please note that the `javafaker` library is being used here to generate the domain seed data.

Next, let's configure the DGS Codegen plugin in the `build.gradle` file as shown in the next code block:

```
generateJava {
    packageName = "com.packt.modern.api.generated"
    generateClient = true
}
```

https://github.com/PacktPublishing/Modern-API-Development-with-Spring-and-Spring-Boot/blob/main/Chapter14/build.gradle

You have configured the following two properties of DGS Codegen using the `generateJava` task, which uses the `com.netflix.graphql.dgs.codegen.gradle.GenerateJavaTask` class:

- `packageName`: The Java package name of the generated Java classes
- `generateClient`: Whether you would like to generate the client or not

The DGS Codegen plugin picks GraphQL schema files from the `src/main/resources/schema` folder directory by default. However, you can modify it using the `schemaPaths` property, which accepts an array. You can add this property in the previous code of `generateTask` along with `packageName` and `generateClient` if you want to change the default schema location, as shown next:

```
schemaPaths = ["${projectDir}/src/main/resources/schema"]
```

You can also configure type mappings as you did for the `org.hidetake.swagger.generator` Gradle plugin while generating the Java code from OpenAPI specs in *step 4* of the *Convert OpenAPI spec to Spring code* section in *Chapter 3, API Specifications and Implementation*. For adding a custom type mapping, you can add the `typeMapping` property to the plugin task as shown next:

```
typeMapping = ["GraphQLType": "mypackage.JavaType"]
```

This property accepts an array; you can add one or more type mappings here. You can refer to the plugin documentation at https://netflix.github.io/dgs/generating-code-from-schema/ for more information.

Let's add the GraphQL schema next.

Adding the GraphQL schema

Netflix's DGS supports both the code-first and design-first approaches. However, you are going to use the design-first approach in this chapter as we have done throughout this book. Therefore, first you'll design the schema using the GraphQL schema language and then use the generated code to implement the GraphQL APIs.

We are going to keep the domain objects minimal to reduce the complexity of business logic and keep the focus on the GraphQL server implementation. Therefore, you'll have just two domain objects – Product and Tag. The GraphQL schema allows the following operation using its endpoint as shown in the following schema file:

```
type Query {
  products(filter: ProductCriteria): [Product]!
  product(id: ID!): Product
}
type Mutation {
  addTag(productId: ID!, tags: [TagInput!]!): Product
  addQuantity(productId: ID!, quantity: Int!): Product
}
type Subscription {
  quantityChanged(productId: ID!): Product
}
```

https://github.com/PacktPublishing/Modern-API-Development-with-Spring-and-Spring-Boot/blob/main/Chapter14/src/main/resources/schema/schema.graphqls

You need to add the schema.graphqls GraphQL schema file at the src/main/resources/schema location. You can have multiple schema files there to create the schema module-wise.

Here, the following root types have been exposed:

- **Query**: The product and products queries for fetching a product by its ID, and a collection of products matched by the given criteria.

- **Mutation**: The addTag mutation would add a tag to the product matched by the given ID. Another mutation, addQuantity, would increase the product quantities. The addQuantity mutation would also be used as an event that would trigger the subscription publication.

- **Subscription**: The quantityChanged subscription would publish the product where the quantity has been updated. The event quantity change would be captured through the addQuantity mutation.

Let's add the object types and input types being used in these root types as shown in the next code block:

```
type Product {
  id: String
  name: String
  description: String
  imageUrl: String
  price: BigDecimal
  count: Int
  tags: [Tag]
}
type Tag {
  id: String
  name: String
}
input ProductCriteria {
  tags: [TagInput] = []
  name: String = ""
  page: Int = 1
  size: Int = 10
}
input TagInput {
  name: String
}
```

https://github.com/PacktPublishing/Modern-API-Development-with-Spring-and-Spring-Boot/blob/main/Chapter14/src/main/resources/schema/schema.graphqls

These are straightforward object and input types. All fields of the `ProductCriteria` input type have been kept optional.

You have also used a `BigDecimal` custom scalar type. Therefore, we need to first declare it in the schema. You can do that by adding `BigDecimal` to the end of the schema file, as shown next:

```
scalar BigDecimal
```

Next, you also need to map it to `java.math.BigDecimal` in the code generator plugin. Let's add it to the `build.gradle` file as shown next (check the highlighted line):

```
generateJava {
  generateClient = true
  packageName = "com.packt.modern.api.generated"
  typeMapping = ["BigDecimal": "java.math.BigDecimal"]
}
```

https://github.com/PacktPublishing/Modern-API-Development-with-Spring-and-Spring-Boot/blob/main/Chapter14/build.gradle

After these changes, your project is ready to generate the GraphQL objects and client. You can run the following command from the project root directory to build the project:

```
gradlew clean build
```

This command would generate the Java classes in the `build/generated` directory.

Before you start implementing the GraphQL root types, let's discuss the custom scalar types in the next subsection.

Adding custom scalar types

You are going to use `BigDecimal` for capturing the monetary values. This is a custom scalar type, therefore you need to add this custom scalar to the code so that the DGS framework can pick it for serialization and deserialization. (This has to be done apart from adding a mapping in the Gradle code generator plugin.)

Create a new Java file called `BigDecimalScaler.java` and add the following code to it:

```
@DgsScalar(name = "BigDecimal")
public class BigDecimalScalar extends
```

```
    GraphqlBigDecimalCoercing {
}
```

https://github.com/PacktPublishing/Modern-API-Development-with-Spring-and-Spring-Boot/blob/main/Chapter14/src/main/java/com/packt/modern/api/scalar/BigDecimalScalar.java

Here, class is marked with the @DgsScalar annotation, which registers this class as a custom scalar with the DGS framework. Ideally, you should implement the graphql.schema.Coercing interface (part of the graphql-java library). This interface lets you implement serialization and parsing for custom scalar types.

However, since the BigDecimal default implementation (GraphqlBigDecimalCoercing) is already there in graphql-java, we'll simply extend it.

The DGS framework also provides custom scalars such as DateTime. These custom scalars can also be added to the DGS framework. The DateTime custom scalar implementation is available at https://github.com/PacktPublishing/Modern-API-Development-with-Spring-and-Spring-Boot/blob/main/Chapter14/src/main/java/com/packt/modern/api/scalar/DateTimeScalar.java, which can be used as a reference for adding other DGS custom scalar types.

Next, let's start implementing the GraphQL root types. First, you are going to implement the GraphQL queries.

Implementing GraphQL queries

Both the queries are straightforward. You pass a product ID to find a product identified by that ID – that's the product query for you. Next, you pass the optional product criteria to find the products based on the given criteria, else products are returned, based on the default values of the fields of product criteria.

In REST, you have done that. You create a controller, pass the call to the service, and the service calls the repository to fetch the data from the database. You are going to use the same design. However, you are going to use ConcurrentHashMap in place of the database to simplify the code. This can also be used in your automated tests.

Let's create a repository class for that, as shown in the next code block:

```
public interface Repository {
    Product getProduct(String id);
```

```
    List<Product> getProducts();
}
```

https://github.com/PacktPublishing/Modern-API-Development-with-Spring-and-Spring-Boot/blob/main/Chapter14/src/main/java/com/packt/modern/api/repository/Repository.java

These are straightforward signatures for fetching the product and collection of products.

Let's implement this interface using `ConcurrentHashMap` as shown in the next code block:

```
@org.springframework.stereotype.Repository
public class InMemRepository implements Repository {
  private final Logger LOG = LoggerFactory.getLogger(
                                 getClass());
  private static final Map<String, Product>
    productEntities = new ConcurrentHashMap<>();
  private static final Map<String, Tag> tagEntities =
    new ConcurrentHashMap<>();
  // rest of the code is truncated
```

Here, you have created two instances of `ConcurrentHashMap` to store the products and tags. Let's add the seed data to these maps using the constructor:

```
  public InMemRepository() {
    Faker faker = new Faker();
    IntStream.range(0, faker.number().numberBetween(20, 50))
      .forEach(number -> {
        String tag = faker.book().genre();
        tagEntities.putIfAbsent(tag,
            Tag.newBuilder().id(UUID.randomUUID().toString())
            .name(tag).build());
    });
    IntStream.range(0, faker.number().numberBetween(4, 20))
      .forEach(number -> {
        String id = String.format("a1s2d3f4-%d", number);
        String title = faker.book().title();
        List<Tag> tags = tagEntities.entrySet().stream()
```

```
      .filter(t -> t.getKey().startsWith(
        faker.book().genre().substring(0, 1)))
      .map(Entry::getValue).collect(toList());
    if (tags.isEmpty()) {
      tags.add(tagEntities.entrySet().stream()
        .findAny().get().getValue());
    }
    Product product = Product.newBuilder().id(id).name(
        title)
      .description(faker.lorem().sentence())
      .count(faker.number().numberBetween(10, 100))
      .price(BigDecimal.valueOf(faker.number()
          .randomDigitNotZero()))
      .imageUrl(String.format("/images/%s.jpeg",
        title.replace(" ", "")))
      .tags(tags).build();
    productEntities.put(id, product);
  });
}
```

This code first generates the tags and then products and stores them in respective maps. This has been done for development purposes only. You should use the database in production applications.

Now, the getProduct and getProducts methods are straightforward, as shown in the next code block:

```
@Override
public Product getProduct(String id) {
  if (Strings.isBlank(id)) {
    throw new RuntimeException("Invalid Product ID.");
  }
  Product product = productEntities.get(id);
  if (Objects.isNull(product)) {
    throw new RuntimeException("Product not found.");
  }
  return product;
}
```

```
@Override
public List<Product> getProducts() {
  return productEntities.entrySet().stream().map(e ->
                         e.getValue()).collect(toList());
}
```

The getProduct method performs the basic validations and returns the product. The getProducts method simply returns the collection of products converted from the map.

Now, you can add the service and its implementation. Let's add the service interface as shown in the next block:

```
public interface ProductService {
  Product getProduct(String id);
  List<Product> getProducts(ProductCriteria criteria);
}
```

https://github.com/PacktPublishing/Modern-API-Development-with-Spring-and-Spring-Boot/blob/main/Chapter14/src/main/java/com/packt/modern/api/services/ProductService.java

These interfaces simply call the repository to fetch the data. Let's add the implementation as shown in the next code block:

```
@Service
public class ProductServiceImpl implements ProductService {
  private final Repository repository;
  public ProductServiceImpl(Repository repository) {
    this.repository = repository;
  }
  @Override
  public Product getProduct(String id) {
    return repository.getProduct(id);
  }
  // continue...
```

https://github.com/PacktPublishing/Modern-API-Development-with-Spring-and-Spring-Boot/blob/main/Chapter14/src/main/java/com/packt/modern/api/services/ProductServiceImpl.java

Here, the repository is injected using constructor injection.

Let's add the getProducts() method also, which also performs filtering based on given filtering criteria, as shown in the next code block:

```
@Override
public List<Product> getProducts(ProductCriteria criteria) {
  List<Predicate<Product>> predicates = new ArrayList<>(2);
  if (!Objects.isNull(criteria)) {
    if (Strings.isNotBlank(criteria.getName())) {
      Predicate<Product> namePredicate = p ->
          p.getName().contains(criteria.getName());
      predicates.add(namePredicate);
    }
    if (!Objects.isNull(criteria.getTags()) &&
        !criteria.getTags().isEmpty()) {
      List<String> tags = criteria.getTags().stream().map(
          ti -> ti.getName()).collect(toList());
      Predicate<Product> tagsPredicate = p ->
          p.getTags().stream().filter(t ->
                  tags.contains(t.getName())).count() > 0;
      predicates.add(tagsPredicate);
    }
  }
  if (predicates.isEmpty()) {
    return repository.getProducts();
  }
  return repository.getProducts().stream()
          .filter(p -> predicates.stream().allMatch(
                  pre -> pre.test(p))).collect(toList());
}
```

This method first checks whether criteria are given or not. If criteria are not given, then it calls the repository and returns all the products.

If criteria are given, then it creates the predicates list. These predicates are then used to filter out the matching products and return back to the calling function.

Now comes the most critical piece of GraphQL query implementation: writing the data fetchers. First, let's write the data fetcher for the `product` query next.

Writing the data fetcher for product

The data fetcher is a critical DSG component to serve the GraphQL requests that fetches the data and DSG internally resolves each of the fields. You mark them with the special @ `DgsComponent` DGS annotation. These are types of Spring components that the DGS framework scans and uses for serving requests.

Let's create a new file called `ProductDatafetcher.java` in the `datafetchers` package for representing a DGS data fetcher component. It will have a data fetcher method for serving the `product` query. You can add the following code to it:

```
DgsComponent
public class ProductDatafetcher {
  private final ProductService productService;
  public ProductDatafetcher(ProductService productService) {
    this.productService = productService;
  }
  @DgsData(
     parentType = DgsConstants.QUERY_TYPE,
     field = QUERY.Product
  )
  public Product getProduct(@InputArgument("id") String id) {
    if (Strings.isBlank(id)) {
      new RuntimeException("Invalid Product ID.");
    }
    return productService.getProduct(id);
  }
}
```

https://github.com/PacktPublishing/Modern-API-Development-with-Spring-and-Spring-Boot/blob/main/Chapter14/src/main/java/com/packt/modern/api/datafetchers/ProductDatafetcher.java

Here, you create a product service bean injection using the constructor. This service bean helps you to find the product based on the given product ID.

Two other important DGS framework annotations have been used in the getProduct method. Let's understand what it does:

- @DgsData: This is a data fetcher annotation that marks the method as the data fetcher. The parentType property represents the type, and the field property represents the type's (parentType) field. Therefore, you can say that method would fetch the field of the given type.

  You have set "Query" as parentType. The field property is set as a "product" query. Therefore, this method works as an entry point for the GraphQL query product call. The @DsgData annotation properties are set using the DgsConstants constants class.

  DgsConstants is generated by the DGS Gradle plugin, which contains all the constant parts of the schema.

- @InputArgument: This annotation allows you to capture the arguments passed by the GraphQL requests. Here, the value of the id parameter is captured and assigned to the id string variable.

You can find the test cases related to this data fetcher method in the *Test automation* section.

Similarly, you can write the data fetcher method for the products query. Let's code it in the next subsection.

Writing the data fetcher for a collection of products

Let's create a new file called ProductsDatafetcher.java in the datafetchers package for representing a DGS data fetcher component. It will have a data fetcher method for serving the products query. You can add the following code to it:

```java
@DgsComponent
public class ProductsDatafetcher {
    private final Logger LOG = LoggerFactory.getLogger(
        getClass());
    private ProductService service;
    public ProductsDatafetcher(ProductService service) {
        this.service = service;
    }
    @DgsData(
        parentType = DgsConstants.QUERY_TYPE,
        field = QUERY.Products
```

```
    )
    public List<Product> getProducts(
                @InputArgument("filter") ProductCriteria
                    criteria) {
      return service.getProducts(criteria);
    }
}
```

https://github.com/PacktPublishing/Modern-API-Development-with-Spring-and-Spring-Boot/blob/main/Chapter14/src/main/java/com/packt/modern/api/datafetchers/ProductsDatafetcher.java

This getProducts() method does not look different from the data fetcher method returned for getProduct() in the second-to-last code block. Here, the parentType and field properties of @DsgData indicate that this method would be used to fetch the collection of products for the "products" query (extra s at the end).

You are done with the GraphQL query implementation. You can now test your changes. You need to build the application before running the test. Let's build the application using the following command:

```
$ gradlew clean build
```

Once the build is done successfully, you can run the following command to run the application:

```
$ java -jar build/libs/chapter14-0.0.1-SNAPSHOT.jar
```

The application should be running on default port 8080 if you have not made any changes in the port settings.

Now, you can open a browser window and open GraphiQL using the following URL: http://localhost:8080/graphiql (part of the DGS framework). Change the host/port accordingly if required.

You can use the following query to fetch the collection of products:

```
{
    products(filter: {name: "His Dark Materials", tags: [{name:
                "Fantasy"}, {name: "Legend"}]}
    ) {
      id
```

```
    name
    price
    description
    tags {
      id
      name
    }
  }
}
```

This would work great. However, what if you have to fetch the tags separately? You might have relations (such as orders having billing information) in objects that may be fetched from separate databases or services or from two separate tables. In that case, you might want to add a field resolver using the data fetcher method.

Let's add a field resolver using the data fetcher method in the next subsection.

Writing the field resolver using the data fetcher method

So far, you don't have a separate data fetcher for fetching the tags. You fetch the products and it also fetches the tags for you because we are using a concurrent map that stores both data together. Therefore, first you need to write a new data fetcher method for fetching the tags for a given product.

Let's add the `tags()` method to the `ProductsDatafetcher` class to fetch the tags, as shown in the next code block:

```
@DgsData(
    parentType = PRODUCT.TYPE_NAME,
    field = PRODUCT.Tags
)
public List<Tags>  tags(String productId) {
  return tagService.fetch(productId);
}
```

https://github.com/PacktPublishing/Modern-API-Development-with-Spring-and-Spring-Boot/blob/main/Chapter14/src/main/java/com/packt/modern/api/datafetch-ers/ProductsDatafetcher.java

Here, the tags() method has a different set of values for the @DsgData properties. The parentType property is not set to a root type like earlier data fetcher methods (set to Query). Instead, it is set to an object type – "Product". The field property is set to "tags".

This method would be called for fetching the tags for each individual product because it is a field resolver for the tags field of the Product object. Therefore, if you have 20 products, this method would be called 20 times to fetch the tags for each of the 20 products. This is an *N+1* problem, which we learned about in the last chapter (*Chapter 13, GraphQL Fundamentals*).

In the *N+1* problem, extra database calls are made for fetching the data for relations. Therefore, given a collection of products, it may hit a database for fetching the tags for each product separately.

You know that you have to use data loaders to avoid the *N+1* problem. Data loaders cache all the IDs of products before fetching their corresponding tags in a single query.

Next, let's learn how to implement a data loader for fixing the *N+1* problem in this case.

Writing a data loader for solving the N+1 problem

You are going to make use of the DataFetchingEnvironment class as an argument in the data fetcher methods. It is injected by the graphql-java library in the data fetcher methods to provide the execution context. This execution context contains information about the resolver, such as the object and its fields. You can also use them in special use cases such as loading the data loader classes.

Let's modify the tags() method in the ProductsDatafetcher class mentioned in the previous code block to fetch the tags without the *N+1* problem, as shown in the next code block:

```
@DgsData(
    parentType = PRODUCT.TYPE_NAME,
    field = PRODUCT.Tags
)
public CompletableFuture<List<Tags>>
            tags(DgsDataFetchingEnvironment env) {
    DataLoader<String, List<Tags>> tagsDataLoader =
            env.getDataLoader(
                TagsDataloaderWithContext.class);
    Product product = env.getSource();
```

```
    return tagsDataLoader.load(product.getId());
}
```

https://github.com/PacktPublishing/Modern-API-Development-with-Spring-and-Spring-Boot/blob/main/Chapter14/src/main/java/com/packt/modern/api/datafetchers/ProductsDatafetcher.java

Here, the modified `tags()` data fetcher method performs the fetch method using a data loader and returns the collection of tags wrapped inside `CompletableFuture`. And it would be called only once even if the number of products are more than 1.

What is CompletableFuture?

`CompletableFuture` is a Java concurrency class that represents the result of asynchronous computation, which is marked as completed explicitly. It can chain multiple dependent tasks asynchronously where the next task would be triggered when the current task's result is available.

You are using `DsgDataFetchingEnvironment` as an argument. It implements the `DataFetchingEnvironment` interface and provides ways to load the data loader class by both its class and name. Here, you are using the data loader class to load the data loader.

The `getSource()` method of `DsgDataFetchingEnvironment` returns the value from the `parentType` property of `@DsgData`. Therefore, `getSource()` returns `Product`.

This modified data fetcher method would fetch the tags for a given list of products. List of products? You are just passing a single product ID. This is correct, the data loader class implements `MappedBatchLoader`, which performs the operation using batches.

The data loader class fetches the tags of the given product (by ID) using the data loader *in batches*. The magic lies in returning `CompletableFuture`. Therefore, though you are passing a single product ID as an argument, the data loader processes it in bunches. Let's implement this data loader class (`TagsDataloaderWithContext`) next to dig into it more.

You can create a data loader class in two ways – with context or without context. Data loaders without context implement `MappedBatchLoader`, which has the following method signature:

```
CompletionStage<Map<K, V>> load(Set<K> keys);
```

On the other hand, data loaders with context implement the
`MappedBatchLoaderWithContext` interface, which has the following method
signature:

```
CompletionStage<Map<K, V>> load(Set<K> keys,
    BatchLoaderEnvironment environment);
```

Both are the same as far as data loading is concerned. However, the data loader with
context provides you with extra information (through `BatchLoaderEnvironment`)
that can be used for various additional features, such as authentication, authorization, or
passing the database details.

Create a new Java file called `TagsDataloaderWithContext.java` in the
`dataloaders` package with the following code:

```java
@DgsDataLoader(name = "tagsWithContext")
public class TagsDataloaderWithContext implements
                MappedBatchLoaderWithContext<String,
                List<Tag>> {
  private final TagService tagService;
  public TagsDataloaderWithContext(TagService tagService) {
    this.tagService = tagService;
  }
  @Override
  public CompletionStage<Map<String, List<Tag>>>
      load(Set<String> keys,
      BatchLoaderEnvironment environment) {
    return CompletableFuture.supplyAsync(() ->
                tagService.getTags(new ArrayList<>(keys)));
  }
}
```

https://github.com/PacktPublishing/Modern-API-Development-with-Spring-and-Spring-Boot/blob/main/Chapter14/src/main/java/com/packt/modern/api/dataloaders/TagsDataloaderWithContext.java

Here, it implements the `load()` method from the `MappedBatchLoaderWithContext` interface. The `BatchLoaderEnvironment` argument exists, which provides the context, but we are not using it as we don't have to pass any additional information to the repository or underlying data access layer. You can find the data loader without context at `https://github.com/PacktPublishing/Modern-API-Development-with-Spring-and-Spring-Boot/blob/main/Chapter14/src/main/java/com/packt/modern/api/dataloaders/TagDataloader.java`. It is similar to what we have written for the data loader with context as we are not using the context.

You could see that it makes use of the tag's service to fetch the tags. Then, it simply returns the completion stage by supplying tags received from the tag service. This operation is performed in batch by the data loader.

You can create a new tag service and its implementation as follows:

```java
public interface TagService {
  Map<String, List<Tag>> getTags(List<String> productIds);
}
```

https://github.com/PacktPublishing/Modern-API-Development-with-Spring-and-Spring-Boot/blob/main/Chapter14/src/main/java/com/packt/modern/api/services/TagService.java

This is the signature of the `getTags` method, which returns the map of product IDs with corresponding tags.

Let's implement this interface as shown in the next code block:

```java
@Service
public class TagServiceImpl implements TagService {
  private final Repository repository;
  public TagServiceImpl(Repository repository) {
    this.repository = repository;
  }
  @Override
  public Map<String, List<Tag>> getTags(List<String>
      productIds) {
    return repository.getProductTagMappings(productIds);
  }
}
```

https://github.com/PacktPublishing/Modern-API-Development-with-Spring-and-Spring-Boot/blob/main/Chapter14/src/main/java/com/packt/modern/api/services/TagServiceImpl.java

Here, the implemented method is straightforward. It passes the call to the repository that fetches the tags based on the passed collection of product IDs.

You can add `getProductTagMappings` to the `Repository` interface as shown in the next line:

```
Map<String, List<Tag>> getProductTagMappings(List<String>
    productIds);
```

Then you can implement this method in the `InMemRepository` class as shown in the next code block:

```
@Override
public Map<String, List<Tag>> getProductTagMappings(
    List<String> productIds) {
  return productEntities.entrySet().stream()
          .filter(e -> productIds.contains(e.getKey()))
          .collect(toMap(e -> e.getKey(),
            e -> e.getValue().getTags()));
}
```

Here, it first creates the stream of the product map's entry set, then filters the products that match the product passed in this method. At the end, it converts filtered products to map with the product ID as `Key` and `Tags` as the value, and then returns it.

Now, if you call the `"product"` GraphQL query, and even if products are fetched with a proper normalized database, it loads the product tags in batches without the *N+1* problem.

You are done with GraphQL query implementation and should be comfortable with implementing queries on your own.

Next, you are going to implement GraphQL mutations.

Implementing GraphQL mutations

As per the GraphQL schema, you are going to implement two mutations – `addTag` and `addQuantity`.

The addTag mutation takes productId and a collection of tags as arguments and returns the Product object. The addQuantity mutation takes productId and quantity to add and returns Product.

Let's add this implementation to the existing ProductDatafetcher class as shown in the following code block:

```java
// rest of the ProductDatafetcher class code
@DgsMutation(field = MUTATION.AddTag)
public Product addTags(@InputArgument("productId") String
    productId,
  @InputArgument(value = "tags", collectionType =
      TagInput.class) List<TagInput> tags) {
  return tagService.addTags(productId, tags);
}

@DgsMutation(field = MUTATION.AddQuantity)
public Product addQuantity(
    @InputArgument("productId") String productId,
    @InputArgument(value = "quantity") int qty) {
  return productService.addQuantity(productId, qty);
}
// rest of the ProductDatafetcher class code
```

https://github.com/PacktPublishing/Modern-API-Development-with-Spring-and-Spring-Boot/blob/main/Chapter14/src/main/java/com/packt/modern/api/datafetchers/ProductDatafetcher.java

Here, these signatures follow the respective mutations written in the GraphQL schema. You are using another DGS framework @DgsMutation annotation, which is a type of @DgsData annotation that is marked on methods to denote them as a data fetcher method. The @DgsMutation annotation by default has the "Mutation" value set to the parentType property. You just have to set the field property in this annotation. Both of these methods have their respective values set to the field property in the @DgsMutation annotation.

If you notice, you will find that the @InputArgument annotation for tags is using another collectionType property that is used for setting the type of input. It is required when the input type is not scalar. If you don't use it, you'll get an error. Therefore, make sure to use the collectionType property whenever you have a non-scalar type input.

These methods use the tag and product services to perform the requested operations. So far, you have not added the tag service in the ProductDatafetcher class. Therefore, you need to add TagService first as shown in the next code block:

```
// rest of the ProductDatafetcher class code
private final TagService tagService;
public ProductDatafetcher(ProductService productService,
                                    TagService tagService) {
    this.productService = productService;
    this.tagService = tagService;
}
// rest of the ProductDatafetcher class code
```

Here, the TagService bean has been injected using the constructor.

Now, you need to implement the addTag() method in the TagService and addQuantity methods in ProductService. Both the interfaces and their implementations are straightforward and pass the call to the repository to perform the operations. You can have a look at the source code in the GitHub code repository (https://github.com/PacktPublishing/Modern-API-Development-with-Spring-and-Spring-Boot/tree/main/Chapter14) to look into these implementations.

Let's add these two methods to the Repository interface as shown in the next code block:

```
Product addTags(String productId, List<TagInput> tags);
Product addQuantity(String productId, int qty);
```

These signatures in the Repository interface also follow the respective mutations written in the GraphQL schema.

Let's implement the addTags() method first in the InMemRepository class as shown in the next code block:

```
@Override
public Product addTags(String productId, List<TagInput> tags) {
```

```
  if (Strings.isBlank(productId)) {
    throw new RuntimeException("Invalid Product ID.");
  }
  Product product = productEntities.get(productId);
  if (Objects.isNull(product)) {
    throw new RuntimeException("Product not found.");
  }
  if (tags != null && !tags.isEmpty()) {
    List<String> newTags = tags.stream()
          .map(t -> t.getName()).collect(toList());
    List<String> existingTags = product.getTags().stream()
          .map(t -> t.getName()).collect(toList());
    newTags.stream().forEach(nt -> {
      if (!existingTags.contains(nt)) {
        product.getTags().add(Tag.newBuilder()
          .id(UUID.randomUUID().toString()).
            name(nt).build());
      }
    });
    productEntities.put(product.getId(), product);
  }
  return product;
}
```

https://github.com/PacktPublishing/Modern-API-Development-with-Spring-and-Spring-Boot/blob/main/Chapter14/src/main/java/com/packt/modern/api/repository/InMemRepository.java

This implementation is straightforward. It performs a couple of validations for the passed product ID. Then it compares the new and existing tags, and adds the new tags to the passed product only if existing tags don't exist. At the end, it updates the concurrent map and returns the updated product.

Let's add the implementation of the addQuantity() method to the InMemRepository class next, as shown in the following code block:

```
@Override
public Product addQuantity(String productId, int qty) {
```

```
if (Strings.isBlank(productId)) {
  throw new RuntimeException("Invalid Product ID.");
}
if (qty < 1) {
  throw new RuntimeException("Quantity arg can't be less
                            than 1");
}
Product product = productEntities.get(productId);
if (Objects.isNull(product)) {
  throw new RuntimeException("Product not found.");
}
product.setCount(product.getCount() + qty);
productEntities.put(product.getId(), product);
return product;
}
```

https://github.com/PacktPublishing/Modern-API-Development-with-Spring-and-Spring-Boot/blob/main/Chapter14/src/main/java/com/packt/modern/api/repository/InMemRepository.java

Here, you first perform the validation for the productId and qty arguments. If everything goes fine, then you increase the quantity of the product, update the concurrent map, and return the updated product.

You are done with the implementation of GraphQL mutations. You can now test your changes. You need to build the application before running the test. Let's build the application using the following command:

```
$ gradlew clean build
```

Once the build is done successfully, you can run the following command to run the application:

```
$ java -jar build/libs/chapter14-0.0.1-SNAPSHOT.jar
```

The application should be running on default port 8080 if you have not made any changes to the port settings.

Now, you can open a browser window and open GraphiQL using the following URL: http://localhost:8080/graphiql (part of the DGS framework). Change the host/port accordingly if required.

You can use the following GraphQL request to perform the addTag mutation:

```
mutation {
  addTag(productId: "a1s2d3f4-0", tags: [
    {
      name:"new Tags..."
    }
  ]) {
    id
    name
    price
    description
    tags {
      id
      name
    }
  }
}
```

Here, you pass productId and tags as arguments. You can use the following GraphQL request to perform the addQuantity mutation:

```
mutation {
  addQuantity(productId: "a1s2d3f4-0", quantity: 10) {
    id
    name
    description
    price
    count
    tags {
      id
      name
    }
  }
}
```

Here, you pass `productId` and `quantity` as arguments. You have learned how to implement GraphQL mutations in the GraphQL server. Let's implement GraphQL subscriptions in the next subsection.

Implementing GraphQL subscriptions

Subscription is another GraphQL root type that sends the object to the subscriber (client) when a particular event occurs.

Let's assume an online shop offers a discount on products when the product's inventory reaches a certain level. You cannot track each and every product's quantity manually and then perform the computation and trigger the discount. This is where you can make use of the subscription.

Each change in the product's inventory (quantity) through the `addQuantity()` mutation should trigger the event and the subscriber should receive the updated product and hence the quantity. Then, the subscriber can place the logic and automate this process.

Let's write the subscription that would send the updated product object to the subscriber. You are going to use Reactive Streams and WebSocket to implement this functionality.

Let's add additional dependencies in `build.gradle` to take care of the auto-configuration of WebSocket and the playground tool to test the subscription functionality. (By default DGS provides the GraphiQL app to explore the documentation and schema and play with queries. However, the bundled GraphiQL tool doesn't work properly for testing the subscription presently. Once it starts working, you don't need to add the playground tool.)

Let's add these dependencies to `build.gradle` as shown in the following code block:

```
dependencies {
    // other dependencies …
    runtimeOnly 'com.netflix.graphql.dgs:graphql-dgs-
        subscriptions-websockets-autoconfigure'
    implementation 'com.graphql-java-kickstart:playground-
        spring-boot-starter:11.0.0'
    // other dependencies …
}
```

Now, you can add the following subscription data fetcher to the `ProductDatafetcher` class as shown in the following code:

```
// rest of the ProductDatafetcher class code
@DgsSubscription(field = SUBSCRIPTION.QuantityChanged)
public Publisher<Product> quantityChanged(
                    @InputArgument("productId") String
                    productId) {
  return productService.gerProductPublisher();
}
// rest of the ProductDatafetcher class code
```

https://github.com/PacktPublishing/Modern-API-Development-with-Spring-and-Spring-Boot/blob/main/Chapter14/src/main/java/com/packt/modern/api/datafetch-ers/ProductDatafetcher.java

Here, you are using another DGS framework annotation, `@DgsSubscription`, which is a type of `@DgsData` annotation that is marked on a method to denote it as a data fetcher method. The `@DgsSubscription` annotation by default has the `Subscription` value set to the `parentType` property. You just have to set the `field` property in this annotation. By setting the field to `quantityChanged`, you are indicating to the DGS framework to use this method when the subscription request for `quantityChanged` is called.

The `Subscription` method returns the `Publisher` instance, which can be sent an unbound number of objects (in this case, `Product` instances) to multiple subscribers. Therefore, the client just needs to subscribe to the product publisher.

You need to add a new method to the `ProductService` interface and its implementation in the `ProductServiceImpl` class. The method signature in the `ProductService` interface and its implementation are straightforward. It passes the call to the repository to perform the operation. You can have a look at the source code in the GitHub code repository at `https://github.com/PacktPublishing/Modern-API-Development-with-Spring-and-Spring-Boot/tree/main/Chapter14`.

Actual work is being performed by the repository. Therefore, you need to make certain changes in the repository as shown in the following steps:

1. First add the following method signature to the `Repository` interface:

    ```
    Publisher<Product> getProductPublisher();
    ```

2. Next, you have to implement the `getProductPublisher()` method in the `InMemRepository` class. This method returns the product publisher as shown in the following code:

    ```
    public Publisher<Product> getProductPublisher() {
        return productPublisher;
    }
    ```

3. Now, we need all the magic to be performed by Reactive Streams. First, let's declare the `FluxSink<Product>` and `ConnectableFlux<Product>` (which is returned by the repository) variables:

    ```
    private FluxSink<Product> productsStream;
    private ConnectableFlux<Product> productPublisher;
    ```

4. Now, we need to initialize these declared instances. Let's do so in `InMemRepository`'s constructor as shown in the following code:

    ```
    Flux<Product> publisher = Flux.create(emitter -> {
        productsStream = emitter;
    });
    productPublisher = publisher.publish();
    productPublisher.connect();
    ```

5. `Flux<Product>` is a product stream publisher that passes the baton to `productsStream (FluxSink)` for emitting the next signals followed by `onError()` or `onComplete()` events. This means `productsStream` should emit the signal when the product quantity gets changed. When `Flux<Product>` calls the `publish()` method, it returns an instance of `connectableFlux`, which is assigned to `productPublisher` (the one that is returned by the subscription).

6. You are almost done with the setup. You just need to emit the signal (product) when the product gets changed. Let's add the following highlighted line to the addQuantity() method before it returns the product, as shown in the following code:

```
product.setCount(product.getCount() + qty);
productEntities.put(product.getId(), product);
productsStream.next(product);
return product;
```

You have completed the subscription quantityChanged implementation. You can test it next.

You need to build the application before running the test. Let's build the application using the following command:

```
$ gradlew clean build
```

Once the build is done successfully, you can run the following command to run the application:

```
$ java -jar build/libs/chapter14-0.0.1-SNAPSHOT.jar
```

The application should be running on default port 8080 if you have not made any changes in the port settings.

The playground tool should be available at http://localhost:8080/playground if the application is running on localhost, else make the appropriate changes in the hostname.

Once the playground app is up, run the following query in it:

```
subscription {
  quantityChanged(productId: "a1s2d3f4-0") {
    id
    name
    description
    price
    count
  }
}
```

This should trigger the application in listening mode. The app will wait for object publications.

Now, you can open another browser window and open GraphiQL using the following URL: `http://localhost:8080/graphiql`. Change the host/port accordingly if required.

Here, you can fire the `addQuantity` mutation by running the following:

```
mutation {
  addQuantity(productId: "a1s2d3f4-0", quantity: 10) {
    id
    name
    price
    count
  }
}
```

Each successful change would publish the updated product to the playground app.

You should know about the instrumentation that helps to implement the tracing, logging, and metrics collection. Let's discuss this in the next subsection.

Instrumenting the GraphQL API

The GraphQL Java library supports the instrumentation of the GraphQL API. This can be used to support metrics, tracing, and logging. The DGS framework also uses it. You just have to mark the instrumentation class with the Spring `@Component` annotation.

The instrumentation bean should implement the `graphql.execution.instrumentation.Instumentation` interface, an easier way to extend the `SimpleInstumentation` class.

Let's add instrumentation that would record the time taken by the data fetcher and complete GraphQL request processing. This metric may help you to fine-tune the performance and identify the fields that take more time to resolve.

Let's create the `TracingInstrumentation.java` file in the `instrumentation` package and add the following code:

```
@Component
public class TracingInstrumentation extends
        SimpleInstrumentation {
```

```java
private final Logger LOG = LoggerFactory.getLogger(
    getClass());
@Override
public InstrumentationState createState() {
  return new TracingState();
}
static class TracingState implements InstrumentationState {
  long startTime;
}
// continue...
```

https://github.com/PacktPublishing/Modern-API-Development-with-Spring-and-Spring-Boot/blob/main/Chapter14/src/main/java/com/packt/modern/api/instrumentation/TracingInstrumentation.java

This class extends `SimpleInstrumentation` and is created as a Spring bean by marking it as `@Component`. First of all, you need to create the instrumentation state by overriding the `createState()` method. Since you are implementing the time metric, you choose `startTime` as the state. A static inner class is added for declaring the `startTime` state.

As a next activity, you would like to initialize the instrumentation state. For that purpose you can override the `beginExecution()` method as shown in the following code:

```java
@Override
public InstrumentationContext<ExecutionResult>
    beginExecution(
      InstrumentationExecutionParameters parameters) {
  TracingState tracingState =
      parameters.getInstrumentationState();
  tracingState.startTime = System.currentTimeMillis();
  return super.beginExecution(parameters);
}
```

This method allows you to set the instrumentation parameters.

The `startTime` state is set. Next, you'll override the `instrumentExecutionResult()` method. This helps you to instrument the execution result such as calculating the total execution time. Let's add the following code to calculate the total execution time:

```
@Override
public CompletableFuture<ExecutionResult>
    instrumentExecutionResult(
    ExecutionResult executionResult,
    InstrumentationExecutionParameters parameters) {
  TracingState tracingState =
  parameters.getInstrumentationState();
  long timeTaken = System.currentTimeMillis() -
    tracingState.startTime;
  LOG.info("Request processing took: {} ms", timeTaken);
  return super.instrumentExecutionResult(
                                    executionResult,
                                    parameters);
}
```

It is a straightforward implementation to calculate the total execution time. It extracts the `startTime` state from the parameters and then uses it to calculate the `timeTaken` value.

So far you have overridden three methods – the initial method (`createState()`), the beginning method (`beginExecution()`) for state initialization, and the end method (`instrumentExecutionResult()`) for final calculations or state recording.

One intermediate method (`instrumentDataFetcher()`) that falls between `beginExecution()` and `instrumentExecutionResult()` is yet to be overridden. It is complex compared to other methods. Therefore, you'll override it after other methods.

Let's add the following code to override the `instrumentDataFetcher()` method:

```
@Override
public DataFetcher<?> instrumentDataFetcher(DataFetcher<?>
        dataFetcher, InstrumentationFieldFetchParameters
        parameters) {
  if (parameters.isTrivialDataFetcher()) {
    return dataFetcher;
  }
```

```
return environment -> {
  long initTime = System.currentTimeMillis();
  Object result = dataFetcher.get(environment);
  String msg = "Instrumentation of datafetcher {} took {}
            ms";
  if (result instanceof CompletableFuture) {
    ((CompletableFuture<?>) result).whenComplete((r, ex)
        -> {
      long timeTaken = System.currentTimeMillis() -
          initTime;
      LOG.info(msg, findDatafetcherTag(parameters),
          timeTaken);
    });
  } else {
    long timeTaken = System.currentTimeMillis() -
        initTime;
    LOG.info(msg, findDatafetcherTag(parameters),
        timeTaken);
  }
  return result;
};
}
```

This method is used for instrumenting the data fetchers. You have added two separate blocks to calculate the data fetching time because values can be returned in two ways by data fetcher methods – a blocking call or an asynchronous call (CompletableFuture). This method would be called for each data fetching call whether it is for the root type or for a field of the object type.

The final piece of instrumentation implementation is the findDatafetcherTag() method. This private method is added to find out the data fetching type of the field/root type.

Let's add it as shown in the following code:

```
private String findDatafetcherTag(
                InstrumentationFieldFetchParameters
parameters) {
  GraphQLOutputType type =
```

```
        parameters.getExecutionStepInfo()
                                 .getParent().getType();
    GraphQLObjectType parent;
    if (type instanceof GraphQLNonNull) {
      parent = (GraphQLObjectType)
                    ((GraphQLNonNull)
                       type).getWrappedType();
    } else {
      parent = (GraphQLObjectType) type;
    }
    return parent.getName() + "." +
        parameters.getExecutionStepInfo().getPath()
           .getSegmentName();
}
```

Here, `GraphQLNonNull` tells us whether the type is a wrapped type or not. Next, let's find out what tool you can use for documenting APIs.

Documenting APIs

You can use GraphQL or a playground tool that provides a graphical interface to explore the GraphQL schema and documentation.

However, if you are looking for a static page, then you can use tools such as graphdoc (`https://github.com/2fd/graphdoc`) for generating the static documentation of GraphQL APIs.

Next, let's learn about GraphQL API testing using the DGS framework.

Test automation

The DGS framework provides you with classes and utilities that you can use to test GraphQL APIs.

Create a new file called `ProductDatafetcherTest.java` inside the `datafetchers` package in the test directory and add the following code:

```
@SpringBootTest(classes = {DgsAutoConfiguration.class,
                  ProductDatafetcher.class,
                  BigDecimalScalar.class})
```

```
public class ProductDatafetcherTest {
  private final InMemRepository repo = new
    InMemRepository();
  private final int TEN = 10;
  @Autowired
  private DgsQueryExecutor dgsQueryExecutor;
  @MockBean
  private ProductService productService;
  @MockBean
  private TagService tagService;
// continue...
```

https://github.com/PacktPublishing/Modern-API-Development-with-Spring-and-Spring-Boot/blob/main/Chapter14/src/test/java/com/packt/modern/api/datafetchers/ProductDatafetcherTest.java

Here, you are using the `@SpringBootTest` annotation to execute the test. By providing limited classes such as `DgsAutoConfiguration`, `ProductDatafetcher`, and `BigDecimalScalar`, you are limiting the Spring context. You should add the classes here that are going to take part in testing.

First of all, you are auto-wiring the `DgsQueryExecutor` class that performs the query execution. After that, you add two Spring-injected mock beans for the `Product` and `Tag` services.

You are ready with the configuration and instances you need to run the tests.

Let's add the setup method that is required before running the tests. You can add the following method for this purpose:

```
@BeforeEach
public void beforeEach() {
  List<Tag> tags = new ArrayList<>();
  tags.add(Tag.newBuilder().id("tag1").name("Tag
                              1").build());
  Product product = Product.newBuilder().id("any")
              .name("mock title").description("mock
                  description")
              .price(BigDecimal.valueOf(20.20)).count(100)
              .tags(tags).build();
```

```
given(productService.getProduct(
  "any")).willReturn(product);
tags.add(Tag.newBuilder().id("tag2").name(
  "addTags").build());
product.setTags(tags);
given(tagService.addTags("any",
       List.of(TagInput.newBuilder().name(
           "addTags").build())))
  .willAnswer(invocation -> product);
}
```

In this method, we used Mockito for stubbing the service methods.

You are done with the setup. Let's run our first test that would fetch the JSON object after running the GraphQL `product` query next.

Testing GraphQL queries

Let's add the following code for testing the `product` query:

```
@Test
@DisplayName("Verify the JSON attrs returned from query
            'product'")
public void product() {
  String name = dgsQueryExecutor.executeAndExtractJsonPath(
        "{ product(id: \"any\") { name }}",
        "data.product.name");
  assertThat(name).contains("mock title");
}
```

Here, you are using the DgsQueryExecutor instance to execute the `product` query and extract the JSON property.

Next, you'll test the `product` query again, but this time for testing the exception.

You can add the following code to test the exception thrown by the `product` query:

```
@Test
@DisplayName("Verify exception for incorrect ID in query
            'product'")
public void productWithException() {
```

```
given(productService.getProduct("any"))
    .willThrow(new RuntimeException("Invalid Product
            ID."));
ExecutionResult result = dgsQueryExecutor.execute(
                    " { product (id: \"any\") {
                        name }}");
verify(productService, times(1)).getProduct("any");
assertThat(result.getErrors()).isNotEmpty();
assertThat(result.getErrors().get(0).getMessage())
    .isEqualTo("java.lang.RuntimeException: Invalid
            Product ID.");
}
```

Here, the product service method is stubbed for throwing the exception. When DgsQueryExecutor runs, the Spring-injected mock bean uses the stubbed method to throw the exception that is being asserted here.

Next, let's query product again, this time to explore GraphQLQueryRequest, which allows you to form the GraphQL query in a fluent way. The GraphQLQueryRequest construction takes two arguments – first the instance of GraphQLQuery, which can be a query/mutation or subscription, and second the projection root type of BaseProjectionNode, which allows you to select the fields.

Let's add the following code to test the product query using GraphQLQueryRequest:

```
@Test
@DisplayName("Verify JSON attrs using GraphQLQueryRequest")
void productsWithQueryApi() {
  GraphQLQueryRequest graphQLQueryRequest = new
      GraphQLQueryRequest(
          ProductGraphQLQuery.newRequest().id(
              "any").build(),
          new ProductProjectionRoot().id().name());
  String name = dgsQueryExecutor.executeAndExtractJsonPath(
          graphQLQueryRequest.serialize(),
          "data.product.name");
  assertThat(name).contains("mock title");
}
```

Here, the `ProductGraphQLQuery` class is part of the auto-generated code by the DGS GraphQL Gradle plugin.

One thing we have not yet tested in previous tests is verifying the sub-fields in the `tags` field of `product`.

Let's verify it in the next test case. Add the following code to verify the tags:

```
@Test
@DisplayName("Verify the Tags returned from the query
    'product'")
void productsWithTags() {
  GraphQLQueryRequest graphQLQueryRequest = new
    GraphQLQueryRequest(
        ProductGraphQLQuery.newRequest().id("any").build(),
        new ProductProjectionRoot().id().name()
            .tags().id().name());
  Product p = dgsQueryExecutor.
    executeAndExtractJsonPathAsObject(
        graphQLQueryRequest.serialize(),
        "data.product", new TypeRef<>() {});
  assertThat(p.getId()).isEqualTo("any");
  assertThat(p.getName()).isEqualTo("mock title");
  assertThat(p.getTags().size()).isEqualTo(2);
  assertThat(p.getTags().get(0).getName()).isEqualTo("Tag 1");
}
```

Here, you can see that you have to use a third argument (`TypeRef`) in the `executeAndExtractJsonPathAsObject()` method if you want to query the sub-fields. If you don't use it, you will get an error.

You are done with GraphQL query testing. Let's move on to testing the mutations in the next subsection.

Testing GraphQL mutations

Testing a GraphQL mutation is no different than testing GraphQL queries.

Let's test the `addTag` mutation as shown in the following code:

```
@Test
@DisplayName("Verify the mutation 'addTags'")
void addTagsMutation() {
  GraphQLQueryRequest graphQLQueryRequest = new
    GraphQLQueryRequest(
      AddTagGraphQLQuery.newRequest().productId("any")
        .tags(List.of(TagInput.newBuilder().name(
            "addTags").build()))
          .build(),new AddTagProjectionRoot().name().
            count());
  ExecutionResult executionResult =
    dgsQueryExecutor.execute(
        graphQLQueryRequest.serialize());
  assertThat(executionResult.getErrors()).isEmpty();
  verify(tagService).addTags("any",
        List.of(
          TagInput.newBuilder().name("addTags").build()));
}
```

Here, the `AddTagGraphQLQuery` class is part of the auto-generated code by the DGS GraphQL Gradle plugin. You fire the request and then validate the results based on the existing configuration and setup.

Similarly, you can test the `addQuantity` mutation. Only the arguments and assertions would change; the core logic and classes would remain the same.

You can add the test as shown in the next code block to test the `addQuantity` mutation:

```
@Test
@DisplayName("Verify the mutation 'addQuantity'")
void addQuantityMutation() {
  given(productService.addQuantity("a1s2d3f4-1", TEN))
      .willReturn(repo.addQuantity("a1s2d3f4-1", TEN));
  GraphQLQueryRequest graphQLQueryRequest = new
    GraphQLQueryRequest(
      AddQuantityGraphQLQuery.newRequest().productId(
          "a1s2d3f4-1")
```

```
                    .quantity(TEN).build(),
        new AddQuantityProjectionRoot().name().count());
    ExecutionResult executionResult =
        dgsQueryExecutor.execute(
          graphQLQueryRequest.serialize());
    assertThat(executionResult.getErrors()).isEmpty();
    Object obj = executionResult.getData();
    assertThat(obj).isNotNull();
    Map<String, Object> data = (Map)((Map)
      executionResult.getData())
                        .get(MUTATION.AddQuantity);
    org.hamcrest.MatcherAssert
          .assertThat((Integer) data.get("count"),
            greaterThan(TEN));
}
```

You are done with GraphQL mutation testing. Let's move on to testing subscriptions in the next subsection.

Testing GraphQL subscriptions

Testing the subscription needs extra effort and care as you can see in the following code, which performs the test for the quantityChanged subscription. It uses the existing addQuantity mutation to trigger the subscription publisher that sends a product object on each call. You capture the product of the first call and store the value of the count field. Then, use it to perform the assertion as shown in the following code:

```
@Test
@DisplayName("Verify the subscription 'quantityChanged'")
void reviewSubscription() {
  given(productService.gerProductPublisher())
      .willReturn(repo.getProductPublisher());
  ExecutionResult executionResult =
    dgsQueryExecutor.execute(
      "subscription { quantityChanged(productId:
        \"a1s2d3f4-0\") { id name price count } }");
  Publisher<ExecutionResult> publisher =
    executionResult.getData();
  List<Product> product = new CopyOnWriteArrayList<>();
```

```java
publisher.subscribe(new Subscriber<>() {
  @Override
  public void onSubscribe(Subscription s) { s.request(2); }
  @Override
  public void onNext(ExecutionResult result) {
    if (result.getErrors().size() > 0) {
      System.out.println(result.getErrors());
    }
    Map<String, Object> data = result.getData();
    product.add(new ObjectMapper().convertValue(
        data.get(SUBSCRIPTION.QuantityChanged),
            Product.class));
  }
  @Override
  public void onError(Throwable t) {}
  @Override
  public void onComplete() {}
});
addQuantityMutation();
Integer count = product.get(0).getCount();
addQuantityMutation();
assertThat(product.get(0).getId())
        .isEqualTo(product.get(1).getId());
assertThat(product.get(1).getCount())
        .isEqualTo(count.intValue() + TEN);
}
```

Here, the core logic lies in the subscription that is done by calling the `publisher.subscribe()` method (check highlighted line). You know that the GraphQL `quantityChanged` subscription returns the publisher. This publisher is received from the data field of the execution result.

The publisher subscribes to the stream by passing an object of `Subscriber`, which is created on the fly. The subscriber's `onNext()` method is used to receive the product sent by the GraphQL server. These objects are pushed into the list. Then, you use this list to perform the assertion.

Summary

In this chapter, you have learned about the different ways of implementing the GraphQL server including federated GraphQL services. You have also explored the complete standalone GraphQL server implementation that performs the following operations:

- Writing the GraphQL schema
- Implementing the GraphQL query APIs
- Implementing the GraphQL mutation APIs
- Implementing the GraphQL subscription APIs
- Writing the data loaders for solving the *N+1* problem
- Adding custom scalar types
- Adding the GraphQL API's instrumentation
- Writing the GraphQL API's test automation using Netflix's DGS framework

You learned about the GraphQL API implementation using Spring and Spring Boot skills that will help you implement GraphQL APIs for your work assignments and personal projects.

Questions

1. Why should you prefer frameworks such as Netflix's DGS in place of the `graphql-java` library to implement GraphQL APIs?
2. What are federated GraphQL services?

Further reading

- GraphQL Java implementation: `https://www.graphql-java.com/` and `https://github.com/graphql-java/graphql-java`
- Netflix DGS documentation: `https://netflix.github.io/dgs/getting-started/`
- *GraphQL and Apollo with Android from Novice to Expert* (video): `https://www.packtpub.com/product/graphql-and-apollo-with-android-from-novice-to-expert-video/9781800564626`

Assessments

This section contains the answers to the questions from every chapter.

Chapter 1 – RESTful Web Services Fundamentals

1. RESTful became popular because it works on top of HTTP protocol, which is the backbone of the internet. You don't need separate protocol implementations such as SOAP. You can use the existing web technologies to implement the REST APIs, with simple application integration compared to other technologies available at the time. REST APIs make application integration simpler compared to other technologies available at the time.

2. RESTful services work on REST, which works on web resources. Resources represent domain models. Actions are defined using HTTP methods, which are performed on web resources. It also allows clients to perform actions based on links available through **Hypermedia as the Engine of Application State HATEOAS** implementation, like a human who can navigate in the browser.

3. RPC is more like functions that perform actions. RPC endpoints are directly formed based on verbs that lead to separate URLs for each action. Whereas REST URLs represent nouns and could be the same for different operations, for example:

```
RPC: GET localhost/orders/getAllOrders
REST: GET localhost/orders


RPC: POST localhost/orders/createOrder
REST: POST localhost/orders
```

4. With **Hypermedia as the Engine of Application State** (**HATEOAS**), RESTful web services provide information dynamically through hypermedia. Hypermedia is the part of the content you receive from a REST call response. This hypermedia content contains links to different types of media such as text, images, and videos. Machines, aka REST clients/browsers, can follow links when they understand the data format and relationship types.

5. Status code 500 should be used for generic server errors. The 502 status code should be used when an upstream server fails. Status code 503 is for unexpected server events such as an overload.

6. Verbs should not be used for forming REST endpoints. Instead, you should use the noun that represents the domain model as a resource. HTTP methods are used to define the actions performed on resources such as POST for creating, GET for retrieving, and so on.

Chapter 2 – Spring Concepts and REST APIs

1. By using the @Scope annotation as shown:

```
@Scope(ConfigurableBeanFactory.SCOPE_PROTOTYPE)
```

2. Beans defined using the Singleton scope instantiate only once per Spring container. The same instance is injected every time it is requested. Whereas a container creates a new instance each time for beans defined with the prototype scope when the injection is done by the Spring container for the requested bean. In short, a container creates a single bean per container for a singleton-scoped bean; whereas a container creates a new instance each time for a new injection for prototype-scoped beans.

3. Session and request scopes only work when a web-aware Spring context is used. Other scopes that also need a web-aware context to work are application and WebSocket scopes.

4. Advice is an action taken by the Aspect at a specific time (JoinPoint). Aspects perform the additional logic (advice) at a certain point (JoinPoint), such as a method being called, an exception being thrown, and so on.

5. The following code would print the method name and argument names before method execution and a message with the return type after the method execution:

```
@Aspect
@Component
```

```
public class TimeMonitorAspect {
  @Around("@annotation(TimeMonitor)")
  public Object logTime(ProceedingJoinPoint joinPoint)
    throws Throwable {
    System.out.println(String.format("Method Name: %s,
        Arg Name: %s", joinPoint.getSignature().
        getName(), Arrays.toString(((CodeSignature)
        joinPoint.getSignature()).getParameterNames())));
    Object proceed = joinPoint.proceed();
    System.out.println(String.format("Method %s contains
        the following return type: %s",
        joinPoint.getSignature().getName(),
        ((MethodSignature) joinPoint.getSignature()).
        getReturnType().toGenericString())));
    return proceed;
  }
}
```

Chapter 3 – API Specifications and Implementation

1. The **OpenAPI Specification (OAS)** is introduced to solve at least a few aspects of REST API's specification and description. It allows you to write REST APIs in YAML or JSON markup languages, which allows you to interact with all stakeholders, including non-technical, for review and discussion in the development phase. It also allows you to generate documentation, models, interfaces, clients, and servers in different languages.

2. The array is defined using the following code:

```
type: array
items:
  type: array
  items:
    type: string
```

3. You need a class annotation, @ControllerAdvice, and method annotation, @ExceptionHandler, to implement the global exception handler.

4. You can use the `--type-mappings` and `--import-mappings rawOptions` in the `swaggerSources` task of the `build.gradle` file.

5. We only generate the models and API interfaces using Swagger Codegen because this allows the complete implementation of controllers by developers only.

Chapter 4 – Writing Business Logic for APIs

1. Repository classes are marked with `@Repository`, which is a specialized `@Component` that makes these classes auto-detected by package-level auto-scanning and makes them available for injection. Spring provides these classes especially for DDD repositories and Java EE **Data Access Object** (**DAO**). This is the layer used by the application for interacting with the database – retrieval and persistence as a central repository.

2. It is possible to change the way models and APIs are generated. You have to copy the template that you want to modify and then place it under the `resources` folder. Then, you have to modify the `swaggerSources` block in the `build.gradle` file by adding an extra configuration parameter for pointing to the template source, such as `templateDir = file("${rootDir}/src/main/resources/templates")`. This is the place where you keep modified templates such as `api.mustache`. This will extend the Swagger Codegen templates. You can find all the templates inside the `openapi` generator JAR file, such as `openapi-generator-cli-4.3.1.jar` in the `\JavaSpring` directory. You can copy the one you want to modify in the `src/main/resource/templates` directory and then play with it. You can make use of the following resources:

   a. **JavaSpring Templates**: `https://github.com/swagger-api/swagger-codegen/tree/master/modules/swagger-codegen/src/main/resources/JavaSpring`

   b. **Mustache Template Variables**: `https://github.com/swagger-api/swagger-codegen/wiki/Mustache-Template-Variables`

c. **An article explaining implementing a similar approach**: `https://arnoldgalovics.com/using-a-custom-template-for-swagger-codegen-with-gradle/`

3. ETag helps to improve the REST/HTTP client performance and user experience by only rerendering the page/section when the underlying API response is updated. It also saves bandwidth by carrying the response body only when required. CPU utilization can be optimized if the ETag is generated based on values retrieved from the database, for example, the version, last modified, and so on.

Chapter 5 – Asynchronous API Design

1. Yes, it is required only if you need vertical scaling. In the cloud, you pay for using the resources, and reactive applications definitely make use of them optimally. It is a new way of achieving scale. You need a small number of threads as compared to non-reactive applications. The cost of connection to a database, I/O, or any external source is the callback, therefore reactive-based applications do not require much memory. However, having said that Reactive programming is superior in terms of vertical scaling, you should continue with your existing or non-reactive applications. Even Spring recommends that. There is no new or old style; both can co-exist. However, when you need scaling for any special component or application, you can go the reactive way. A few years back, Netflix replaced the Zuul API gateway with the reactive API gateway Zuul2. They got scale and efficiency. However, they still have/use non-reactive applications.

2. You always have some pros and cons. Reactive is not an exception. Reactive code is not easy to write compared to the imperative style. It is very difficult to debug because it does not use a single thread. However, if you have developers who are proficient in the reactive paradigm, it does not stand as claimed.

3. Mono/Flux streams are subscribed by WebFlux internal classes when the controller sends the Mono/Flux stream. These classes convert them into HTTP packets. HTTP protocol does support event streams. However, for other media types such as JSON, Spring WebFlux subscribes Mono/Flux streams and waits till `onComplete()` or `onError()` is triggered. Then, it serializes the whole list of elements, or a single element in the case of Mono, in one HTTP response. You can learn more about it in the *Reactive core* section.

Chapter 6 – Security (Authorization and Authentication)

1. The security context stores the principal using `SecurityContextHolder` and is always available in the same thread of execution. The security context allows you to extract the principal during the flow execution and use it wherever you want. This is where a security annotation such as `@PreAuthorize` makes use of it for validation. The principal is the currently logged-in user. It can either be an instance of `UserDetails` or a string carrying a username. You can use the following code to extract it:

    ```
    Object principal = SecurityContextHolder
                        .getContext().getAuthentication().
    getPrincipal();

    if (principal instanceof UserDetails) {
      String username = ((UserDetails)principal).
    getUsername();
    } else {
      String username = principal.toString();
    }
    ```

2. This is a subjective question. However, there is a recommendation to use the signing of tokens (JWS) if JWT doesn't contain sensitive and private information such as a date of birth or credit card information. In such cases, you should make use of JWE for encrypting the information. If you want to use both together, then the preferred way is to use encryption for information carried by the token and then sign it with keys.

3. You can follow these guidelines, and add to them if you come across any new ones:

*   Make sure that JWT always has issuer and audience validations.

*   Make sure that JWT validation does not allow a none algorithm (without the algorithm mentioned in JWT). Instead, make sure that you have verification in place that checks the specific algorithm (whatever you configured) and a key.

*   Keep an eye on the **National Vulnerability Database (NVD)**.

- Don't use a weak key (secret). Instead use the asymmetric private/public keys with SHA 256, SHA 384, and SHA 512.

- Use a minimum key size of 2,048 for normal cases and 3,072 for business cases.

- A private key should be used for authentication and the verification server should use a public key.

- Make sure clients use the security guidelines to store the tokens and web applications should use HTTPS for communication with servers.

- Make sure the web application is tested thoroughly for **cross-site scripting** (**XSS**) attacks. It is always best to use the **content security policy** (**CSP**).

- Keep a short expiration time and use a refresh token to refresh an access token.

- Keep a tab on OWASP security guidelines and new threats.

Chapter 7 – Designing the User Interface

1. Props are special objects that you use to pass the values/objects/functions from the parent component to a child component, whereas state belongs to a component – it could be global or local to the component. From a functional component perspective, you use the `useState` hook for local state and `useContext` for global state.

2. In general, events are objects generated by the browser on input such as `keydown` or `onclick`. React uses `SyntheticEvent` to ensure that the browser's native events work identically across all browsers. `SyntheticEvent` wraps on top of the native event. You have used the code `onChange={ (e) => setUserName(e.target.value) }` in the login component. Here, e is `SyntheticEvent` and `target` is one of its attributes. The event onChange is binded in JSX that calls `setUserName` when the input value is changed. You can also use the same JavaScript way to bind events such as `window.addEventListener("click", handleClick)`. Ideally, you would like to do this in the `useEffect` hook, however, the event should be removed as a part of the cleanup. That can also be done in `useEffect` when you return the arrow function that removes the binding, for example, `return () => { window.removeEventListener("click", handleClick); }`. You can find this example in the `Header.js` file in the `src/components` directory.

3. In JavaScript, higher-order functions take a function as an argument and/or return a function such as an array's function (map, filter, and so on). On similar lines, in React, **higher-order components (HOCs)** are a pattern to use composition with an existing component and return a new component. Basically, you write a new function that takes a component as an argument and returns it. A HOC allows you to reuse the existing component and its logic.

4. In the `ecomm-ui` application, the components `ProductCard` and `ProductDetail` are similar in nature and you can use a HOC to reuse the logic.

Chapter 8 – Testing APIs

1. Unit testing is done to test the smallest code unit, such as a method. Whereas integration testing is performed where either different layers are involved or multiple modules are involved. In this chapter, integration testing has been done for the entire application, which involves all the layers of the application, including the database, whereas unit testing is performed class-wise for each of the methods. In the context of this chapter, unit testing is white-box testing, whereas API integration testing is a kind of black-box testing because you verify the API's functional requirement.

2. Having a separate unit and integration test, including their source location, allows you to manage tests easily. You can also have a configurable build setup that would perform unit testing during development or on-demand because unit tests are faster. You can run only unit tests by using the command `gradlew clean build -x integrationTest`, whereas on merge request builds, you can execute the integration tests to verify the merge request. The default build (`gradlew clean build`) would execute both unit and integration tests.

3. When you use `Mockito.mock()` or `@Mock`, it creates the complete fake object of the given class, then you can stub its method based on the test requirement. Whereas `Mockito.spy()` or `@Spy` creates the real object, on which you can stub the required methods. If stubbing is not done on the spy object, then its real methods will be called during the test.

Chapter 9 – Deployment of Web Services

1. You can create VMs using virtualization that are created on top of the host system which shares its hardware, whereas containerization creates containers that are executed as an isolated process on top of the hardware and its OS. Containers are lightweight and take only a few MB (exceptionally GB). VMs are heavyweight and take many GB. Containers run faster than VMs. Containers are more portable than VMs:

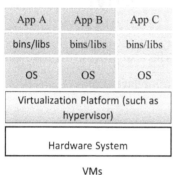
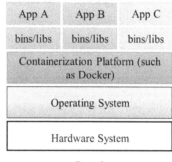

2. Kubernetes is a container orchestration system and is used for managing application containers. It keeps track of running containers. It shuts down containers when they are not being used and restarts orphaned containers. A Kubernetes cluster is also used for scale. It can provision resources such as CPU, memory, and storage automatically when required.

3. kubectl is a **command-line interface (CLI)** that is used for running commands against a Kubernetes cluster. You can manage Kubernetes resources using kubectl. You have used the `apply` and `create` kubectl commands in this chapter.

Chapter 10 – gRPC Fundamentals

1. **RPC** stands for **Remote Procedure Call**. A client can call an exposed procedure on a remote server, which is just like calling a local procedure but it gets executed on a remote server. An RPC is best suited for inter-service communication in connected systems.

2. gRPC is based on client-server architecture, whereas this is not true for REST. gRPC also supports full-duplex streaming communication in contrast to REST. gRPC performs better than REST as it uses the static paths and single source of the request payload.

3. A REST response error depends on HTTP status codes, whereas gRPC has formalized the set of errors to make it well-aligned with APIs. gRPC has also been built for supporting and handling call cancellations, load balancing, and failovers. For more information, please refer to the subsection *REST versus gRPC*.

4. You should use the Server Stream RPC methods because you would like to receive the latest messages from the server, such as tweets.

Chapter 11 – gRPC-Based API Development and Testing

1. Because, unlike HTTP libraries, gRPC libraries also provide the following features:

    a. Interaction with flow control at the application layer

    b. Cascading call cancellation

    c. Load balancing and failover

2. Yes, we can. You can use the metadata as shown in the next code block. However, making use of com.google.rpc.Status allows you to use the details (with a type of Any) object, which can capture more information:

    ```
    Metadata.Key<SourceId.Response> key =
        ProtoUtils.keyForProto(SourceId.Response.
    getDefaultInstance);

    Metadata metadata = new Metadata();
    metadata.put(key, sourceIdResponse);
    respObserver.onError(Status.INVALID_ARGUMENT
        .withDescription("Invalid Source ID")
        .asRuntimeException(metadata));
    ```

3. com.google.rpc.Status can include details of type Any, which can be used to provide more error details. io.grpc.Status does not have a field that contains the error details. You have to rely on another class' metadata to provide the error-related details, which is just like containing information not only specific to the error.

Chapter 12 – Logging and Tracing

1. Trace IDs and span IDs are created when the distributed transaction is initiated. A trace ID is generated for the main API call by the receiving service using Spring Cloud Sleuth. A trace ID is generated only once for each distributed call. Span IDs are generated by all the services participating in the distributed transaction. A trace ID is a correlation ID that will be common across the service for a call that requires a distributed transaction. Each service will have its own span ID for each of the API calls.

2. Yes, a broker such as Kafka, RabbitMQ, or Reddis allows robust persistence of logs and removes the risk of losing log data in unavoidable circumstances. It also performs better and can handle sudden spikes of data.

3. A tracer such as Spring Cloud Sleuth (which performs *instrumentation*) does two jobs – (1) records the time and metadata of the call being performed, and (2) propagates the trace IDs to other services participating in the distributed transaction. Then, it pushes the tracing information to Zipkin using *Reporter* once the scan completes. The Reporter uses *Transport* such as HTTP and Kafka to publish the data in Zipkin.

    The *Collector* in Zipkin collects the data sent by the transporters from the running services and passes it to the storage layer. The storage persists the data. Persisted data is exposed by the Zipkin APIs. The Zipkin UI calls these APIs to show the information graphically:

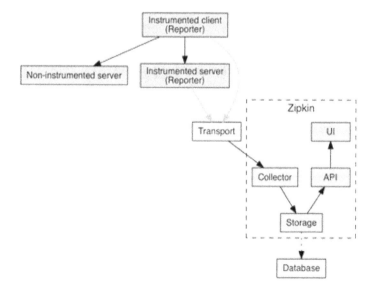

Chapter 13 – GraphQL Fundamentals

1. It depends on the use cases. However, it performs much better for mobile apps and web-based UI applications.

2. Fragments should be used while sending a request from the GraphQL client when the response contains an interface or union.

3. You can use a variable in a GraphQL query/mutation as shown next. You are going to modify the GraphQL request sent in *step 6* of the *Testing GraphQL Query and Mutation* section:

```
mutation removeStar ($repoId: String) {
  addStar(input: {
    starrableId: $repoId
  }) {
    clientMutationId
  }
}
```

4. Here, you can see that the `$repoId` variable is used. You have to declare that variable in the named mutation and then you use it in the mutation's argument as shown in the following code block:

```
{
    "repoId": "MDEwOlJlcG9zaXRvcnkyOTMyOTU5NDA="
}
```

Chapter 14 – GraphQL API Development and Testing

1. You should prefer a framework such as Netflix DGS in place of the `graphql-java` library to implement the GraphQL APIs because it bootstraps the development and avoids writing boilerplate code.

2. Apart from the ease of development, the framework uses `graphql-java` internally, therefore it keeps itself in sync with the GraphQL specification's Java implementation. It also supports developing federated GraphQL services.

3. It also provides plugins, the Java client, and testing utilities that help you to automate the development. The Netflix DGS framework is well tested and has been used by Netflix in production for quite some time.

4. A federated GraphQL service contains a *single distributed graph* exposed using a gateway. Clients call the gateway, which is an entry point to the system. A data graph will be distributed among multiple services and each service can maintain its own development and release cycle independently. Having said that, federated GraphQL services would still follow the OneGraph principle. Therefore, the client would query a single endpoint for fetching any part of the graph.

Other Books You May Enjoy

If you enjoyed this book, you may be interested in these other books by Packt:

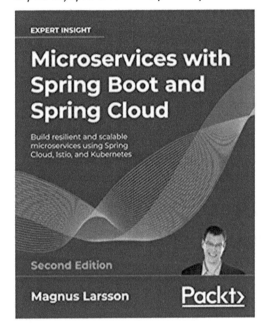

Microservices with Spring Boot and Spring Cloud - Second Edition

Magnus Larsson

ISBN: 978-1-80107-297-7

- Build cloud-native production-ready microservices with this comprehensively updated guide
- Understand the challenges of building large-scale microservice architectures
- Learn how to get the best out of Spring Cloud, Kubernetes, and Istio in combination

Hands-On Full Stack Development with Spring Boot 2 and React - Second Edition

Juha Hinkula

ISBN: 978-1-83882-236-1

- Unleash the power of React Hooks to build interactive and complex user interfaces
- Build scalable full stack applications designed to meet demands of modern users
- Understand how the Axios library simplifies CRUD operations

Full Stack Development with JHipster - Second Edition

Deepu K Sasidharan , Sendil Kumar N

ISBN: 978-1-83882-498-3

- Build full stack applications with modern JavaScript frameworks such as Angular, React, and Vue.js

- Explore the JHipster microservices stack, which includes Spring Cloud, Netflix OSS, and the Elastic Stack

- Learn advanced local and cloud deployment strategies using Docker and Kubernetes

Packt is searching for authors like you

If you're interested in becoming an author for Packt, please visit `authors.packtpub.com` and apply today. We have worked with thousands of developers and tech professionals, just like you, to help them share their insight with the global tech community. You can make a general application, apply for a specific hot topic that we are recruiting an author for, or submit your own idea.

Leave a review - let other readers know what you think

Please share your thoughts on this book with others by leaving a review on the site that you bought it from. If you purchased the book from Amazon, please leave us an honest review on this book's Amazon page. This is vital so that other potential readers can see and use your unbiased opinion to make purchasing decisions, we can understand what our customers think about our products, and our authors can see your feedback on the title that they have worked with Packt to create. It will only take a few minutes of your time, but is valuable to other potential customers, our authors, and Packt. Thank you!

Index

www.ingramcontent.com/pod-product-compliance
Lightning Source LLC
Chambersburg PA
CBHW081450050326
40690CB00015B/2746